D0384943

Two Heads

A Graphic Exploration of How Our Brains Work with Other Brains

UTA FRITH, CHRIS FRITH, AND ALEX FRITH

ILLUSTRATIONS BY DANIEL LOCKE

SCRIBNER

New York London Toronto Sydney New Delhi

SCRIBNER
AN IMPRINT OF SIMON & SCHUSTER, INC.
1230 AVENUE OF THE AMERICAS
NEW YORK, NY 10020

FIRST SCRIBNER HARDCOVER EDITION APRIL 2022

INTERIOR DESIGN BY MELISSA GANDHI

MANUFACTURED IN THE UNITED STATES OF AMERICA

1 3 5 7 9 10 8 6 4 2

LIBRARY OF CONGRESS CATALOGING-IN-PUBLICATION DATA IS AVAILABLE.

ISBN 978-1-5011-9407-8
ISBN 978-1-5011-9409-2 (EBOOK)

CHRIS: FOR UTA

UTA: FOR CHRIS

ALEX: FOR CHRIS AND UTA,
WHO TAUGHT ME THAT SCIENCE WAS DEEPLY WORTHWHILE
AND HARDER THAN IT LOOKS.

DAN: FOR HANNAH, POLLY, AND FELIX.
AND FOR MY GOOD FRIEND GORDON
WHO HAD ONE OF THE BEST BRAINS I'VE EVER KNOWN.

...are better than one

Contents

PSSST!
HEY YOU!
DO YOU WANT TO KNOW A SECRET?
ALEX FRITH, AUTHOR
HERE IT IS: NO ONE UNDERSTANDS HOW THE BRAIN WORKS.
OH, SURE, PEOPLE HAVE WRITTEN MANY BOOKS ABOUT THE TOPIC, BUT THEY DON'T REALLY KNOW.

I DON'T KNOW EITHER. I'M NOT EVEN A NEUROSCIENTIST. BUT I HAVE SPENT MY LIFE AROUND TWO VERY GOOD ONES.

LOOK, HERE'S ME, PICKING UP METAPHORICAL CRUMBS OF KNOWLEDGE.
CRUMBS GATHERED FROM THE DINNER TABLE I GREW UP EATING AT.

THESE ARE MY PARENTS, PROFESSORS UTA AND CHRIS FRITH, THEY ARE NEUROSCIENTISTS. HAVE BEEN FOREVER! (AS FAR AS I'M CONCERNED.)
EVEN AS A BABY, I PICKED UP BITS ABOUT THE LATEST DEVELOPMENTS IN NEUROSCIENCE.*
*YES, SOME GOSSIP ABOUT PEOPLE, TOO – BUT HONESTLY, MUCH OF THEIR CONVERSATION REALLY WAS ABOUT THE SCIENCE.

I LIKE TO THINK I UNDERSTOOD MOST OF THAT INTEL. THIS BOOK IS MY ATTEMPT TO EXPLAIN IT ALL.
IN FACT, IT'S OUR ATTEMPT. AFTER ALL, TWO HEADS ARE BETTER THAN ONE.
DAN LOCKE, ARTIST
RIGHT, RIGHT. NOW, LET'S GET OUT OF THE WAY AND LET THE PROFS DO THE HARD WORK...

LOOK, HERE THEY ARE NOW. LET'S FOLLOW THEM INTO THEIR HOME, IN NORTH WEST LONDON, WHERE THEY'VE LIVED FOR MORE THAN 30 YEARS.

UTA FRITH, NÉE AURNHAMMER (1941 – ?)
CHRIS FRITH, (1942 – ?)
WELCOME TO OUR HOME. I EXPECT YOU'VE ALREADY NOTICED SOMETHING: LIKE ALL GOOD ACADEMICS, WE SURROUND OURSELVES WITH CLUTTER, ESPECIALLY ART.
THAT'S PART OF THE REASON WHY THIS BOOK IS A COMIC.
INDEED. NOW, THIS DOOR LEADS TO MY STUDY. COME WITH ME AND WE CAN BEGIN OUR JOURNEY INTO THE WONDERS OF THE HUMAN BRAIN, THE MOST SOCIAL OF ALL OUR ORGANS.
INVITATION TO A PARTY (IT'S IN THE EPILOGUE)
R.S.V.P
New Year's Day
CAN YOU SPOT ALL THE BRAINS HIDDEN IN THIS PROLOGUE?
MORE CARPETS THAN FLOOR SPACE

WALLS COVERED WITH ART...
...MOSTLY ABSTRACT
HIGH-END CHRISTMAS DECORATIONS
BUT FIRST, A QUICK VISIT TO MY STUDY, UP THESE STAIRS...
MORE WINE BOTTLES THAN CAN REALISTICALLY BE DRUNK
LADDER TO REACH BOOKS
TOWERING BOOKSHELVES LOOM JUST OUT OF SIGHT...

PROFESSOR FRITH'S* STUDY
I'VE BEEN A RESEARCH PSYCHOLOGIST SINCE THE 1960S. FOR MANY YEARS I STUDIED SCHIZOPHRENIA, BEFORE TURNING MY ATTENTION TO BRAIN SCANNING.
WORLD'S MOST COMFORTABLE READING CHAIR
PILE OF COMICS RECOMMENDED BY SON (READ)
STANDING DESK
PILE OF WORK-RELATED PAPERS TO REVIEW (NOT READ)
PROFESSOR FRITH'S* STUDY
COLLECTION OF ASSORTED ARTIFACTS AND PLATES
SINCE I BECAME A RESEARCH PSYCHOLOGIST IN THE 1960S, I'VE SPENT A LOT OF TIME STUDYING AUTISM...
...A CONDITION PRACTICALLY DEFINED BY THE FACT THAT IT HINDERS PEOPLE'S ABILITY TO RELATE TO AND COMMUNICATE WITH OTHERS.
THE ROLE THAT COGNITIVE PROCESSES PLAY IN SOCIAL INTERACTIONS.
LATELY, I'VE BEEN COMING AT THE QUESTION OF HUMAN INTERACTIONS FROM ANOTHER ANGLE, BY WORKING WITH TEAMS OF PEOPLE TO STUDY HOW WE COOPERATE WITH EACH OTHER.
SHELVES FILLED WITH ART HISTORY BOOKS
*MAKE A GENDER-BASED NAME DISTINCTION AT YOUR PERIL!

SHELVES FULL OF 1960S SCI-FI AND DETECTIVE NOVELS
SELF-PORTRAIT OF THE SCIENTIST AS A YOUNG MAN
NOWADAYS I'M CALLED A NEUROSCIENTIST. I'VE BECOME INVOLVED WITH THE CHALLENGE OF TRYING TO WORK OUT WHAT CONSCIOUSNESS IS AND HOW IT ARISES...
STILL PREFERS 1990S-STYLE BALL-TRACKER MOUSE
SITTING DESK
...A QUESTION THAT HAS LONG ELUDED BRAIN SCIENTISTS.
TRAIN SET (FOR GRANDCHILDREN, HONEST)
BUT THAT'S NOT WHAT THIS BOOK IS ABOUT.
SPARE BIRD FEED
THIS BOOK IS ALL ABOUT PEOPLE, AND HOW WE RELATE TO EACH OTHER – OR, TO GET TECHNICAL, SOCIAL COGNITION.
LITERARY NOVELS IN ENGLISH AND GERMAN
YES, THAT IS A LAOCOON. GOOD EYE.
RING RING
GRAND DESK
OH! EXCUSE ME. THAT'S PROBABLY OUR SON CALLING.
GRAND PIANO (PLAYED WITH SURPRISING FREQUENCY)

RING
RING

YES, HELLO, ALEX. YES, I'VE SEEN YOUR EMAIL BUT HAVEN'T HAD TIME TO REPLY.*
*DISTRACTED BY CHECKING TWITTER.

NOW, HOW'S TWO HEADS GOING?

TO BE HONEST, I'M TRYING TO WORK OUT WHERE TO START.

HOW ABOUT THE MIND/ BRAIN PROBLEM?
YOUR FATHER AND I HAVE BEEN DISCUSSING IT LATELY.

THE PROBLEM IS THAT IT'S VERY HARD TO DEFINE THE IDEA OF "MIND."

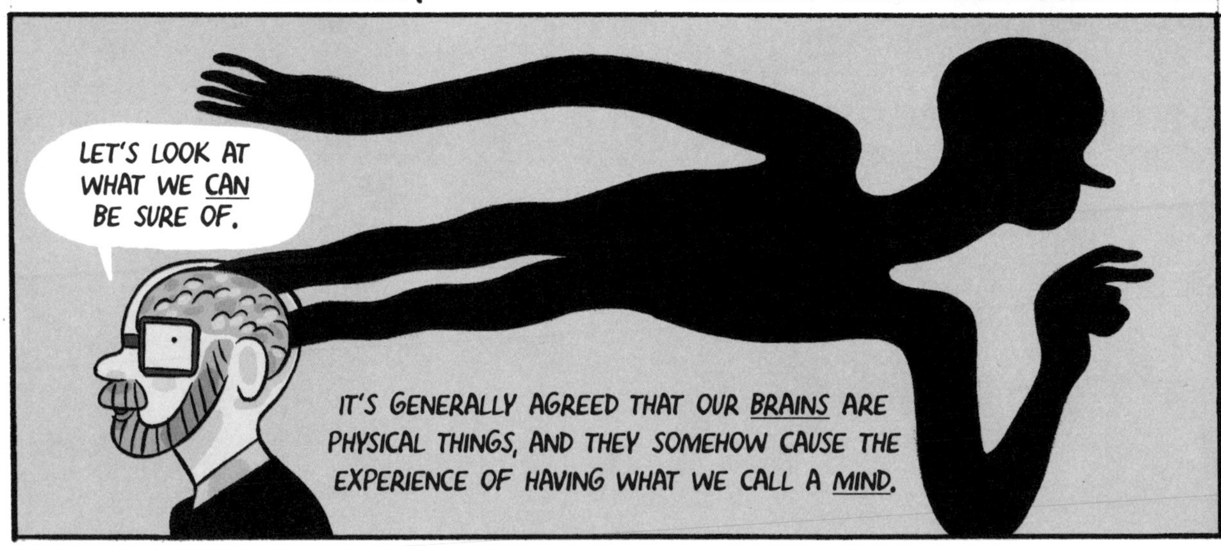
LET'S LOOK AT WHAT WE CAN BE SURE OF.
IT'S GENERALLY AGREED THAT OUR BRAINS ARE PHYSICAL THINGS, AND THEY SOMEHOW CAUSE THE EXPERIENCE OF HAVING WHAT WE CALL A MIND.

MORE THAN THAT, THE MIND IS AFFECTED BY THE WHOLE BODY, NOT JUST THE BRAIN.

FOR INSTANCE, IMAGINE YOU'VE JUST STUBBED A TOE.

1) YOUR BRAIN SIMPLY REGISTERS PAIN, BUT CARRIES ON WITH ITS OTHER BUSINESSES.

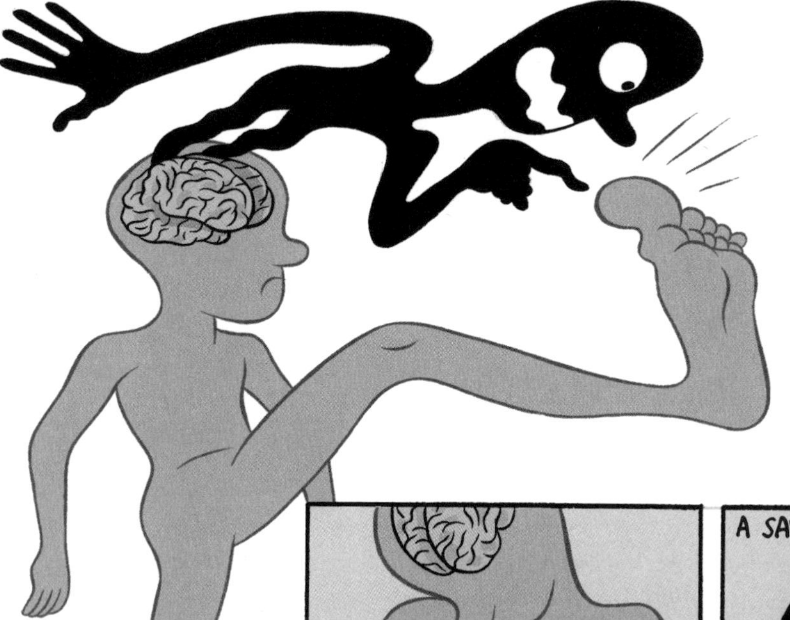

2) YOUR MIND IS, FOR A MOMENT, ABSOLUTELY OVERWHELMED BY THE SENSATION, AND YOU FEEL YOU CAN BARELY FUNCTION.

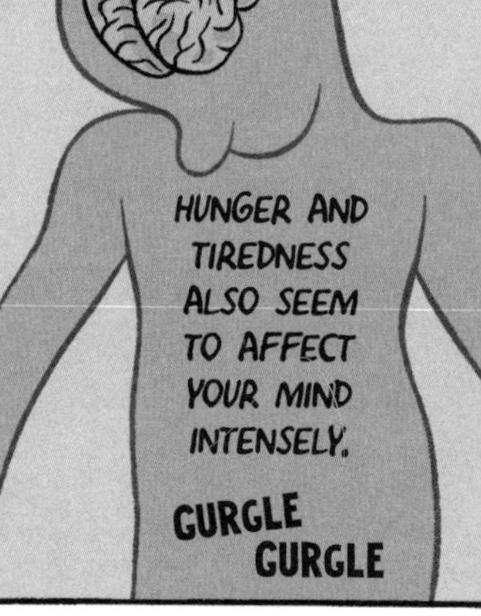
HUNGER AND TIREDNESS ALSO SEEM TO AFFECT YOUR MIND INTENSELY.
GURGLE
GURGLE

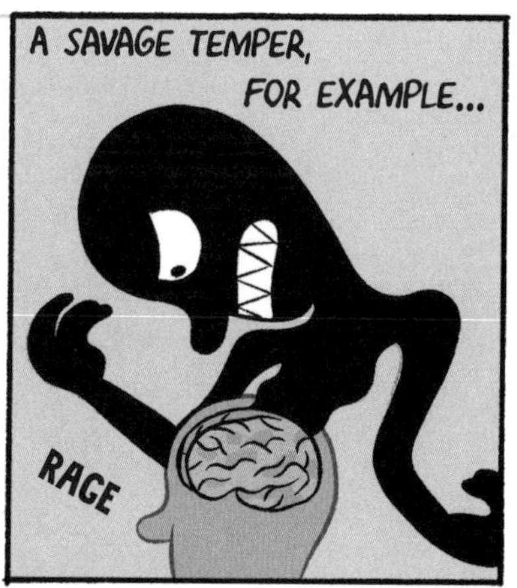
A SAVAGE TEMPER, FOR EXAMPLE...
RAGE

...CAN BE SOOTHED BY SITTING DOWN TO A CUP OF TEA AND A SLICE OF CAKE.
A CASE OF THE NEEDS OF THE BODY RULING OUR STATE OF MIND.

BUT THAT'S NOT THE END OF IT.
EVERYDAY EXPERIENCE TELLS US THAT OUR MINDS CAN CONTROL OUR BODIES AND OUR BRAINS, TOO.
WE HAVE ALL SORTS OF TOOLS TO INVESTIGATE THE WAYS IN WHICH BRAIN AND BODY AFFECT THE MIND. BUT, SO FAR, NO ONE TRULY UNDERSTANDS HOW THE MIND WORKS.

WE HAVEN'T EVEN REACHED THE ROOT OF THE MIND/BRAIN PROBLEM YET.
IN FACT, YOU COULD SAY WE'VE MADE IT WORSE.

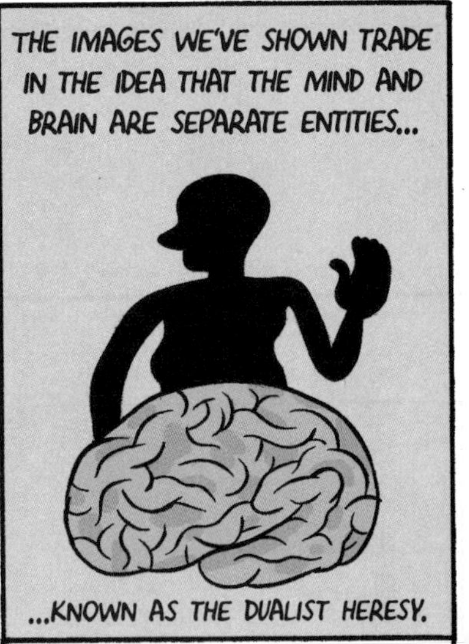
THE IMAGES WE'VE SHOWN TRADE IN THE IDEA THAT THE MIND AND BRAIN ARE SEPARATE ENTITIES...
...KNOWN AS THE DUALIST HERESY.

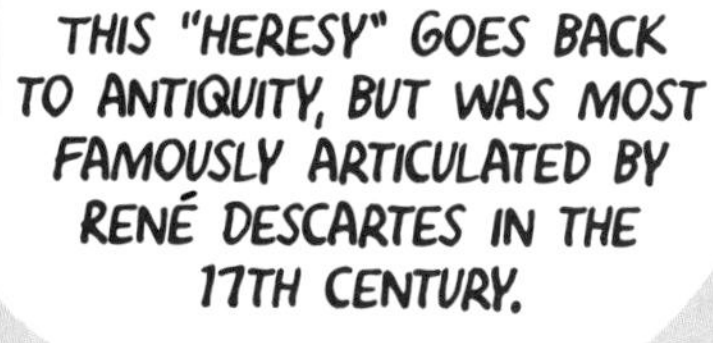

THIS "HERESY" GOES BACK TO ANTIQUITY, BUT WAS MOST FAMOUSLY ARTICULATED BY RENÉ DESCARTES IN THE 17TH CENTURY.

IT IS CERTAIN THAT I AM REALLY DISTINCT FROM MY BODY, AND CAN EXIST WITHOUT IT.
You think I am Descartes therefore I am.

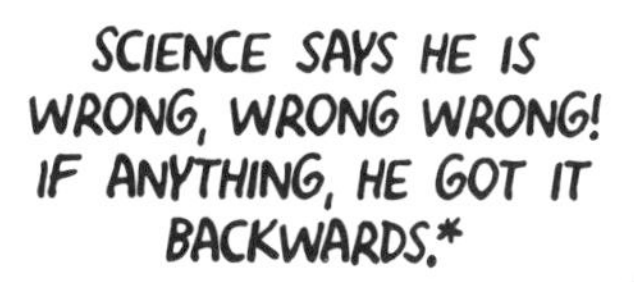

SCIENCE SAYS HE IS WRONG, WRONG WRONG! IF ANYTHING, HE GOT IT BACKWARDS.*

ALL AVAILABLE EVIDENCE SAYS THAT THE SENSATION YOU CALL "YOU" – ROUGHLY EQUIVALENT TO YOUR MIND – ABSOLUTELY COULD NOT AND WOULD NOT EXIST WITHOUT YOUR BRAIN.
NOW, BEFORE WE GET INTO OUR STORY ABOUT HOW BRAINS COOPERATE WITH EACH OTHER, WE'RE GOING TO GO OVER SOME BRAIN BASICS NICE AND QUICKLY.
THE THINGS WE DO KNOW, THAT IS. DON'T WORRY, WE'LL KEEP IT SIMPLE...

*ALTHOUGH I WILL SAY, IN A FOOTNOTE, THAT DESCARTES WASN'T LITERALLY TALKING ABOUT HUMAN PHYSIOLOGY, HE WAS FOCUSED INSTEAD ON A SEMANTIC ARGUMENT ABOUT "THINGS IN THEMSELVES."
BE WARNED – PLENTY MORE FOOTNOTES TO COME IN THIS BOOK!

AND I DON'T WANT TO BE TOO HARSH ON OLD RENÉ EITHER – THE DUALIST HERESY COMES NATURALLY TO ALL OF US.

Chapter 1
ONE OF OUR FAVORITE BRAIN ANALOGIES IS CALLED "THE COOK AND THE DINER."
YOUR CONSCIOUS BRAIN IS LIKE A DINER, WHO CAN ONLY EAT THANKS TO A COOK – THE UNCONSCIOUS BRAIN.
YOU NEED THIS COMBINATION TO HAVE A MIND.
What your brain is, what it can do,

THIS RELATIONSHIP HAPPENS IN EVERY INDIVIDUAL'S BRAIN. (NOT TO MENTION IN SOME HOUSEHOLDS.)
YOU MIGHT THINK WE'RE SAYING THAT THE DINER IS IN CHARGE...
...BUT THAT'S NOT THE CASE. AS WE'LL SEE, EACH PART OF YOUR BRAIN IS MADE OF THE SAME STUFF, AND IT ALL WORKS IN THE SAME WAY.
how it works, and all that.

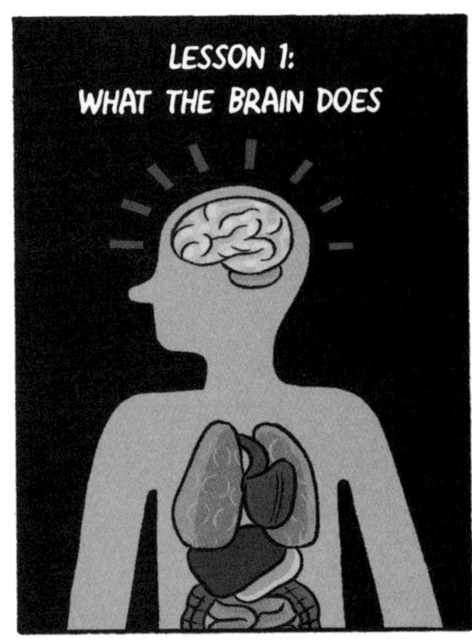
LESSON 1:
WHAT THE BRAIN DOES

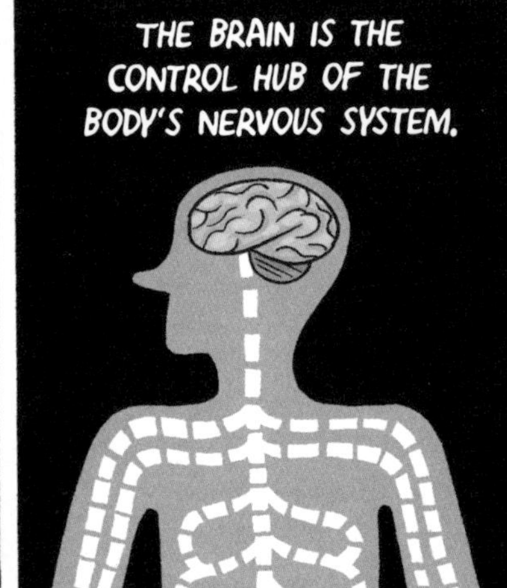
THE BRAIN IS THE CONTROL HUB OF THE BODY'S NERVOUS SYSTEM.

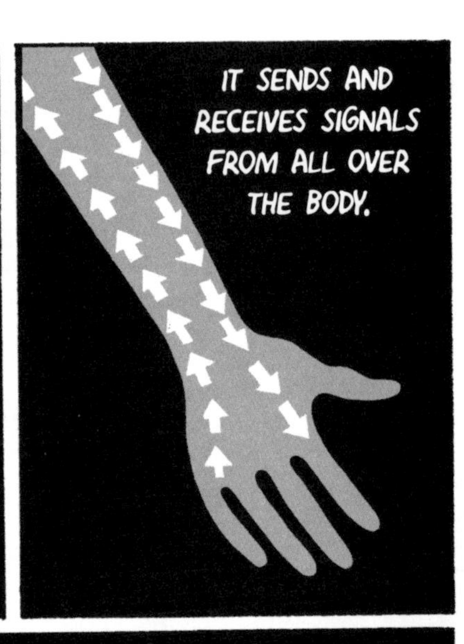
IT SENDS AND RECEIVES SIGNALS FROM ALL OVER THE BODY.

THE VAST MAJORITY OF THESE SIGNALS ARE CONTAINED IN CHEMICAL MESSENGERS KNOWN AS NEUROTRANSMITTERS.

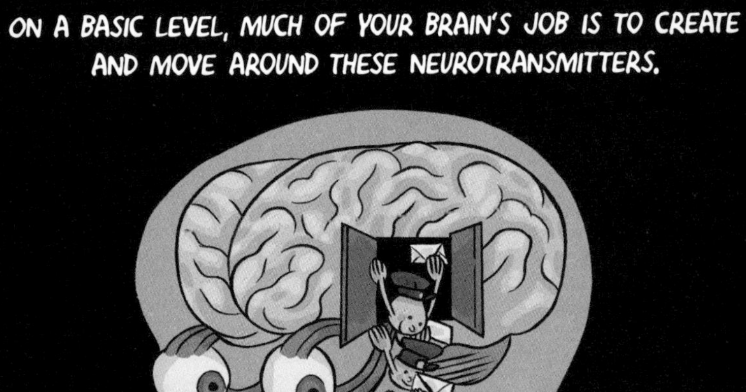
ON A BASIC LEVEL, MUCH OF YOUR BRAIN'S JOB IS TO CREATE AND MOVE AROUND THESE NEUROTRANSMITTERS.

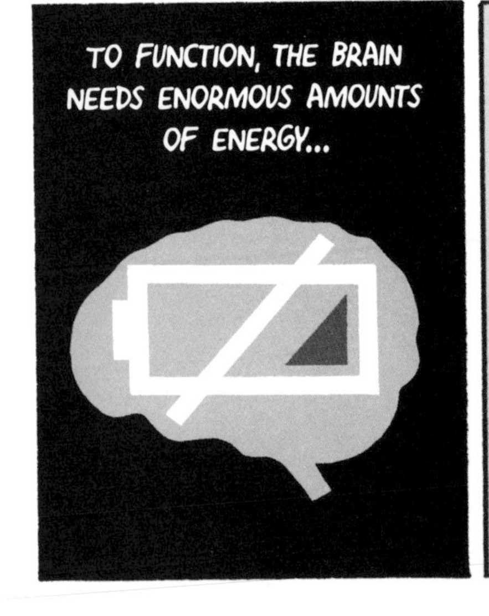
TO FUNCTION, THE BRAIN NEEDS ENORMOUS AMOUNTS OF ENERGY...

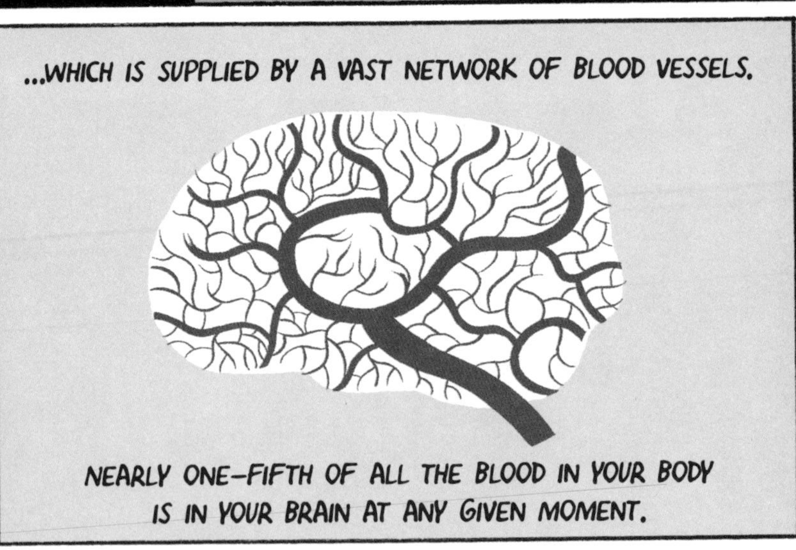
...WHICH IS SUPPLIED BY A VAST NETWORK OF BLOOD VESSELS.
NEARLY ONE-FIFTH OF ALL THE BLOOD IN YOUR BODY IS IN YOUR BRAIN AT ANY GIVEN MOMENT.

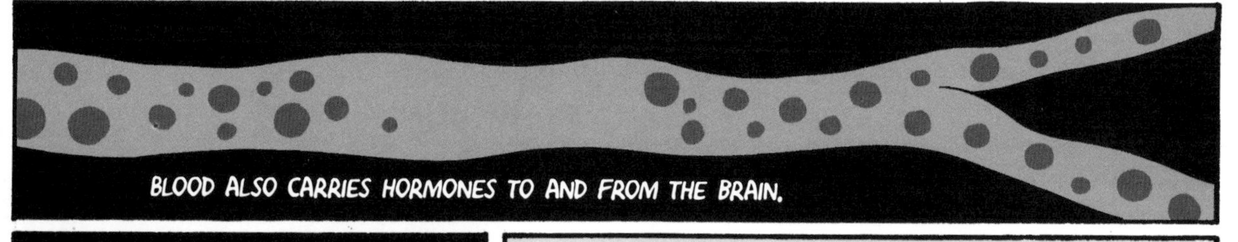
BLOOD ALSO CARRIES HORMONES TO AND FROM THE BRAIN.

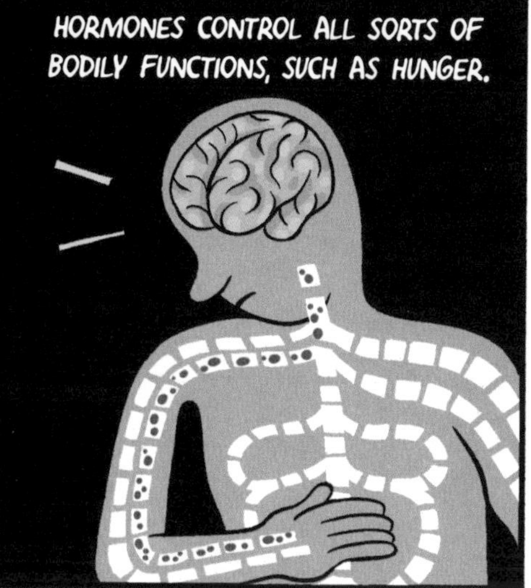
HORMONES CONTROL ALL SORTS OF BODILY FUNCTIONS, SUCH AS HUNGER.

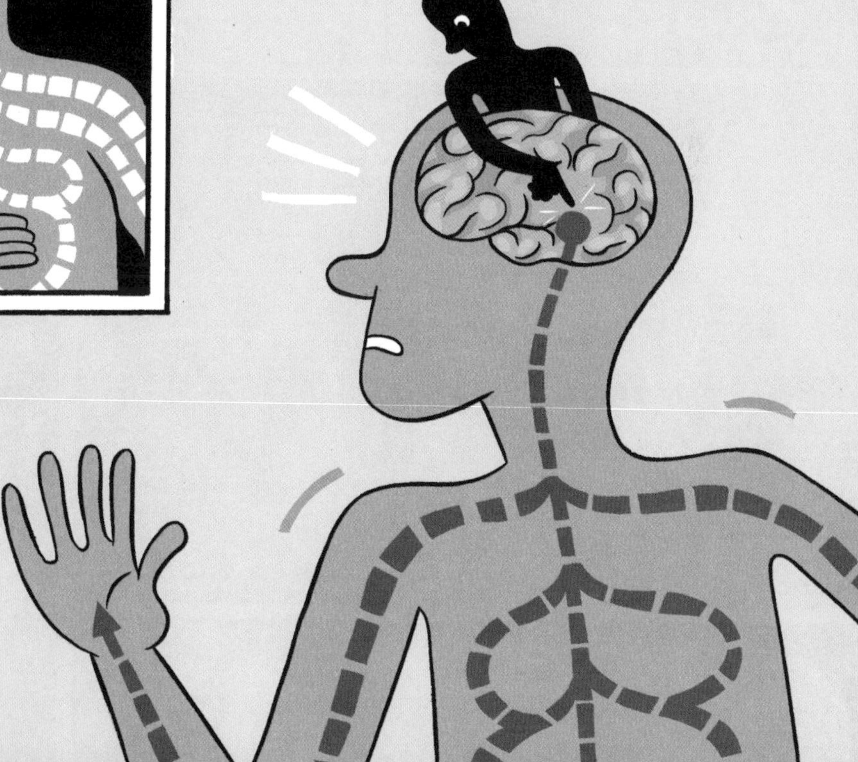
HORMONES OFTEN DETERMINE OUR EMOTIONS, TOO. SOMETIMES, THE TRIGGER TO RELEASE THEM SEEMS TO START IN YOUR "MIND." FOR EXAMPLE, WHEN YOU PERCEIVE A PROBLEM...

...YOUR BRAIN STARTS TO RELEASE HORMONES THAT RAISE YOUR STRESS LEVELS.

THIS IS YOUR BRAIN WORKING PROPERLY (UNPLEASANT THOUGH IT CAN BE).

THE QUESTION REMAINS, IS IT YOUR MIND OR YOUR BRAIN THAT CAUSES THE FEELING OF BEING STRESSED OUT? ANSWER: UNKNOWN.

AND INDEED, SOME WOULD SAY THE QUESTION ITSELF IS FAULTY – REMEMBER, OUR BEST GUESS AT THE MOMENT IS THAT THE MIND AND BRAIN ARE NOT SEPARATE ENTITIES!

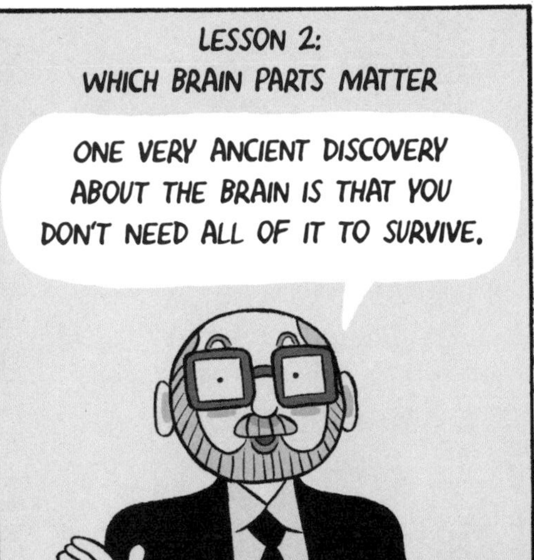

LESSON 2:
WHICH BRAIN PARTS MATTER
ONE VERY ANCIENT DISCOVERY ABOUT THE BRAIN IS THAT YOU DON'T NEED ALL OF IT TO SURVIVE.

YOU CAN LITERALLY HAVE MORE THAN HALF YOUR BRAIN REMOVED AND LIVE OUT A PRETTY NORMAL LIFE.

ONE PATIENT HAD ONLY BEEN ABLE TO SAY A SINGLE WORD...
TAN! TAN!
...THE OTHER JUST FIVE WORDS.
OUI. NON. TOIS. TOUJOURS. LELO.

HERE'S A SIDE-ON VIEW OF THE BRAIN, SHOWING ROUGHLY WHAT SOME PARTS OF THIS SIDE OF THE BRAIN DO. PLEASE NOTE, SOME PARTS OF THE BRAIN ON THE OTHER SIDE CONTROL VERY SIMILAR FUNCTIONS.*

BROCA SPECULATED THAT THE PART OF THEIR BRAINS THAT HAD DEVELOPED A HOLE (CAUSED BY EPILEPTIC SEIZURES, IN BOTH CASES) WAS THE PART RESPONSIBLE FOR ARTICULATING WORDS.

THUS WAS BORN ONE OF THE MAJOR PROJECTS OF NEUROSCIENCE – STILL ONGOING – TO MAP OUT THE BRAIN IN MINUTE DETAIL, DETERMINING WHICH PARTS (IF ANY) ARE RESPONSIBLE FOR WHICH ACTIVITIES.

IN FACT, MUCH OF WHAT WE KNOW ABOUT THE BRAIN COMES FROM STUDYING PEOPLE WHO WERE MISSING PARTS OF THEIRS.

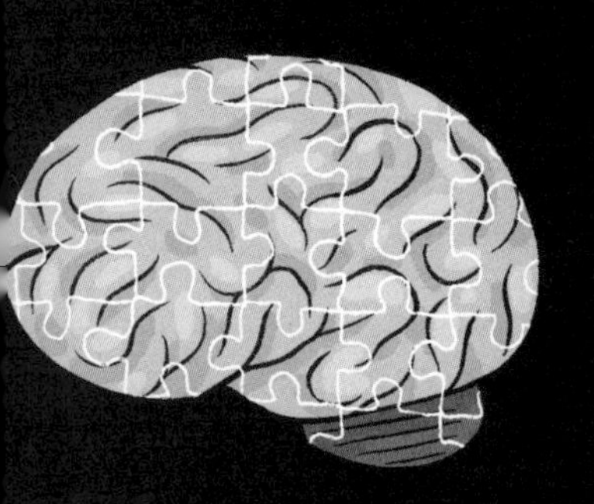

IN THE 1860S, FRENCH PHYSICIAN PIERRE PAUL BROCA FOUND NEAR-IDENTICAL HOLES IN THE BRAINS OF TWO OF HIS PATIENTS.

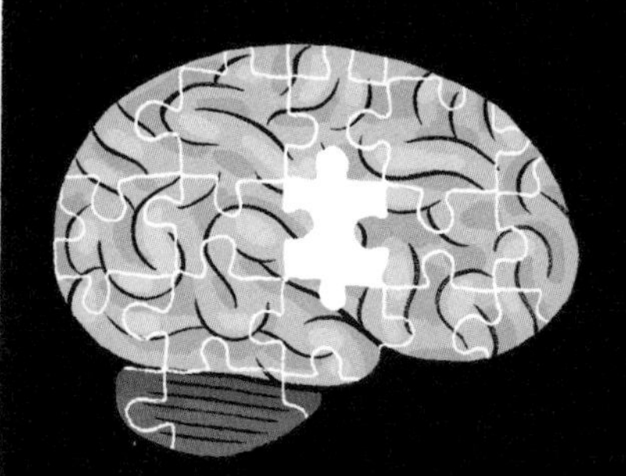

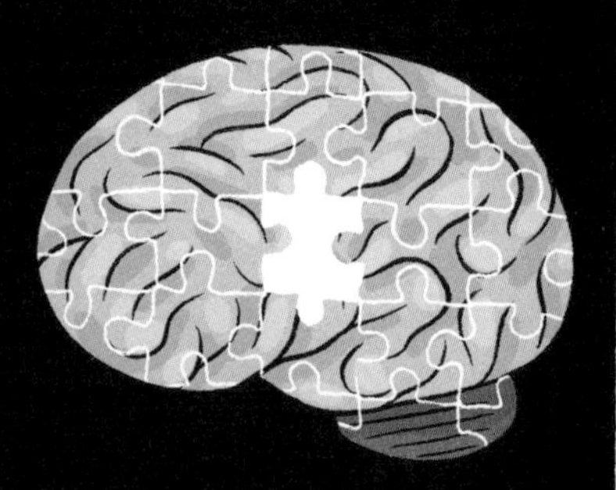

(FOUND DURING AUTOPSIES AFTER THEIR DEATHS, NATURALLY!)

SOME OF THESE BRAIN REGIONS ARE ONLY BROADLY UNDERSTOOD. BUT OTHERS ARE PINPOINTED WITH A SURPRISING DEGREE OF ACCURACY – AND THEY'RE THE SAME FOR ALL HUMAN BRAINS.

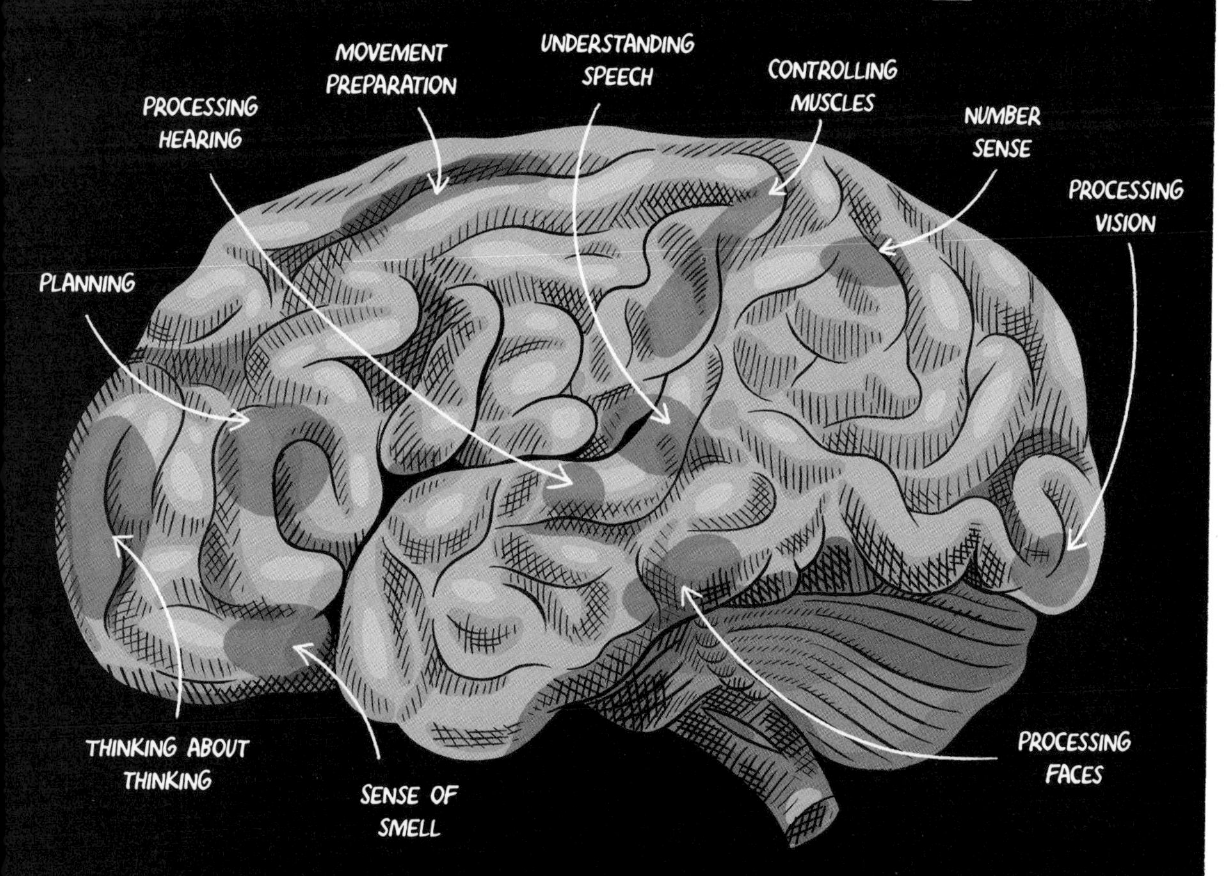
MOVEMENT PREPARATION
UNDERSTANDING SPEECH
CONTROLLING MUSCLES
PROCESSING HEARING
NUMBER SENSE
PROCESSING VISION
PLANNING
THINKING ABOUT THINKING
SENSE OF SMELL
PROCESSING FACES

*BORING BUT NECESSARY FOOTNOTE: THIS IS ONLY THE OUTER SURFACE OF THE BRAIN. ALL THE BITS THAT KEEP US ALIVE ARE DEEP IN THE CENTER. PLUS, WE RESERVE THE RIGHT TO GO OUT OF DATE AS PEOPLE GATHER MORE DATA.

LESSON 3:
WHAT THE BRAIN IS MADE OF
BY WEIGHT, BRAINS ARE MOSTLY FAT AND BLOOD. BUT IF YOU ZOOM IN CLOSE ENOUGH WITH A MICROSCOPE, YOU'LL MAKE OUT LOTS AND LOTS OF KEY ITEMS, NOTABLE FOR THEIR HAIRLIKE STRANDS.
THESE ARE BRAIN CELLS, MORE PROPERLY CALLED NEURONS.
NEURONS SEND SIGNALS TO EACH OTHER WITHIN YOUR BRAIN – AND ACROSS YOUR WHOLE BODY – USING YOUR SPINE. THE SIGNALS THEMSELVES ARE ELECTRIC...
...A FACT DISCOVERED BY ITALIAN BIOLOGIST LUIGI GALVANI TWO CENTURIES AGO.

I PROVED THAT MUSCLES CAN BE CONTROLLED USING ELECTRICITY AFTER I WIRED UP A DEAD FROG'S LEGS TO A BATTERY.*

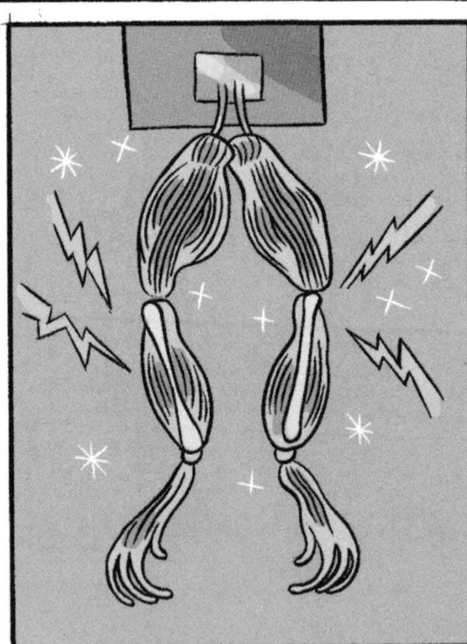

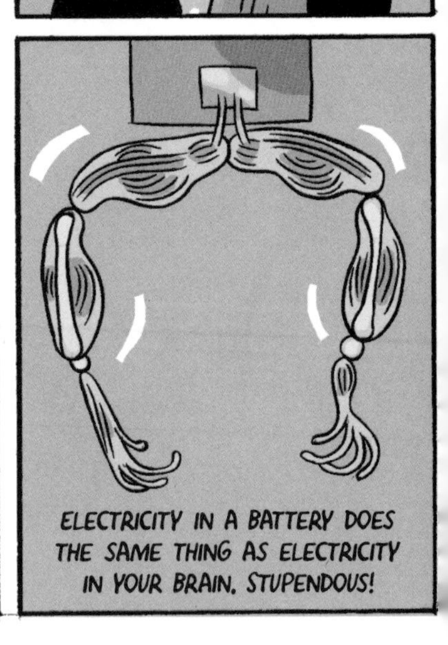
ELECTRICITY IN A BATTERY DOES THE SAME THING AS ELECTRICITY IN YOUR BRAIN. STUPENDOUS!

*NO ETHICS COMMITTEES IN THE 18TH CENTURY!

ELECTRICITY IS THE MEDIUM, BUT DON'T FORGET ABOUT NEUROTRANSMITTERS. THEY ARE WHAT TELL ONE NEURON TO STIMULATE THE NEXT NEURON, AND IN WHICH DIRECTION.

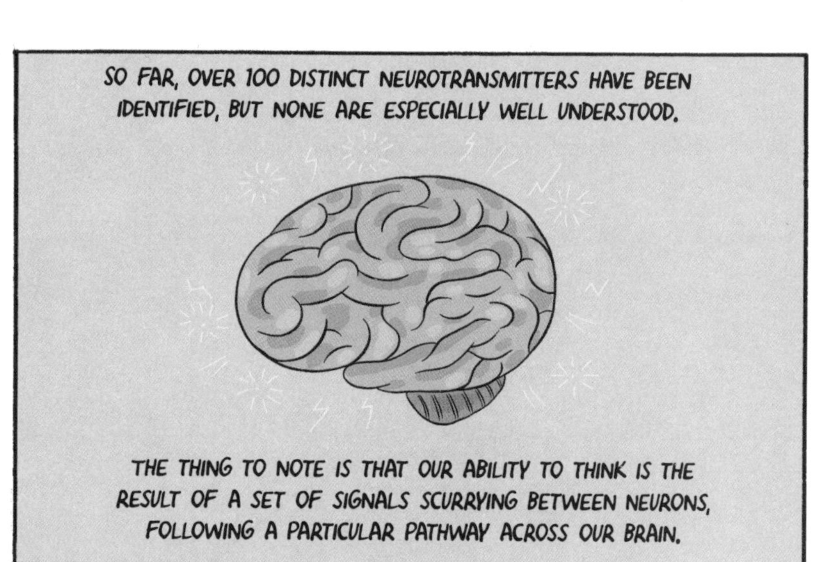
SO FAR, OVER 100 DISTINCT NEUROTRANSMITTERS HAVE BEEN IDENTIFIED, BUT NONE ARE ESPECIALLY WELL UNDERSTOOD.
THE THING TO NOTE IS THAT OUR ABILITY TO THINK IS THE RESULT OF A SET OF SIGNALS SCURRYING BETWEEN NEURONS, FOLLOWING A PARTICULAR PATHWAY ACROSS OUR BRAIN.

THIS IS TRUE OF ALL THOUGHTS. WHEN YOU PRACTICE AN ACTIVITY, SUCH AS TENNIS, ON ONE LEVEL YOU ARE REFINING AND MAINTAINING A SPECIFIC NEURAL PATHWAY.
WHICH MEANS YOU CAN IMPROVE YOUR TENNIS...
...JUST BY THINKING ABOUT IT.*
*ALTHOUGH DO NOTE, NEITHER PROFESSOR HAS EVER ACTUALLY PLAYED TENNIS.

AS A BABY, YOUR BRAIN CONSTANTLY WIRES UP NEW PATHWAYS AND CONNECTIONS ACROSS THE BRAIN. THESE GET OVERGROWN, SO YOU THEN HAVE TO PRUNE DOWN EXCESS CONNECTIONS.

SOMETHING YOU DO A LOT OF AS AN INFANT...
...AND AGAIN AS AN ADOLESCENT.

FOR THE REST OF YOUR LIFE, YOU AND YOUR BRAIN ARE CONSTANTLY REFINING THOSE CONNECTIONS – AS WELL AS MAKING NEW ONES.
INDEED, WHEN A BRAIN STOPS MAKING NEW CONNECTIONS, IT'S A SIGN OF DEMENTIA.

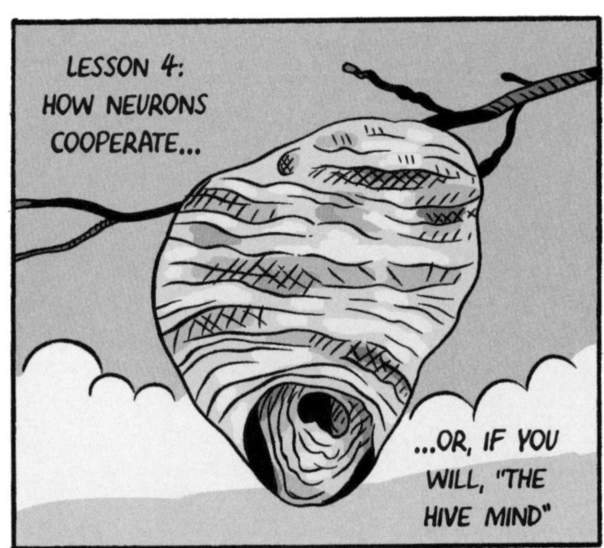
LESSON 4:
HOW NEURONS
COOPERATE...
...OR, IF YOU
WILL, "THE
HIVE MIND"

HOW EXACTLY DO BILLIONS OF NEURONS
CONSPIRE TO MAKE YOUR BRAIN WORK?
WELL, IT'S A LITTLE LIKE BEES.

WHEN A
COLONY OF
BEES NEEDS
A NEW NEST,
THE HIVE
SENDS OUT A
FEW HUNDRED
SCOUTS.

EACH SCOUT IS LOOKING FOR A GOOD-QUALITY SITE.
ROOMY,
A DECENT HEIGHT
OFF THE GROUND...

THE SCOUTS RETURN TO THE HIVE
WITH A REPORT TO EACH OTHER IN
THE FORM OF A WAGGLE-DANCE.

THE PATTERN OF THE DANCE
SHOWS WHICH DIRECTION TO GO
TO FIND THE NEW SITE.
NORTH,
BABY, NORTH.
THE LONGER THE DANCE, THE BETTER
THAT SCOUT THINKS ITS DISCOVERY IS.

I'M TELLING YOU, IT'S REALLY GREAT!

SCOUTS WITH
LONG DANCES
PERSUADE
OTHER SCOUTS
WHO ONLY
PERFORMED
SHORT DANCES
TO JOIN THEM.
OK, OK, I'LL GO YOUR WAY
AND SEE WHAT I CAN FIND.

INSTEAD OF DANCING, SOME DISRUPT ANY BEES WHO ARE DANCING TO PROMOTE A DIFFERENT SITE.
HEY! I'M DANCIN' HERE.

OVER TIME, MORE AND MORE SCOUTS ARE PERSUADED TO GO AND SEE A NEW SITE FOR THEMSELVES. WHEN THEY RETURN, THEY'LL JOIN ONE TEAM, AND HELP TO EITHER DANCE OR DISRUPT.
JOIN US! JOIN US!
NO, YOU JOIN US!

EVENTUALLY, A CRITICAL MASS OF SCOUTS ALL DANCE FOR ONE TEAM. THIS PERSUADES THE ENTIRE HIVE TO FLY TO THEIR CHOSEN SITE TO SET UP THEIR NEW HOME.*
OK, DONE HERE. LET'S MOVE OUT, YOU BEES.
*WE'LL GET INTO HOW THIS SUBTLE FORM OF PERSUASION WORKS – STATISTICALLY SPEAKING AT LEAST – IN A LATER CHAPTER.

IT'S POSSIBLE THAT THE HIVE IS WRONG, AND THE OTHER SITE WAS A BETTER OPTION.
WHY DIDN'T THEY LISTEN TO US??
BUT OBSERVERS REPORT THAT IT USUALLY IS THE BEST SITE.

IT'S A QUESTION OF PERSUADING ENOUGH BEES TO AGREE, RATHER THAN AIMING TO MAKE THE LITERAL "BEST" DECISION.
WHAT EVEN IS "BEST" WHEN YOU THINK ABOUT IT?

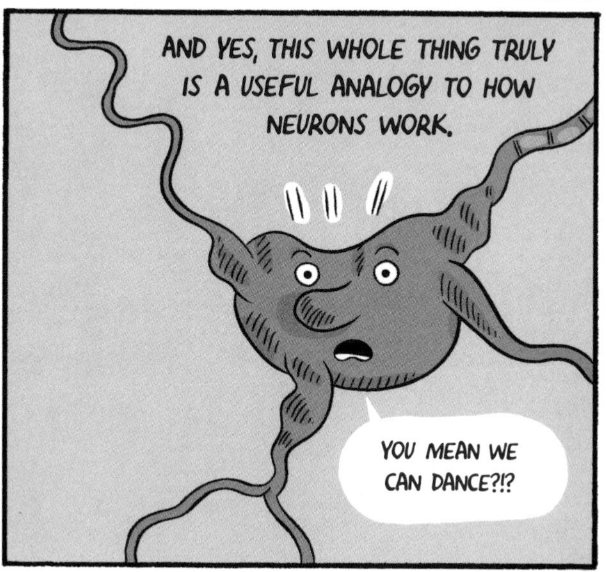
AND YES, THIS WHOLE THING TRULY IS A USEFUL ANALOGY TO HOW NEURONS WORK.
YOU MEAN WE CAN DANCE?!?

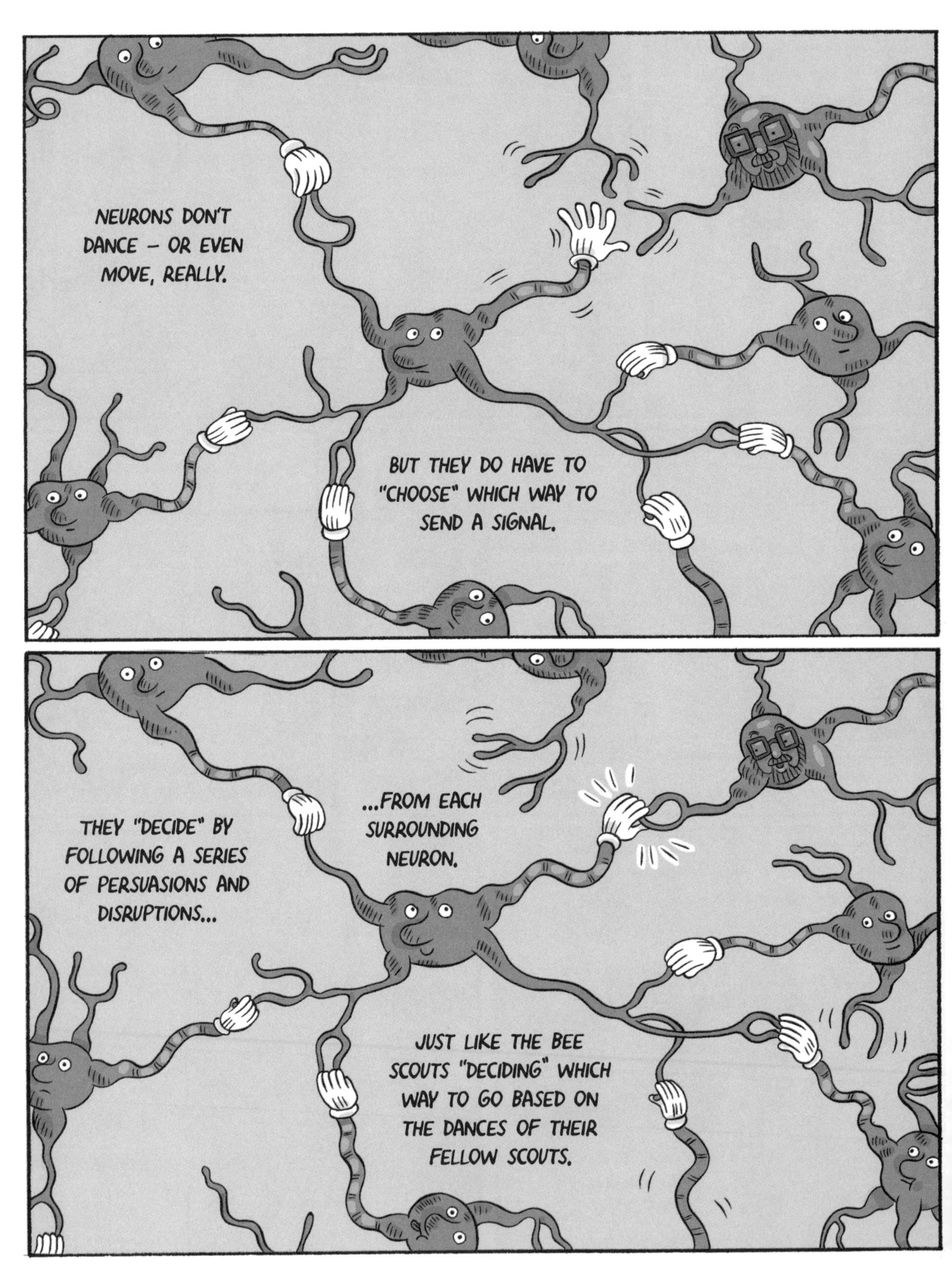
NEURONS DON'T DANCE – OR EVEN MOVE, REALLY.
BUT THEY DO HAVE TO "CHOOSE" WHICH WAY TO SEND A SIGNAL.
THEY "DECIDE" BY FOLLOWING A SERIES OF PERSUASIONS AND DISRUPTIONS...
...FROM EACH SURROUNDING NEURON.
JUST LIKE THE BEE SCOUTS "DECIDING" WHICH WAY TO GO BASED ON THE DANCES OF THEIR FELLOW SCOUTS.

LESSON 5: HOW YOU USE YOUR BRAIN – AND HOW YOUR BRAIN USES YOU

I THINK, THEREFORE YOU ARE.

ONE OF THE BEST UNDERSTOOD BRAIN FUNCTIONS IS VISION. HERE'S WHAT'S GOING ON WHEN YOU LOOK AT THE WORLD.

ALTHOUGH YOUR EYES ARE AT THE FRONT, THEY'RE WIRED UP TO THE BACK OF THE BRAIN, WHERE THE VISUAL CORTEX IS.

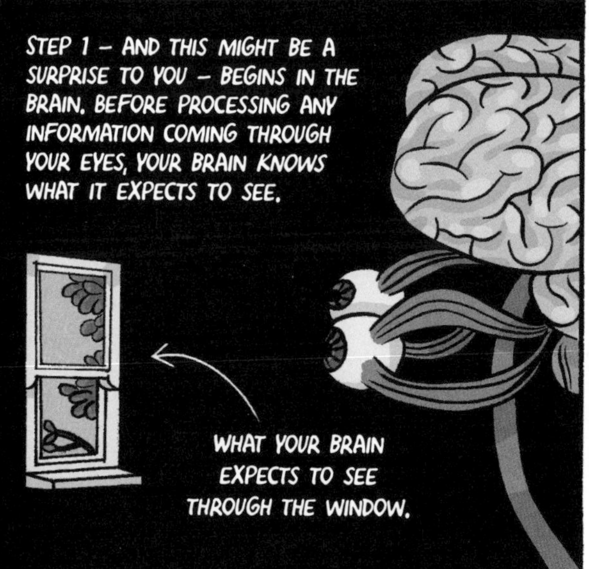
STEP 1 – AND THIS MIGHT BE A SURPRISE TO YOU – BEGINS IN THE BRAIN. BEFORE PROCESSING ANY INFORMATION COMING THROUGH YOUR EYES, YOUR BRAIN KNOWS WHAT IT EXPECTS TO SEE.
WHAT YOUR BRAIN EXPECTS TO SEE THROUGH THE WINDOW.

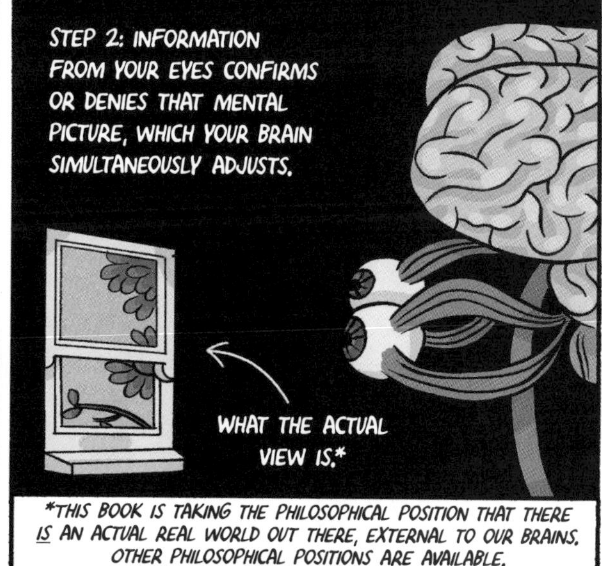
STEP 2: INFORMATION FROM YOUR EYES CONFIRMS OR DENIES THAT MENTAL PICTURE, WHICH YOUR BRAIN SIMULTANEOUSLY ADJUSTS.
WHAT THE ACTUAL VIEW IS.*
*THIS BOOK IS TAKING THE PHILOSOPHICAL POSITION THAT THERE IS AN ACTUAL REAL WORLD OUT THERE, EXTERNAL TO OUR BRAINS. OTHER PHILOSOPHICAL POSITIONS ARE AVAILABLE.

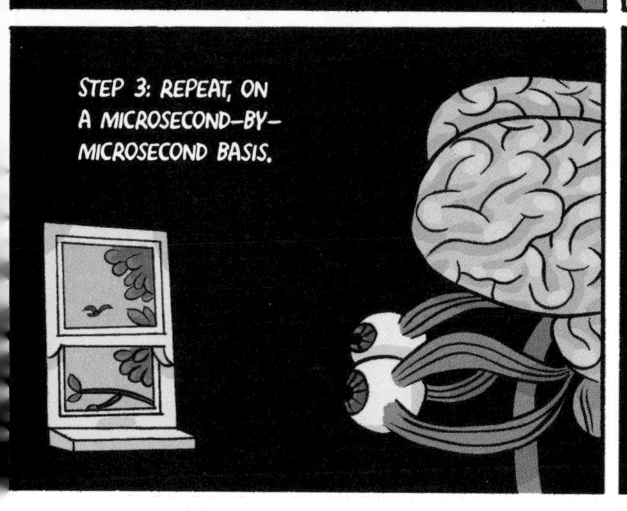
STEP 3: REPEAT, ON A MICROSECOND-BY-MICROSECOND BASIS.

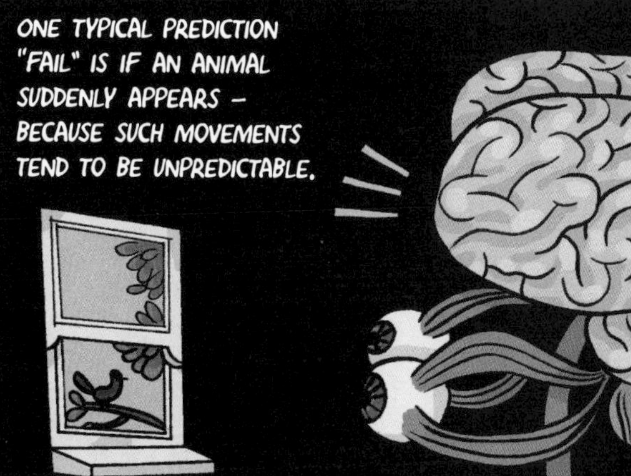
ONE TYPICAL PREDICTION "FAIL" IS IF AN ANIMAL SUDDENLY APPEARS – BECAUSE SUCH MOVEMENTS TEND TO BE UNPREDICTABLE.

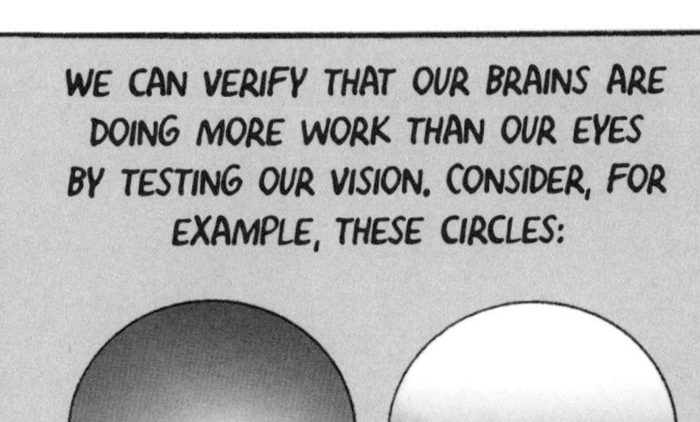
WE CAN VERIFY THAT OUR BRAINS ARE DOING MORE WORK THAN OUR EYES BY TESTING OUR VISION. CONSIDER, FOR EXAMPLE, THESE CIRCLES:

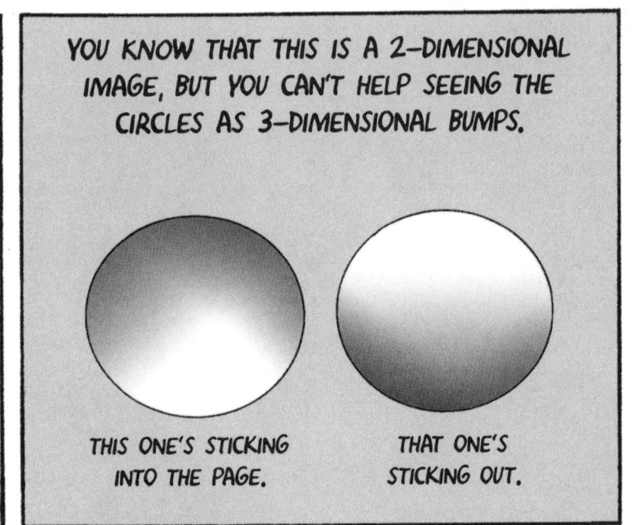
YOU KNOW THAT THIS IS A 2-DIMENSIONAL IMAGE, BUT YOU CAN'T HELP SEEING THE CIRCLES AS 3-DIMENSIONAL BUMPS.
THIS ONE'S STICKING INTO THE PAGE.
THAT ONE'S STICKING OUT.

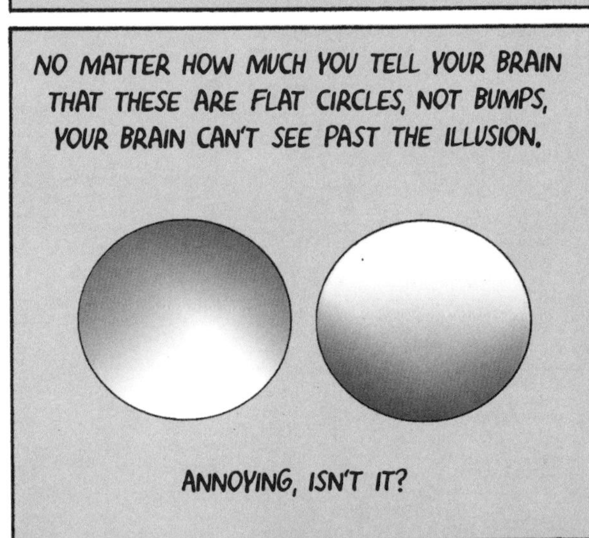
NO MATTER HOW MUCH YOU TELL YOUR BRAIN THAT THESE ARE FLAT CIRCLES, NOT BUMPS, YOUR BRAIN CAN'T SEE PAST THE ILLUSION.
ANNOYING, ISN'T IT?

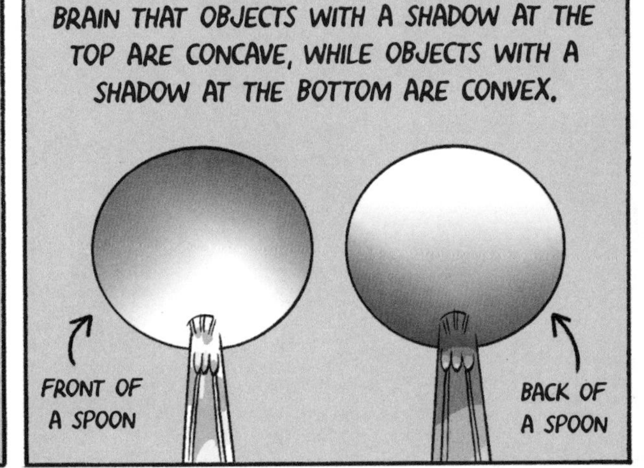
A LIFETIME OF EVIDENCE HAS PERSUADED YOUR BRAIN THAT OBJECTS WITH A SHADOW AT THE TOP ARE CONCAVE, WHILE OBJECTS WITH A SHADOW AT THE BOTTOM ARE CONVEX.
FRONT OF A SPOON
BACK OF A SPOON

TRY TURNING THE PAGE UPSIDE DOWN.
YOU STILL CAN'T TELL YOUR BRAIN TO SEE THE CIRCLES AS "FLAT".

I THINK THIS IS SOMETHING WIRED INTO OUR BRAINS VERY EARLY IN LIFE, BASED ON LIVED EXPERIENCE.
AND I THINK THIS IS SOMETHING WIRED INTO OUR BRAINS IN OUR DNA, THE PRODUCT OF EVOLUTION.
THIS IS ABOUT AS DRAMATIC AS OUR FAMILY ARGUMENTS GET.

YOUR BRAIN WORKS BY MAKING ASSUMPTIONS ABOUT THE WORLD AND THEN USING YOUR SENSES TO CONFIRM, OR, LESS OFTEN, DENY THOSE ASSUMPTIONS. WHEN FACED WITH AN OPTICAL ILLUSION LIKE THE CIRCLES, THE ASSUMPTIONS YOU MAKE ARE SO STRONG THAT YOUR SENSES AREN'T ENOUGH TO DENY THEM! BUT OVERALL, OUR BEST DESCRIPTION FOR THE WAY IT WORKS IS TO SAY:

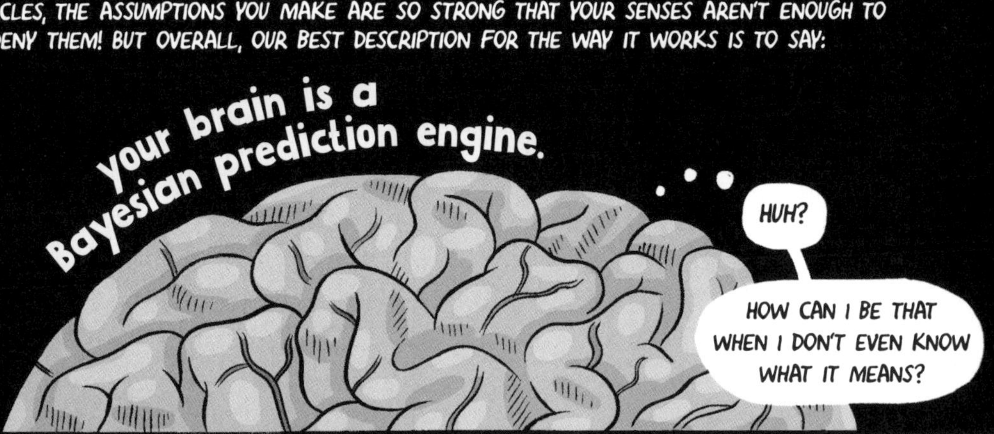
your brain is a Bayesian prediction engine.
...
HUH?
HOW CAN I BE THAT WHEN I DON'T EVEN KNOW WHAT IT MEANS?

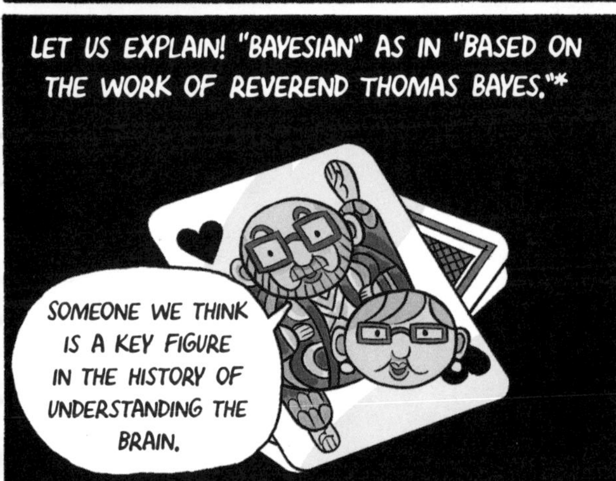
LET US EXPLAIN! "BAYESIAN" AS IN "BASED ON THE WORK OF REVEREND THOMAS BAYES."*
SOMEONE WE THINK IS A KEY FIGURE IN THE HISTORY OF UNDERSTANDING THE BRAIN.

BAYES WAS AN 18TH-CENTURY NONCONFORMIST MINISTER WHO, IN FACT, NEVER GAVE THE BRAIN ITSELF ANY PARTICULAR THOUGHT. HE WAS INTERESTED IN PROBABILITY.
PICK A BALL FROM MY BAG.

*THERE ARE NO KNOWN PORTRAITS OF BAYES, SO THE LIKELIHOOD THAT ANY OF OUR PICTURES HERE REPRESENT BAYES HIMSELF IS P(A/B) ETC., ETC. (LITTLE PROBABILITY JOKE FOR YOU THERE.)

HUH, A WHITE BALL. WHAT ARE THE CHANCES OF THAT?

HIS WORK ON PROBABILITY GOT HIM ELECTED AS A "FELLOW" OF THE UK'S ROYAL SOCIETY – MOSTLY, ONE SUSPECTS, BECAUSE MANY FELLOWS WERE KEEN GAMBLERS.

SHHH! IMPORTANT EXPERIMENT IN PROGRESS.

BAYES BEGAN TO BECOME POPULAR AGAIN IN THE 1950S. TODAY HE IS EMBRACED ACROSS THE INTERNET.

WHAT MATTERS IS THAT BAYES IS COOL, AND IF YOU DON'T KNOW BAYES YOU AREN'T COOL.*

*WE'RE QUOTING NOTED BAYES ENTHUSIAST ELIEZER S. YUDKOWSKY.

BAYES DEVELOPED A WAY TO MEASURE PROBABILITIES THAT TAKES ACCOUNT OF PEOPLE'S INTUITION AND BELIEFS, AS WELL AS LOOKING AT THE PURE ARITHMETIC OF PROBABILITY.

MY CALCULATOR TELLS ME THERE'S A 1 IN 13 CHANCE THIS CARD WILL BE AN ACE.

BUT MY CALCULATOR DOESN'T KNOW THAT THE HAND BELONGS TO JAMES BOND, IN THE PIVOTAL FINAL SCENE OF A MOVIE.

THE BAYESIAN PROBABILITY, TAKING INTO ACCOUNT THIS KEY PIECE OF KNOWLEDGE, IS 1 IN 1.

IF YOU'RE PLAYING AN ACTUAL GAME (NOT MERELY ACTING OUT A PLOT POINT), BAYES DOESN'T HELP YOU CALCULATE THE ODDS.

THE AIM OF THE GAME SHOWN HERE IS TO GET CLOSE TO, BUT NOT MORE THAN, 21. MATHEMATICS CAN HELP YOU CALCULATE THE LIKELIHOOD OF DRAWING A GOOD OR BAD CARD.

WHERE BAYES CAN HELP IS BY POINTING OUT THAT, IN GENERAL, YOU WILL LOSE MORE GAMES OF CARDS THAN YOU WILL WIN.

THE KEY TO BAYES IS TAKING ACCOUNT OF GENERAL FACTORS BEYOND ANY SPECIFIC SITUATION. IN TECHNICAL LANGUAGE, THESE GENERAL FACTORS ARE CALLED PRIORS.
IN THE EXAMPLE OF CASINO-STYLE CARD GAMES, THE MOST SALIENT PRIOR IS: ONLY THE HOUSE GETS RICH BY GAMBLING.

ONCE YOU TAKE THIS INTO ACCOUNT, IT IS ARGUABLY NOT WORTH CALCULATING THE ODDS OF ANY GIVEN CARD GAME.

BAYES CREATED A WAY TO BUILD THE IDEA OF PRIORS INTO A MATHEMATICAL FORMULA USED IN A BRANCH OF MATH KNOWN AS BAYESIAN STATISTICS.

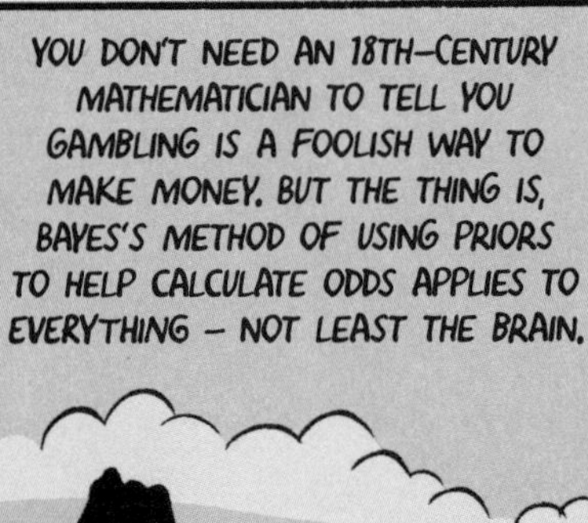
YOU DON'T NEED AN 18TH-CENTURY MATHEMATICIAN TO TELL YOU GAMBLING IS A FOOLISH WAY TO MAKE MONEY. BUT THE THING IS, BAYES'S METHOD OF USING PRIORS TO HELP CALCULATE ODDS APPLIES TO EVERYTHING – NOT LEAST THE BRAIN.

WE'VE BEEN BAYESIANS SINCE WE EVOLVED!

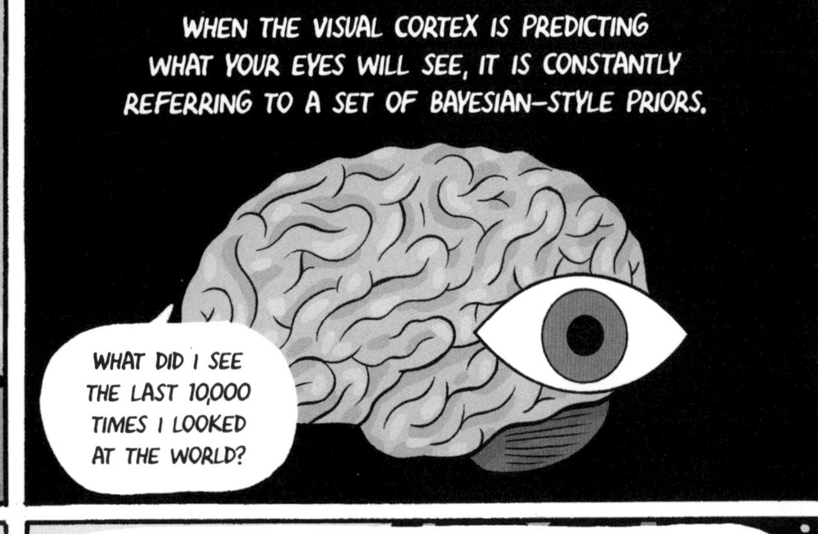
WHEN THE VISUAL CORTEX IS PREDICTING WHAT YOUR EYES WILL SEE, IT IS CONSTANTLY REFERRING TO A SET OF BAYESIAN-STYLE PRIORS.
WHAT DID I SEE THE LAST 10,000 TIMES I LOOKED AT THE WORLD?

NEUROSCIENTISTS ARE BEGINNING TO UNDERSTAND HOW THIS WORKS ON A NEURON-BY-NEURON BASIS.
PROFESSOR KARL FRISTON

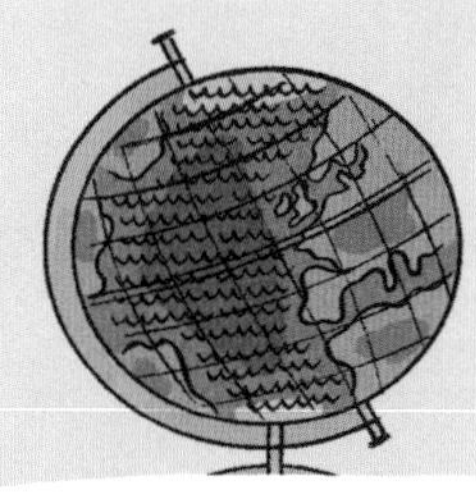

WHAT NEURONS DO IS: COMPARE THE PREDICTED MODEL OF THE WORLD AROUND YOU TO THE DATA THAT COMES IN, AND ADJUST TO MAKE THE DIFFERENCE BETWEEN THE TWO AS SMALL AS POSSIBLE.
IT TURNS OUT THAT BAYESIAN STATISTICS IS THE BEST WAY TO DO THIS WITH THE LEAST AMOUNT OF EFFORT. AS WE SAID, THE BRAIN IS A BAYESIAN PREDICTION ENGINE.

BAYESIAN THINKING IS REALLY JUST A MATHEMATICAL WAY TO TAKE ACCOUNT OF HOW PEOPLE ACTUALLY BEHAVE.

FOR INSTANCE, YOU'VE PROBABLY GOT TWO PRIOR BELIEFS ABOUT THIS BOOK THAT WILL AFFECT THE WAY YOU READ IT:
1. YOU MIGHT EXPECT TO LEARN SOME ACTUAL SCIENCE.

2. BECAUSE IT'S A COMIC, YOU CAN REASONABLY EXPECT TO CHUCKLE A FEW TIMES.

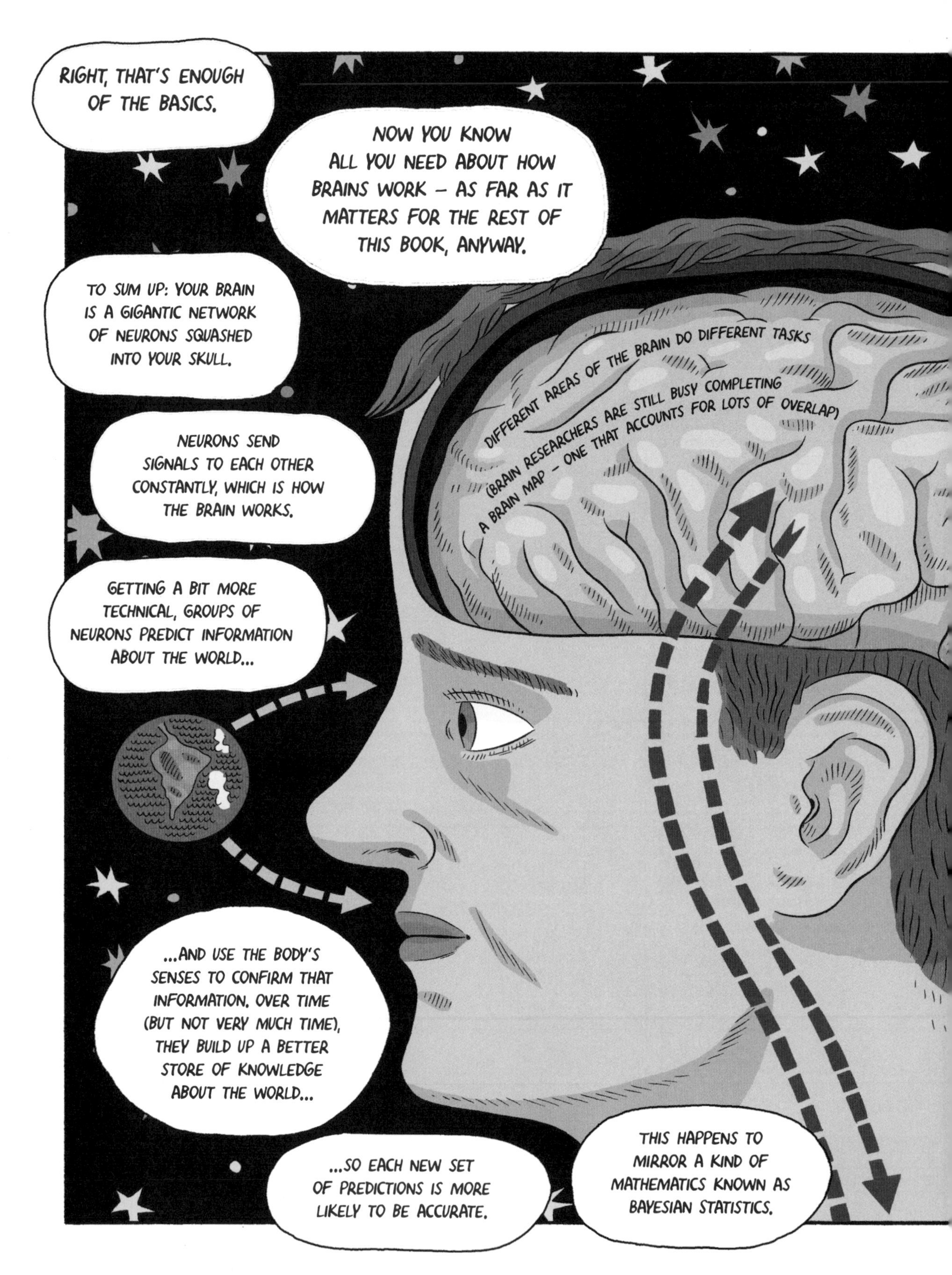
RIGHT, THAT'S ENOUGH OF THE BASICS.
NOW YOU KNOW ALL YOU NEED ABOUT HOW BRAINS WORK – AS FAR AS IT MATTERS FOR THE REST OF THIS BOOK, ANYWAY.
TO SUM UP: YOUR BRAIN IS A GIGANTIC NETWORK OF NEURONS SQUASHED INTO YOUR SKULL.
DIFFERENT AREAS OF THE BRAIN DO DIFFERENT TASKS
(BRAIN RESEARCHERS ARE STILL BUSY COMPLETING A BRAIN MAP – ONE THAT ACCOUNTS FOR LOTS OF OVERLAP)
NEURONS SEND SIGNALS TO EACH OTHER CONSTANTLY, WHICH IS HOW THE BRAIN WORKS.
GETTING A BIT MORE TECHNICAL, GROUPS OF NEURONS PREDICT INFORMATION ABOUT THE WORLD...
...AND USE THE BODY'S SENSES TO CONFIRM THAT INFORMATION. OVER TIME (BUT NOT VERY MUCH TIME), THEY BUILD UP A BETTER STORE OF KNOWLEDGE ABOUT THE WORLD...
...SO EACH NEW SET OF PREDICTIONS IS MORE LIKELY TO BE ACCURATE.
THIS HAPPENS TO MIRROR A KIND OF MATHEMATICS KNOWN AS BAYESIAN STATISTICS.

SOME OF THE THINGS WE FIND ESPECIALLY INTERESTING ABOUT ALL THIS:
NEURONS, INDIVIDUALLY, HAVE NO FREE WILL OR ABILITY TO MAKE CHOICES...
...BUT THE WAY IN WHICH THEY WORK TOGETHER GIVES RISE TO THE PHENOMENON WE CALL THE MIND.
WHAT DOES THAT MEAN FOR US AS INDIVIDUALS? IF THE SMALLEST UNITS OF OUR BRAINS HAVE NO FREE WILL, DOES THAT MEAN OUR BRAINS AND MINDS HAVE NONE EITHER?
DOES THE WAY OUR NEURONS COOPERATE HAVE ANYTHING TO DO WITH THE WAY INDIVIDUAL PEOPLE COOPERATE WITH EACH OTHER?
WE'LL GET INTO THAT SOON ENOUGH.
BUT FOR NOW, WE'RE GOING TO TELL YOU A LITTLE MORE ABOUT OURSELVES, AND HOW WE FELL IN LOVE WITH NEUROSCIENCE.
AND EACH OTHER.

OCCASIONALLY I WONDER – HAVE I WASTED MY LIFE?
EVEN SUCH A LITTLE THING AS A WORKING DEFINITION OF CONSCIOUSNESS IS EVER OUT OF MY GRASP!
HUMANITY NEEDS PEOPLE LIKE US! IF WE DIDN'T LOCK OURSELVES INTO PSYCHOLOGY LABS ALL DAY, WE'D NEVER SOLVE PROBLEMS SUCH AS ELIMINATING PREJUDICE!
BUT DID WE CHOOSE THIS LIFE, OR DID THIS LIFE CHOOSE US?

Chapter 2

DON'T YOU REMEMBER THAT DAY, ALL THOSE DECADES AGO, WHEN A STRANGE CREATURE FLEW THROUGH OUR WINDOW...
A FLYING BRAIN! IT'S AN OMEN. WE SHALL BECOME... NEUROSCIENTISTS.

The Friths: who they are, and how they came to be.*

*WITH APOLOGIES TO BILL FINGER & BOB KANE

AS IT HAPPENS, THE QUESTION OF DESTINY IS PART OF BRAIN SCIENCE.
ALTHOUGH WE NOW KNOW THAT OUR BRAINS GRADUALLY BUILD UP USEFUL BAYESIAN PRIORS, BASED ON LIFE EXPERIENCES, WE DON'T KNOW HOW IT STARTS OFF. ONE OF THE GREAT PURSUITS OF CONTEMPORARY NEUROSCIENCE IS TO UNCOVER SO-CALLED INNATE PRIORS.

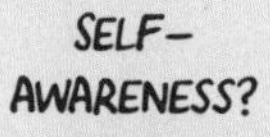

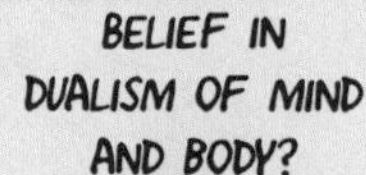

IN OTHER WORDS, WHICH BASIC FACTS OF THE WORLD ARE HARDWIRED INTO OUR BRAINS FROM THE START?
LANGUAGE SKILLS?
SELF-AWARENESS?
COLOR?
MOTOR CONTROL OF MUSCLES?
BELIEF IN DUALISM OF MIND AND BODY?

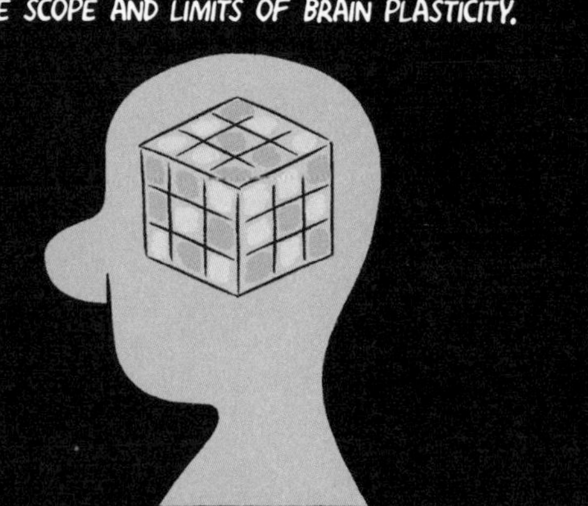
IT'S QUITE A PUZZLE, NOT LEAST BECAUSE THE BRAIN HAS EXTREME PLASTICITY. IN FACT, ONE OF THE GREAT PROJECTS OF NEUROSCIENCE IS UNCOVERING THE SCOPE AND LIMITS OF BRAIN PLASTICITY.

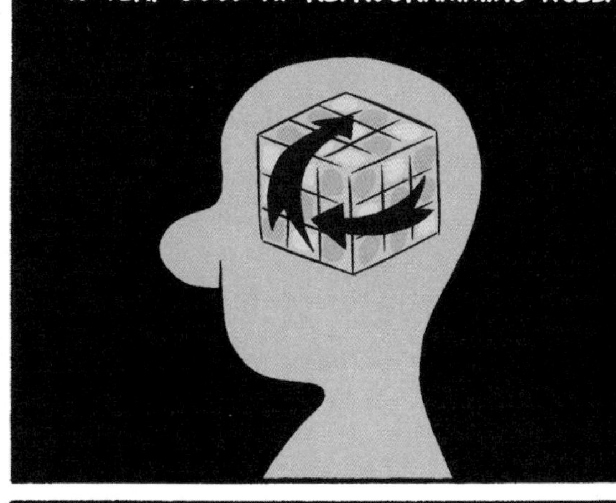
WHAT "PLASTICITY" MEANS IS THAT THE BRAIN IS VERY GOOD AT REPROGRAMMING ITSELF.

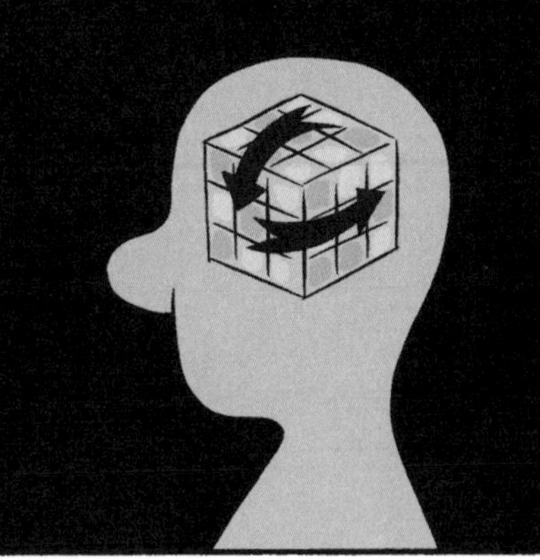
HOWEVER DEEPLY HELD ANY GIVEN PRIOR IS, WITH ENOUGH REAL-WORLD EXPERIENCE, THE BRAIN CAN OVERWRITE THOSE PREEXISTING PRIORS.

ONE WAY TO FIND OUT HOW BRAINS CAN CHANGE IS TO ASK THEIR OWNERS SIMPLE THINGS, SUCH AS HOW THEIR TASTES HAVE ADJUSTED OVER TIME:
SHE'S ALWAYS AFTER MORE ROCK 'N' ROLL RECORDS...
UTA ON A SCOOTER WITH HER PAPA IN 1955.

BUT NOW I CAN UNDERSTAND THE LYRICS, AND I'M NOT SO IMPRESSED ANYMORE. WHEN I WAS INVITED ONTO DESERT ISLAND DISCS,* I MADE NO ROOM FOR ROCK 'N' ROLL.

*A LONG-RUNNING RADIO SHOW IN THE UK, IN WHICH GUESTS ARE ASKED TO SHARE THEIR FAVORITE PIECES OF MUSIC WHILE TELLING THEIR LIFE STORIES.

HANDEL: ORGAN CONCERTO OPUS 7 NO. 1 IN B-FLAT MAJOR
ENGELBERT HUMPERDINCK: "EIN MANNLEIN STEHT IM WALDE" (FROM HANSEL AND GRETEL)
MOZART: SONATA FOR 4 HANDS IN B-FLAT MAJOR
SCHUBERT: STRING QUARTET NO. 13 IN A MINOR
KURT WEILL: "THE CANNON SONG" (FROM THE THREEPENNY OPERA)
JONATHAN HARVEY: MORTUOS PLANGO, VIVOS VOCO
FRED FRITH: "SOME CLOUDS DON'T"
STRAUSS: DER ROSENKAVALIER

EVERYONE IS USED TO THEIR TASTE CHANGING – CAN THE SAME BE TRUE OF ALL THE OTHER THINGS YOUR BRAIN CONTROLS?

IN FACT, THE BRAIN ISN'T INFINITELY PLASTIC.

IF YOU LISTENED TO THAT DESERT ISLAND DISCS EPISODE, YOU'LL HAVE NOTICED THAT I HAVE AN UNMISTAKABLY GERMAN ACCENT.

DESPITE HAVING LIVED AND WORKED IN LONDON SINCE MY EARLY 20S.

ON THE ONE HAND, IT FRUSTRATES ME, BECAUSE I WANT TO FIT IN.

WHERE ARE YOU FROM?*

*A QUESTION I AM INVARIABLY ASKED ON MEETING PEOPLE (WHEN IN ENGLAND, ANYWAY).

ON THE OTHER HAND, IT HAS SURELY GIVEN ME A SMALL AMOUNT OF MEDIA CACHET.

DER LANGUICH OF ZIENCE IS ENGLISH MIT DER ACZENT.

INCIDENTALLY, THESE THEMES OF "FITTING IN" AND PERSONAL REPUTATION ARE GOING TO PLAY A BIG PART AT THE END OF THE BOOK.

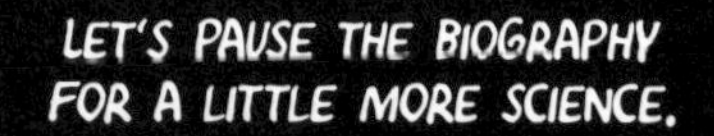
LET'S PAUSE THE BIOGRAPHY FOR A LITTLE MORE SCIENCE.

HOW CAN WE PROVE, PHYSICALLY, THAT THE BRAIN CAN AND DOES CHANGE WITH ITS OWNER'S SITUATION?

HERE'S ELEANOR MAGUIRE, ONE OF CHRIS'S FORMER POSTDOCS,* TO GIVE A CLEAR-CUT EXAMPLE.
*POSTDOC: SOMEONE WHO HAS COMPLETED A PHD AND, TYPICALLY, JOINS A RESEARCH TEAM BEFORE STRIKING OUT ON THEIR OWN.

THE SECRET IS TO USE BRAIN SCANNING MACHINES. ONE THING THEY'RE VERY GOOD AT IS MEASURING DIFFERENCES BETWEEN BRAINS.

I ARRANGED FOR A NUMBER OF LONDON TAXI DRIVERS TO COME IN AND BE SCANNED, ALONGSIDE LOTS OF NON-TAXI DRIVERS.

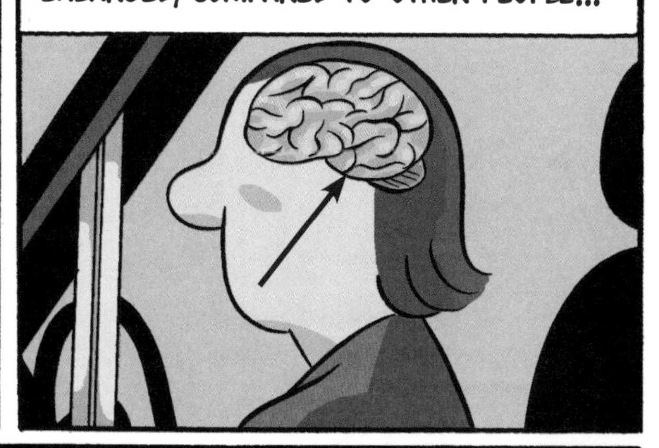
WE DISCOVERED THAT IN MOST TAXI DRIVERS ONE PART OF THE HIPPOCAMPUS WAS ENLARGED, COMPARED TO OTHER PEOPLE...

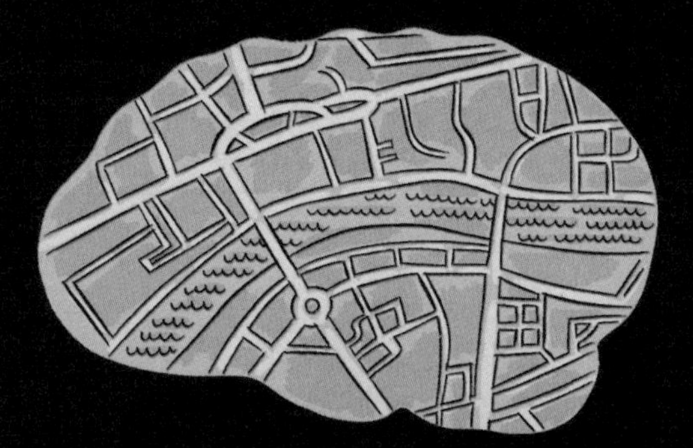
...TO MAKE SPACE FOR THE SPECIALIZED WORKING MEMORY OF LONDON'S JUMBLED STREETS (CALLED "THE KNOWLEDGE").
(YOU HAVE TWO HIPPOCAMPUSES, NEAR THE BASE OF THE BRAIN. THEY'RE INVOLVED WITH STORING MEMORY.)
WE ALSO FOUND THAT ANOTHER PART OF THE HIPPOCAMPUS WAS REDUCED IN SIZE. MOST INTERESTING OF ALL, AFTER A DRIVER RETIRED, BOTH REGIONS "SNAPPED BACK" TO TYPICAL SIZES.

THIS VITAL AREA OF RESEARCH WON THE TEAM THE IG NOBEL PRIZE FOR MEDICINE IN 2003.
A SET OF PARODY PRIZES AWARDED TO UNUSUAL RESEARCH TOPICS THAT MAKE PEOPLE LAUGH.*

I'M STILL NOT SURE WHETHER THE IG NOBEL PRIZE HELPED OUR REPUTATION OR NOT, BUT IT DOES CALL FOR UNPICKING PART OF THE DARK SIDE OF MODERN SCIENCE...
...THE POLITICS OF GETTING YOUR NAME ATTACHED TO A PAPER (PRIZE-WINNING OR NOT).

IN THE ~~GOOD~~ BAD OLD DAYS, WHEN WE STARTED OUT, OFTEN A SINGLE PERSON WOULD HAVE THEIR NAME ON A PAPER. IT MIGHT BE THE PERSON WHO DID THE WORK – OR, SIMPLY, THE TEAM LEADER.
BY THE TIME WE WERE SENIOR ENOUGH TO HAVE OUR OWN TEAMS OF RESEARCHERS, CREDITING HAD BECOME A LITTLE FAIRER. SOME MIGHT SAY TOO FAIR SINCE NAMES ON PAPERS STARTED TO LOOK MORE LIKE THE END CREDITS OF A FILM.

HERE'S THE FULL LIST OF CREDITS, IN ORDER, ON THE FIRST TAXI DRIVER PAPER:
AND INDEED, ALL THESE PEOPLE WERE INVOLVED. HERE'S THE BREAKDOWN OF WHO DID WHAT:
Eleanor Maguire
David Gadian
Ingrid Johnsrude
Catriona Good
John Ashburner
Richard Frackowiak
Chris Frith
HER IDEA; SHE RECRUITED AND SCANNED PEOPLE, ANALYZED THE DATA, AND WROTE THE PAPER
PREPARED THE SCAN RESULTS TO MAKE THE DATA USABLE
HELPED DEVISE THE CONCEPT OF THE EXPERIMENT, AND HELPED GET THE PAPER PUBLISHED
JOINTLY DEVELOPED THE SYSTEM USED TO ANALYZE THE DATA
SET UP (AND SECURED CONTINUED FUNDING FOR) THE LAB AND ITS EQUIPMENT
HEAD OF THE RESEARCH TEAM, HELPED DEVELOP THE IDEA AND PROVIDED ENCOURAGEMENT ALONG THE WAY. (LAST BILLING IS, CURRENTLY, THE MOST RESPECTED POSITION.)

AS WITH ALL JOBS IN LIFE, IT'S ALL ABOUT A CERTAIN NUMBER OF PEOPLE COOPERATING.
THIS IS GOING TO BE A MAJOR THEME, TOO.

*BUT IT HAS TO BE A GENUINE AND INDEED USEFUL PIECE OF SCIENTIFIC RESEARCH, TOO.

IF THE PAPER IS REFERRED TO BY OTHERS, THE STANDARD VERSION IS: "MAGUIRE ET AL 2001." *
WE'RE GOING TO BREAK THAT RULE A LITTLE FOR THIS COMIC, SINGLING OUT INDIVIDUALS WHO MAY NOT ALWAYS BE THE LEAD AUTHOR ON THE OFFICIAL PAPER.
AND YES, QUITE OFTEN THOSE INDIVIDUALS WILL BE OUR COLLEAGUES AND FRIENDS – BUT ALWAYS PEOPLE HEAVILY INVOLVED IN BRINGING A NEW PIECE OF RESEARCH TO LIFE.

*EARNING AN "ET AL" AFTER YOUR NAME IS A SURE SIGN YOU'RE IN THE BIG LEAGUES.
FULL CREDITS TO ALL PAPERS IN THIS BOOK ARE GIVEN AS ENDNOTES AT THE BACK OF THE BOOK.

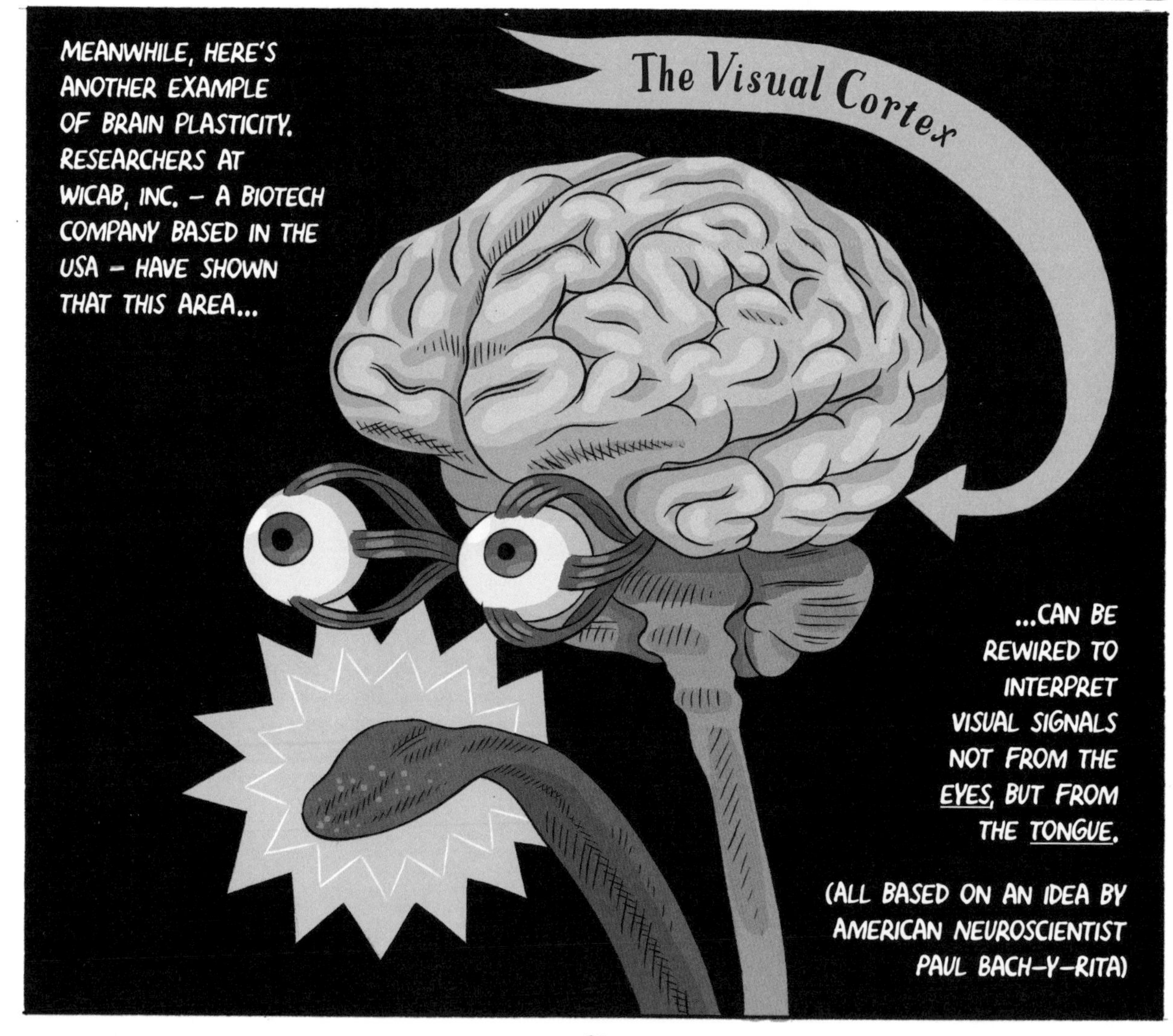
MEANWHILE, HERE'S ANOTHER EXAMPLE OF BRAIN PLASTICITY. RESEARCHERS AT WICAB, INC. – A BIOTECH COMPANY BASED IN THE USA – HAVE SHOWN THAT THIS AREA...
The Visual Cortex
...CAN BE REWIRED TO INTERPRET VISUAL SIGNALS NOT FROM THE EYES, BUT FROM THE TONGUE.
(ALL BASED ON AN IDEA BY AMERICAN NEUROSCIENTIST PAUL BACH-Y-RITA)

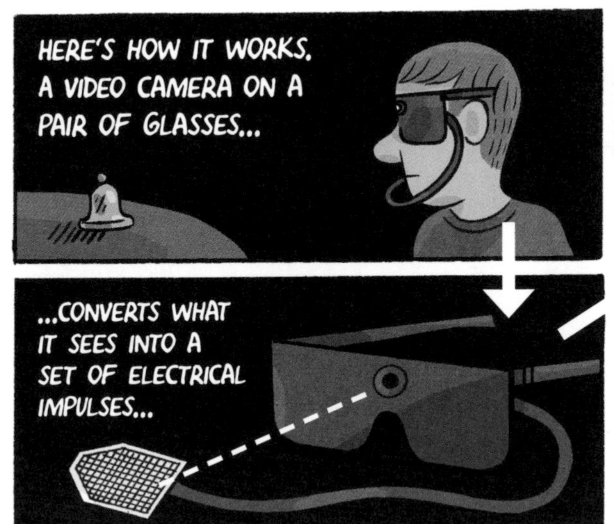
HERE'S HOW IT WORKS. A VIDEO CAMERA ON A PAIR OF GLASSES...
...CONVERTS WHAT IT SEES INTO A SET OF ELECTRICAL IMPULSES...

...AND SENDS THOSE IMPULSES ALONG A WIRE DOWN INTO THE MOUTH...

...WHERE A STRIP OF SMALL BUMPS SITS ON TOP OF THE TONGUE.

THE ELECTRICAL IMPULSES ACTIVATE A PATTERN OF BUMPS, MATCHING WHAT THE CAMERA SEES.

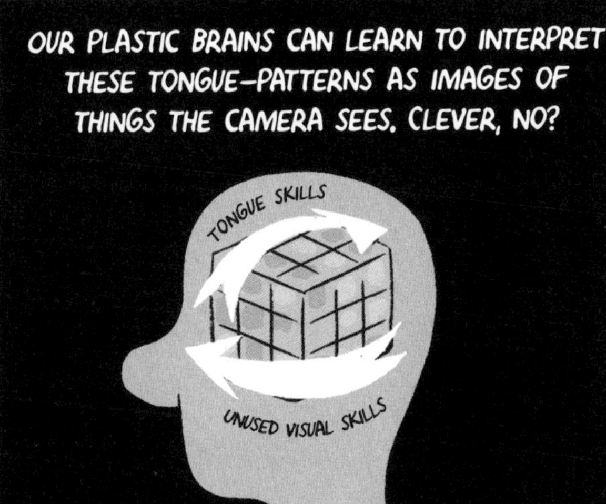
OUR PLASTIC BRAINS CAN LEARN TO INTERPRET THESE TONGUE-PATTERNS AS IMAGES OF THINGS THE CAMERA SEES. CLEVER, NO?
TONGUE SKILLS
UNUSED VISUAL SKILLS

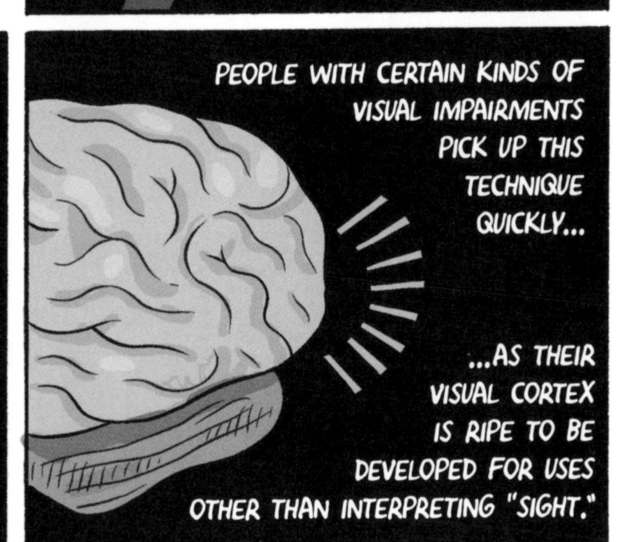
PEOPLE WITH CERTAIN KINDS OF VISUAL IMPAIRMENTS PICK UP THIS TECHNIQUE QUICKLY...
...AS THEIR VISUAL CORTEX IS RIPE TO BE DEVELOPED FOR USES OTHER THAN INTERPRETING "SIGHT."

QUICK REMINDER FROM CHAPTER 1: BRAINS RELY ON CHEMICAL TRANSMITTERS THAT STIMULATE ELECTRICAL CONNECTIONS BETWEEN NEURONS. THIS CREATES PATHWAYS ACROSS THE BRAIN THAT ULTIMATELY RESULTS IN THE EXPERIENCE OF HAVING A MIND.

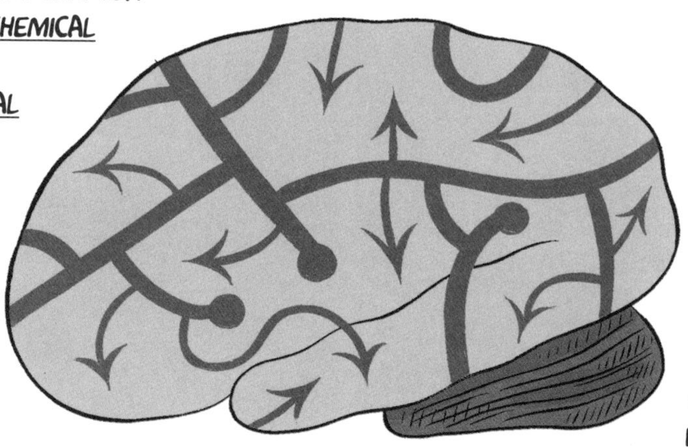

IN FACT, ONE OF THE MOST CONSISTENT WAYS IN WHICH RESEARCHERS HAVE LEARNED HOW THE BRAIN WORKS IS BY INVESTIGATING CASES WHEN IT DOESN'T. WITH SO MUCH COMPLEXITY, IT'S NO SURPRISE TO LEARN THAT THINGS CAN GO WRONG.

I HAD MY FIRST TASTE OF THIS DURING A SERIES OF LECTURES I ATTENDED AS AN UNDERGRADUATE IN 1963 AT THE UNIVERSITÄT DES SAARLANDES (IN SOUTHWEST GERMANY).

HOORAY! IT'S FRIDAY. TIME FOR MY FAVORITE LECTURE OF THE WEEK.

THE PROFESSOR OF PSYCHIATRY* WOULD PRESENT PATIENTS FROM HIS OWN CLINIC TO A PACKED LECTURE HALL OF MEDICAL STUDENTS.

*PSYCHIATRY: MEDICAL STUDY OF MENTAL DISORDERS.
PSYCHOLOGY: NONMEDICAL STUDY OF HUMAN BEHAVIOR.

HE ASKED HIS PATIENTS QUESTIONS, AND ENCOURAGED US TO ASK QUESTIONS, TOO.

HERE WE HAVE A DEPRESSIVE, A SCHIZOPHRENIC, AND AN OBSESSIVE...*

I HAVE NO MOTIVATION TO DO ANYTHING.

A VOICE TELLS ME I AM A BAD PERSON.

I KEEP WASHING MY HANDS BUT THEY NEVER FEEL CLEAN.

*I'M AFRAID THAT IN THOSE DAYS, PEOPLE WERE ROUTINELY LABELED BY THEIR DIAGNOSIS.

HOW OFTEN DO YOU WASH YOUR HANDS?
48 TIMES THIS MORNING.
WHAT ABOUT YOUR ARMS?
NO, I DON'T CARE ABOUT MY ARMS.
WHAT?! THIS CLEARLY ISN'T ABOUT KEEPING CLEAN.

THE STORY TOLD BY THE WOMAN WHO OBSESSIVELY WASHED HER HANDS REALLY STRUCK ME.

SHE SEEMED SO NORMAL ON THE OUTSIDE: NICE, INTELLIGENT, WELL DRESSED.

YET SHE WAS CLEARLY A PRISONER IN HER OWN MIND – A SLAVE TO HER THOUGHTS, NOT IN CHARGE OF THEM.

IT BECAME CLEAR TO ME THAT I WAS MORE ENGAGED BY PSYCHIATRY THAN I WAS BY MY PRIMARY STUDIES.*

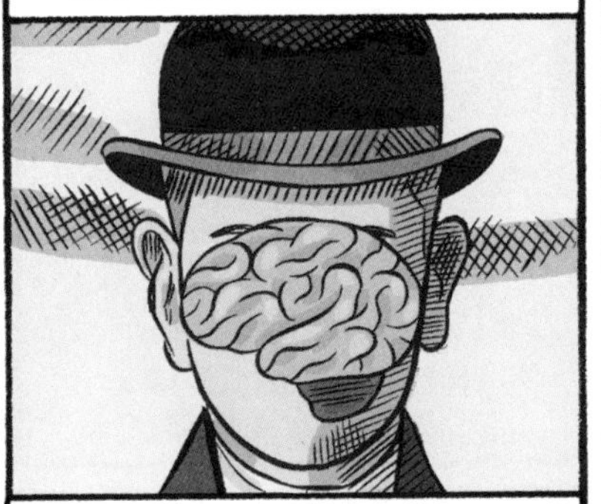

*ART HISTORY AND OLD BULGARIAN. NO, REALLY – ASK ME ABOUT ST. CYRIL ONE DAY.

I DECIDED TO USE MY COMING SUMMER TO "DO A PRAKTIKUM" – THE TERM WE USED AT THE TIME FOR INTERNSHIPS – AT THE INSTITUTE OF PSYCHIATRY IN LONDON.
LITTLE KNOWING WHAT WAS WAITING FOR ME THERE...

AROUND THE SAME TIME, I WAS IN MY FINAL YEAR AT CAMBRIDGE, STUDYING PSYCHOLOGY.
PROFESSOR OLIVER ZANGWILL (A CURIOUS CHAP, WHO NEVER LOOKED ANYONE IN THE EYE)...

...DESCRIBED A FAMOUS CASE ABOUT A WOMAN WITH AMNESIA RECORDED BY SWISS DOCTOR ÉDOUARD CLAPARÈDE IN 1911.

IT'S A STORY THAT GRABBED MY ATTENTION, AND MADE ME QUESTION HOW RELIABLE MY MIND AND BRAIN WERE.

CLAPARÈDE MET THE AMNESIAC PATIENT EVERY MORNING OVER SEVERAL DAYS. SHE NEVER RECOGNIZED HIM.

ONE MORNING, HE HID A PIN IN HIS HAND BEFORE SHAKING HER HAND IN GREETING.

THE NEXT DAY, THE LADY ONCE AGAIN CLEARLY DID NOT RECOGNIZE THE DOCTOR...

...BUT SHE REFUSED TO SHAKE HIS HAND.

SOMETIMES PEOPLE HIDE PINS IN THEIR HANDS.

TWO THINGS STRUCK ME: 1) A PERSON'S MIND LIKES TO THINK IT IS IN CONTROL OF THE BRAIN'S PROCESSES – FOR EXAMPLE, CHOOSING TO SHAKE HANDS; 2) BUT SOMETIMES THE MIND ISN'T IN CONTROL. I BADLY WANTED TO KNOW HOW ON EARTH THIS WAS POSSIBLE. A CAREER IN PSYCHOLOGY BECKONED...

...AND IN 1963, I WAS DEEP INTO A GRADUATE COURSE IN CLINICAL PSYCHOLOGY,* BASED AT THE LONDON INSTITUTE OF PSYCHIATRY. IN THEORY, I WAS LEARNING HOW TO DIAGNOSE AND EVEN TREAT PEOPLE WITH VARIOUS BRAIN/MIND-BASED DISORDERS, AT CANE HILL HOSPITAL IN COULSDON, LONDON, WHERE I MET MANY PEOPLE DIAGNOSED WITH SCHIZOPHRENIA.

HERE ARE A FEW THINGS THEY REPORTED.

*CLINICAL PSYCHOLOGY = PSYCHOLOGY IN SERVICE OF DIAGNOSING AND TREATING PEOPLE.

THE FORCE MOVED MY LIPS. I BEGAN TO SPEAK. THE WORDS WERE MADE FOR ME.

IT'S JUST AS IF I WERE BEING STEERED AROUND, BY WHOM OR WHAT I DON'T KNOW.

MY GRANDFATHER HYPNOTIZED ME AND NOW HE MOVES MY FOOT UP AND DOWN.

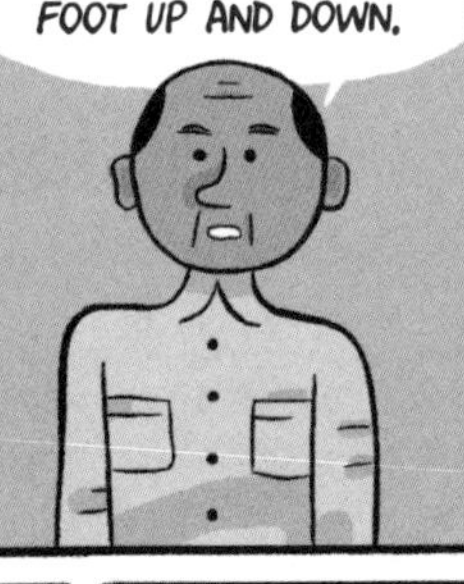

HER FACE CHANGES INTO THAT OF A RABBIT WITH EARS AND WHISKERS WHILE SHE WATCHES IN THE MIRROR.

I LEARNED VERY QUICKLY THAT I HAD NO FUTURE IN CLINICAL WORK.

WITH ALL DUE RESPECT TO PSYCHOLOGISTS,

I NEVER DEVELOPED THE CONFIDENCE TO BE ABLE TO PERSUADE SOMEONE THAT...

A) THERE'S SOMETHING WRONG WITH YOU.
B) I CAN MAKE YOU BETTER!
PRETTY SURE THERE'S SOMETHING WRONG WITH YOU, BUSTER!

MOSTLY, I WAS FASCINATED BY THE DIVERSE EFFECTS THAT THE BRAIN COULD PRODUCE. I WANTED TO UNDERSTAND HALLUCINATIONS AND DELUSIONS.

AND I ENJOYED DEVISING AND RUNNING ELEGANT EXPERIMENTS. IN PSYCHOLOGY, WHEN THERE ARE SO MANY FACTORS TO TAKE INTO ACCOUNT, ACHIEVING ELEGANCE IS OFTEN DELICIOUSLY DIFFICULT.

I CAME TO A VERY SIMILAR CONCLUSION. I KNEW HELP WAS NEEDED. EXISTING TREATMENTS OFTEN DIDN'T WORK. RESEARCH WAS THE WAY FORWARD.

I HOPED THAT BY STUDYING PSYCHOLOGY I COULD ULTIMATELY FIND OUT MORE ABOUT MY OWN MIND, AND ABOUT MYSELF. ALTHOUGH I WAS TOLD AT THE TIME THAT IF I REALLY WANTED TO KNOW MYSELF I SHOULD WRITE NOVELS. AND OF COURSE THERE WAS – AND STILL IS – A STIGMA AGAINST PSYCHOLOGY AS A WASTE OF TIME...

Psychology is a non-science!
WHAT DOES PSYCHOLOGY TELL US THAT WE DON'T ALREADY KNOW?
YOU CAN NEVER BELIEVE WHAT PEOPLE SAY THEY ARE THINKING.
YOU CAN'T PREDICT WHAT PEOPLE ARE LIKE.

I CAN NOW CATEGORICALLY SAY THIS IS WRONG ON ALL COUNTS!

BUT NEITHER OF US WAS THINKING ABOUT THE BRAIN ITSELF AT THIS TIME.

I WAS SOMEWHAT DISTRACTED BY THE INTRIGUING YOUNG GERMAN WOMAN WHO APPEARED ONE SUMMER AT THE INSTITUTE OF PSYCHIATRY.

I FOUND AN EXCUSE TO TALK TO HER WHEN SHE WAS STRUGGLING WITH A CALCULATOR.
GUESS WHO STILL PROVIDES I.T. SUPPORT IN OUR HOUSE TODAY.

BY THIS TIME I HAD GIVEN IN TO AN OVERWHELMING INTEREST IN HUMAN BEHAVIOR...

...AND THE CONTINUED ROMANTIC ATTRACTIONS OF LONDON.
(ALTHOUGH I WAS DISAPPOINTED BY THE LACK OF PEA-SOUPER FOGS AS PROMISED BY TOO MANY NOVELS.)
EUSTON
OXFORD St.
MARBLE ARCH
BY THE END OF THE SUMMER, I MANAGED TO SNEAK ONTO THE SAME PSYCHOLOGY COURSE CHRIS HAD JUST COMPLETED. I NEVER WENT BACK TO THE UNIVERSITY OF THE SAARLAND AGAIN.

I WORKED ON MY PHD BY DAY, AND TAUGHT UTA ENGLISH BY NIGHT.

4 YEARS LATER, I COMPLETED MY PHD (A NORMAL AMOUNT OF TIME). UTA HAD ALREADY COMPLETED THE COURSE AND HER PHD, ALL IN FLAWLESS ENGLISH.
Dr
Dr

LUCKILY, WE CONTINUED TO SHARE A LOVE NOT ONLY FOR EACH OTHER...
...BUT ALSO FOR PSYCHOLOGY.

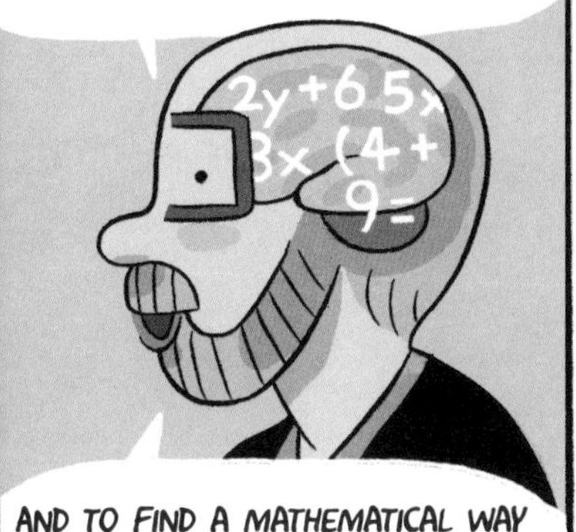
I HOPED ONE DAY TO EXPLAIN THE DISCONNECT BETWEEN A PERSON'S EXPERIENCES AND THEIR ACTIONS.
AND TO FIND A MATHEMATICAL WAY TO DESCRIBE HUMAN BEHAVIOR.

I HOPED TO EXPLAIN WHY PEOPLE, WHO ALL HAVE VERY SIMILAR BRAINS...
...HAVE SUCH DIFFERENT ABILITIES.

SADLY, THESE GOALS ELUDED US. BUT WE ENDED UP DISCOVERING SOMETHING ABOUT BRAINS THAT WE WEREN'T EVEN LOOKING FOR:
BRAINS ARE INCREDIBLY SOCIAL. THEY THRIVE ON AND ARE DESIGNED TO WORK, LEARN, AND PLAY ALONGSIDE OTHER BRAINS. THAT'S WHAT WE'LL EXPLORE IN THE REST OF THIS BOOK.

Chapter 3

How brains know what they know.
OUR FIRST CHILD, MARTIN, WAS BORN IN 1975, WHICH ALREADY FEELS TOO LONG AGO FOR ME TO REMEMBER.
LOOKING AT MY INFANT SON, IT STRUCK ME AS IMPOSSIBLE TO IMAGINE HAVING A NEWBORN BRAIN, WHICH HAS SO MUCH LEARNING AND DEVELOPING TO DO!
IN FACT, ONE REASON WHY WE CAN'T RECALL OUR EARLIEST YEARS IS THAT OUR BRAINS HAVE TO LEARN HOW LEARNING ITSELF WORKS.

THERE IS STATISTICAL LEARNING BY EXTRACTING PATTERNS.
BRICK. BRICK. BRICK.

THERE IS ASSOCIATION LEARNING FROM REWARD AND PUNISHMENT.
BON
YUM
DON'T STEAL SWEETS!

AND THERE IS A SHORTCUT – LEARNING BY COPYING!*

IN THE VERY EARLY STAGES, THE IDEA OF COPYING IS CURIOUS.

AFTER ALL, YOU HAVE TO RECOGNIZE OTHER PEOPLE BEFORE YOU CAN COPY THEM, SURELY?

*A SHORTCUT READILY AVAILABLE TO OUR SECOND SON, ALEX, BORN THREE YEARS AFTER MARTIN.

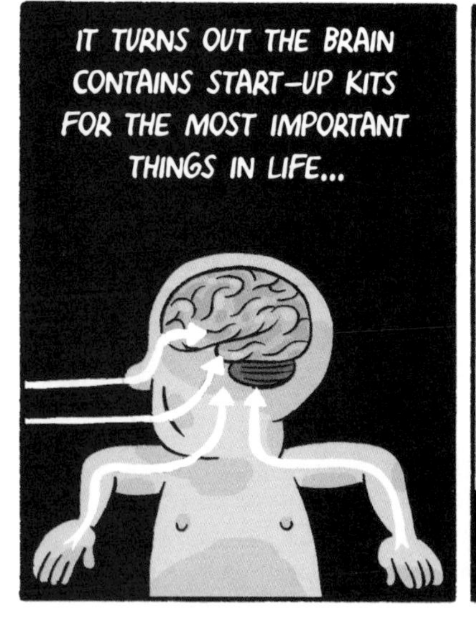
IT TURNS OUT THE BRAIN CONTAINS START-UP KITS FOR THE MOST IMPORTANT THINGS IN LIFE...

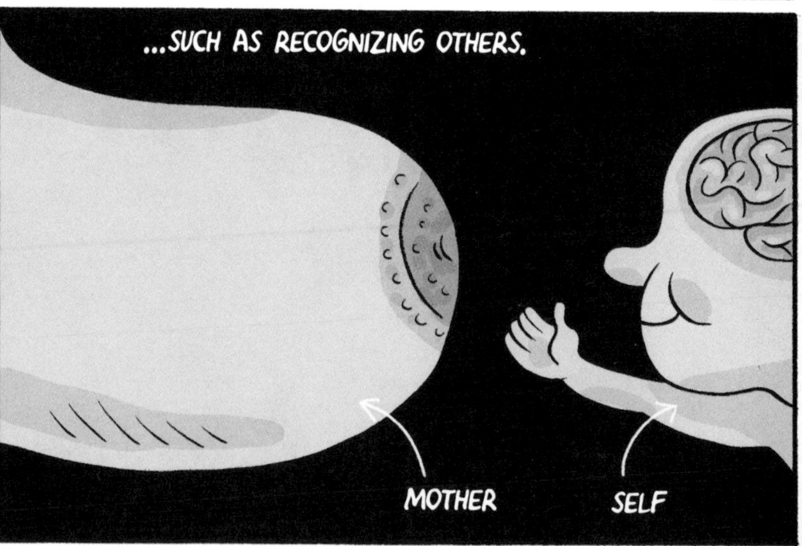
...SUCH AS RECOGNIZING OTHERS.
MOTHER
SELF

NEWBORN INFANTS SEE THINGS OUT OF FOCUS, AND NOT IN COLOR.

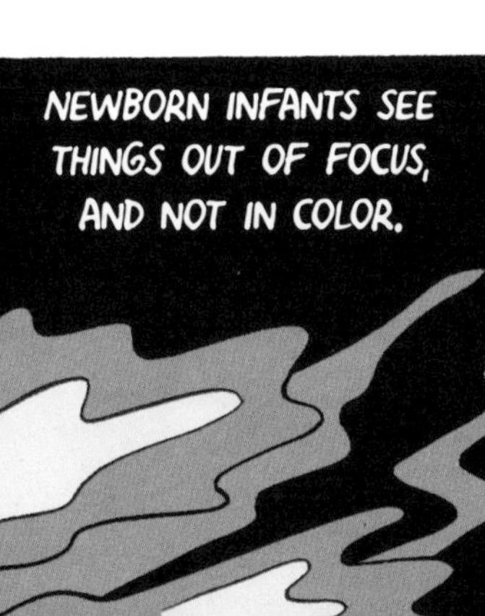

SEEING IN COLOR, AND IN FOCUS,* CAN TAKE SEVERAL MONTHS.

*SURPRISINGLY, SEEING THINGS BLURRED AT FIRST IS KNOWN TO HELP CHILDREN RECOGNIZE FACES MORE EASILY.

ALL THE THE SENSES NEED TO BE FINE-TUNED. THIS CAN TAKE YEARS.

BUT NEWBORNS CAN, IN FACT, RECOGNIZE HUMAN FACES AS SOON AS THEY ARE BORN.

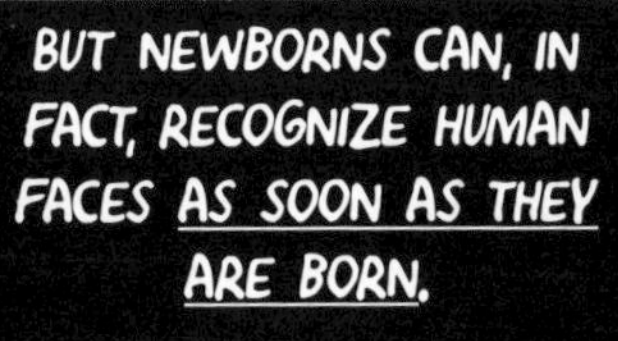

THE TRICK IS IN THE ARRANGEMENT OF DETAILS ON A FACE. WE KNOW THE NEWBORNS CAN RECOGNIZE THE IDEA OF A FACE BECAUSE THEY PREFER TO LOOK AT THESE

TWO STYLIZED IMAGES...

...RATHER THAN THESE.

THERE IS A REGION IN THE BRAIN THAT IS SPECIALIZED FOR THE PERCEPTION OF FACES.

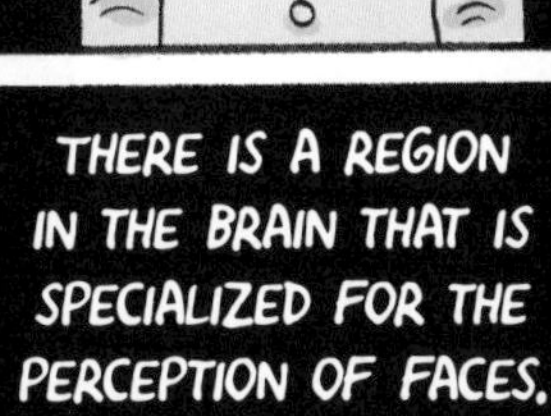

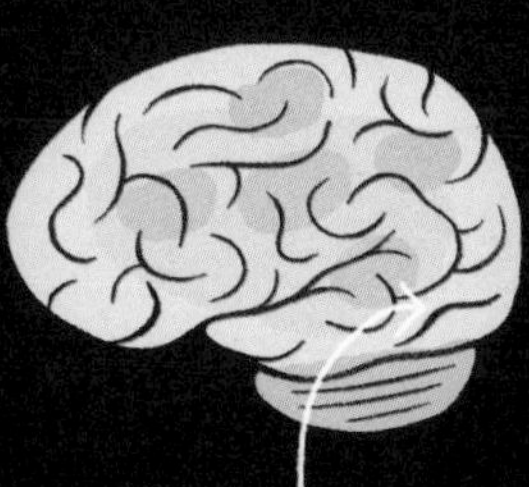
THE FUSIFORM FACE AREA

THIS ABILITY IS PRESENT WITHIN 9 MINUTES OF BIRTH.
FROM THAT POINT, WE'RE PREDISPOSED TO SEE FACES EVERYWHERE, FROM CLOUDS TO SHROUDS TO PIECES OF TOAST.

ONE UNSOLVED MYSTERY IS WHICH ACTIONS ARE DIRECTLY COPIED FROM OTHERS, AND WHICH ARE JUST THINGS THAT EVERYBODY DOES.

AS A CHILD, I LOVED TO EXPLORE THE WORLD BY MYSELF.
NOT REALIZING I WAS DOING WHAT ALL CHILDREN DO.

AS A CHILD, I LOVED TO EXPLORE THE WORLD BY MYSELF, NOT REALIZING I WAS DOING WHAT ALL CHILDREN DO.

SOME ACTIONS ARE DEFINITELY ABOUT ONE PERSON COPYING ANOTHER.
STOP COPYING ME!
STOP COPYING ME!

(UTA TUSSLING WITH HER LITTLE SISTER, ELKE)

WE COPY THOSE WE LIKE BECAUSE WE WANT TO BE LIKED. WE LEARN THAT THERE IS UNCONSCIOUS COPYING (GOOD) AND CONSCIOUS COPYING (BAD).
I'M AN INDIVIDUAL
I'M AN INDIVIDUAL
I'M AN INDIVIDUAL

YOU'VE GOT TO MAKE YOUR OWN MISTAKES.

YOU CAN'T LEARN ANYTHING PROPERLY WITHOUT DOING IT FOR YOURSELF.

WRONG! LEARNING FROM OTHERS IS FAST LEARNING – AND YOU CAN LEARN FROM THE MISTAKES OF OTHERS.

LEARNING THROUGH COPYING TURNS OUT TO BE MORE EFFICIENT FOR HUMAN BRAINS, ANIMAL BRAINS, AND EVEN MACHINE BRAINS.

THIS LAST CASE IS ESPECIALLY INTERESTING BECAUSE IT'S SOMETHING PEOPLE CAN TEST IN CONTROLLED CONDITIONS.
VR

RESEARCHERS WORKING TO DEVELOP ARTIFICIAL INTELLIGENCE (AI) DESIGN A VARIETY OF PROGRAMS TO HELP MACHINES LEARN.

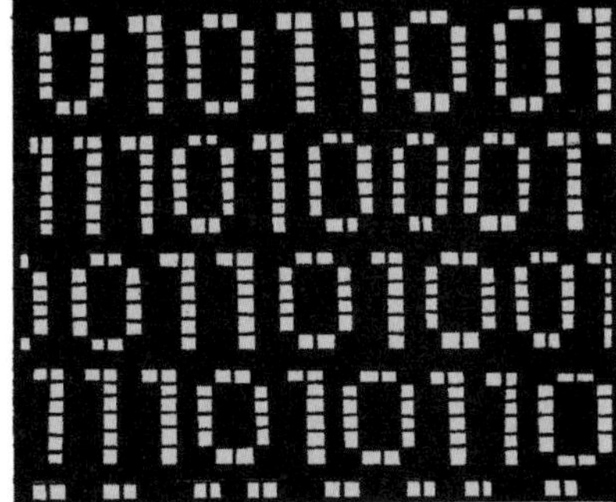

FOR EXAMPLE, COMPETING AI PROGRAMS CAN BE GIVEN THE TASK OF LEARNING HOW TO PLAY SIMPLE GAMES.
THE OBJECT OF THIS GAME IS TO MOVE THE PADDLE AT THE BOTTOM TO HIT THE BALL, SMASHING THE BRICKS.

PRESS A KEY

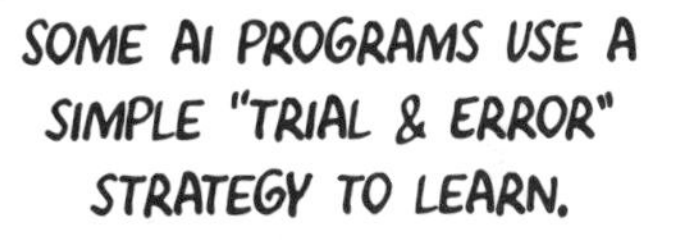
SOME AI PROGRAMS USE A SIMPLE "TRIAL & ERROR" STRATEGY TO LEARN.

THE AI BEGINS...

...BY TRYING OUT EVERY MOVE...

GAME OVER

...ONE AFTER ANOTHER...

...AFTER ANOTHER...
GAME OVER

...
GAME OVER

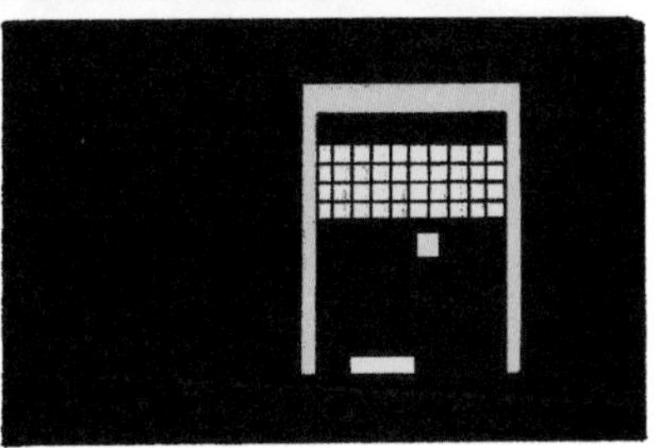

AS A PROPORTION OF POSSIBLE MOVES, THE VAST MAJORITY ARE DOOMED TO END THE GAME IMMEDIATELY.
THE AI DOESN'T MIND FAILING. IT IS HAPPY TO START AGAIN, ARMED WITH TINY INCREMENTS OF NEW INFORMATION.
EVENTUALLY, IT WILL LEARN TO PUT THE PADDLE WHERE THE BALL IS, SO THAT THE GAME CONTINUES.
BEAR IN MIND, THE AI DOESN'T KNOW THE POINT OF THE GAME. ITS ONLY GOAL IS TO KEEP PLAYING IT. ULTIMATELY, IT WILL LEARN THAT IT CAN PLAY FOR LONGER BY GETTING BETTER AT THE GAME.

IF THE AI WERE A HUMAN BRAIN...

...YOU COULD SAY THAT IT HAS NO INNATE PRIORS.

OTHER AIS ARE PROGRAMMED TO WATCH AND COPY PEOPLE...

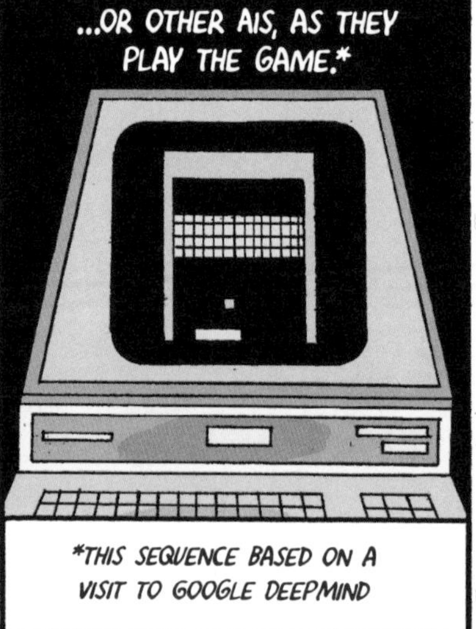
...OR OTHER AIS, AS THEY PLAY THE GAME.*
*THIS SEQUENCE BASED ON A VISIT TO GOOGLE DEEPMIND

ACROSS MANY EXPERIMENTS, REGARDLESS OF THE GAME BEING PLAYED, THE AIS THAT LEARN BY COPYING GET BETTER AT THE GAME MUCH, MUCH FASTER THAN THE TRIAL AND ERROR MODELS.

THE SAME TURNS OUT TO BE TRUE OF PEOPLE.

YOU MIGHT THINK THAT THE "REAL" WAY TO LEARN HOW TO GET GOOD AT DOING SOMETHING IS TO PRACTICE UNTIL YOU KNOW EVERY MOVE INSIDE OUT.

IN FACT, IT'S MUCH MORE EFFICIENT TO COPY OTHER PEOPLE. IDEALLY, LOTS OF OTHER PEOPLE.

WE ARE SPECIFICALLY TALKING ABOUT COPYING PEOPLE'S ACTIONS AND DECISION-MAKING.

THERE'S NO WAY AROUND THE NEED TO PRACTICE AN ACTIVITY TO BECOME ADEPT AT EXECUTING IT.

COPYING A PROFESSIONAL VIOLIN* PLAYER CAN TELL YOU WHAT YOU'RE SUPPOSED TO DO WITH THE INSTRUMENT...

...BUT UNTIL YOUR NEURAL PATHWAYS ARE FINELY TUNED, YOU LIKELY WON'T BE ABLE TO COPY THE PROFESSIONAL VERY WELL.

*THIS IS IN FACT A VIOLA, NOT A VIOLIN – THE INSTRUMENT CHRIS PLAYED IN AN ORCHESTRA FOR MANY YEARS.

COPYING MAY BE "CHEATING," BUT IT'S WHAT OUR BRAINS ARE WIRED UP TO DO.

AND IT'S NOT JUST PEOPLE. ANIMALS ALSO LEARN BY COPYING EACH OTHER.

FEMALE FRUIT FLIES HAVE THE POWER TO CHOOSE OR REJECT POTENTIAL SUITORS.

YOUNGER FEMALES WILL WATCH THE OLDER ONES AT WORK.
HOW CAN I KEEP THE WRONG MEN AWAY?
FOLLOW ME AND SEE HOW IT'S DONE!

SEE THAT GUY WITH THE BROKEN WING?

HE'S NO GOOD!

FEMALE FRUIT FLIES HEAD-BUTT AWAY UNWANTED SUITORS.

YOUNG FLIES NOTICE WHAT MAKES A SUITOR UNDESIRABLE...

...AND IN TURN USE THEIR OWN HEADS TO TURN THE SAME TYPES AWAY.

BUT WE DON'T COPY JUST ANYBODY. OUR BRAINS USE A TWO-STAGE PROCESS TO SET UP THE COPYING MECHANISM.

1. RECOGNIZE OTHER AGENTS – THAT'S TO SAY, PEOPLE WHO ARE NOT US.

2. IDENTIFY THE MOST RELIABLE AND COMPETENT OF THOSE "OTHER AGENTS"...
...TYPICALLY SOMEONE SIMILAR TO US.*

SO HOW DO WE RECOGNIZE OTHER SELVES? PART OF THE ANSWER (WHICH WE DON'T FULLY KNOW) IS ABOUT CONTROL AND OBSERVATION OF MOVEMENT.
1. WE CAN MAKE OUR LIMBS MOVE JUST BY THINKING IT. BUT WE CAN'T MAKE OTHER PEOPLE'S LIMBS MOVE BY THINKING ABOUT IT – WE HAVE TO ASK THEM!

2. THEN THERE'S THE WAY OUR BRAINS REACT TO OTHER THINGS MOVING.
WE CAN PERCEIVE AND IGNORE REPETITIVE, PREDICTABLE MOVEMENTS, SUCH AS THE HANDS OF A CLOCK.

WE ENJOY BUT CAN STILL IGNORE RANDOM YET INORGANIC MOTION.

BUT IF IT'S AN ANIMAL MOVING...

...WE PAY ATTENTION.

*LOTS MORE ON THE BRAIN'S BIASES TOWARDS PEOPLE WHO ARE LIKE US – AND AGAINST PEOPLE WHO AREN'T – LATER ON IN THE BOOK.

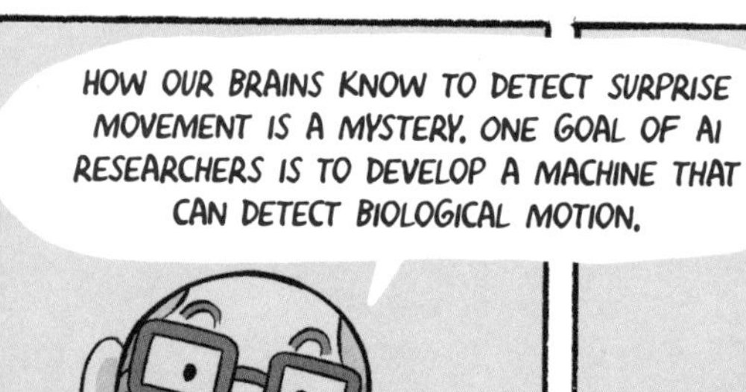
HOW OUR BRAINS KNOW TO DETECT SURPRISE MOVEMENT IS A MYSTERY. ONE GOAL OF AI RESEARCHERS IS TO DEVELOP A MACHINE THAT CAN DETECT BIOLOGICAL MOTION.

HEY, THIS IS MY BIT! DON'T INTERRUPT.

BRAINS DON'T JUST WATCH THINGS MOVING – THEY INSTINCTIVELY JUDGE HOW INTERESTING IT IS.

IMAGINE A CAT WANDERING ALONG A ROAD.

ALL OF A SUDDEN, IT STOPS.

THE CAT IS LOOKING AND LISTENING FOR SOMETHING. IT MAY WELL BE OF NO CONSEQUENCE FOR YOU, BUT YOUR BRAIN IS PRIMED TO NOTICE WHERE THE CAT IS LOOKING.

THE DRIVE IS EVEN MORE POWERFUL IF IT'S ANOTHER HUMAN YOU SEE LOOKING AT SOMETHING.

COMPARED TO MOST MAMMALS, WE HUMANS ARE, GENERALLY, INCREDIBLY RELIANT ON VISION.
WHENEVER YOU SEE ANOTHER PERSON'S FACE, YOU ARE INSTINCTIVELY DRAWN TO NOTICE WHERE THEIR EYES ARE LOOKING.

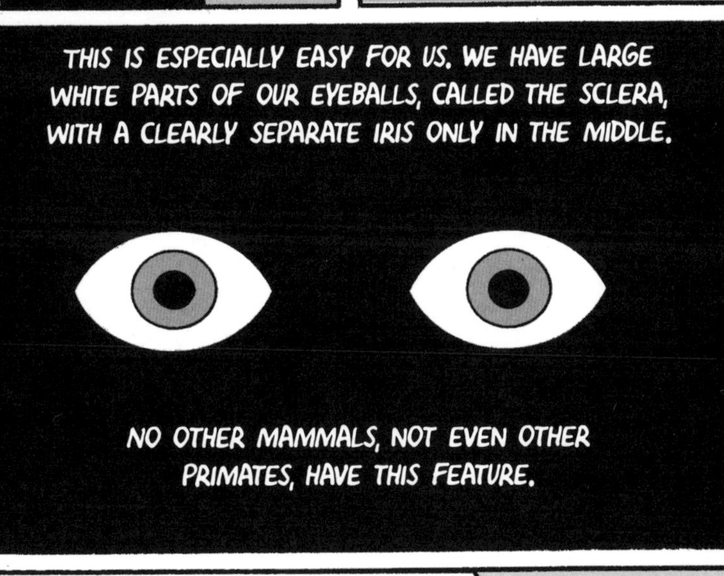
THIS IS ESPECIALLY EASY FOR US. WE HAVE LARGE WHITE PARTS OF OUR EYEBALLS, CALLED THE SCLERA, WITH A CLEARLY SEPARATE IRIS ONLY IN THE MIDDLE.
NO OTHER MAMMALS, NOT EVEN OTHER PRIMATES, HAVE THIS FEATURE.

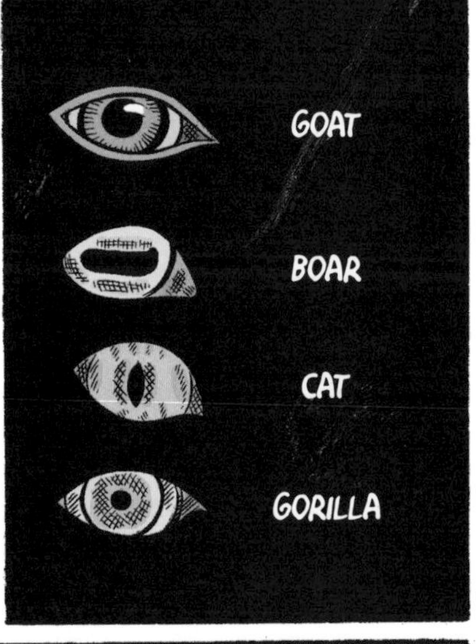
GOAT
BOAR
CAT
GORILLA

THERE'S AN OBVIOUS EVOLUTIONARY BENEFIT TO LARGE SCLERA – IF YOU THINK SOCIALLY. ONE PERSON NEED ONLY LOOK AT ANOTHER PERSON'S EYES TO SEE WHERE THEY ARE LOOKING. IN TURN, THIS MAY HELP THEM...
...BECOME AWARE OF SOME HIDDEN REWARD...
...OR AVOID A HIDDEN DANGER.

FOLLOWING SOMEONE'S EYE GAZE IS ONE OF THE MOST BASIC CLUES TO WHAT A PERSON MIGHT BE THINKING ABOUT.
I BET HE'S HUNGRY.

IT'S SO OBVIOUS, AND SOMETHING MOST PEOPLE TAKE FOR GRANTED SO MUCH, THAT IT'S EASY TO FORGET THIS IS ARGUABLY A FORM OF TELEPATHY.

IT'S ALSO SOMETHING HUMANS LEARN ABOUT INCREDIBLY YOUNG.
TERESA FARRONI

I RAN A STUDY THAT SHOWED THAT 4-MONTH-OLD INFANTS CAN CLEARLY FOLLOW ANOTHER PERSON'S EYE GAZE.

BY THE TIME A CHILD IS 9 MONTHS OLD, THEY CAN EVEN TELL WHETHER TWO PEOPLE ARE LOOKING AT EACH OTHER OR NOT.

AND THEY EXPECT YOU TO LOOK AT THEM WHEN YOU'RE TALKING, TOO.

IN ADULTS, THIS "TELEPATHY" GOES FAR BEYOND READING OTHER PEOPLE'S GAZES.

IT CAN EVEN BECOME A BURDEN, WHEN ONE PERSON EXPECTS ANOTHER TO BE ABLE TO READ THEIR MIND MORE LITERALLY.

OBVIOUSLY I COULD FILL MY OWN SAKE CUP, BUT A CUSTOM WE PICKED UP IN JAPAN* SAYS IT'S THE DONE THING TO FILL SOMEONE ELSE'S GLASS, BUT NEVER YOUR OWN.
*THE PROFESSORS' OLDEST SON HAS LIVED IN TOKYO SINCE 2004.

IN THIS SORT OF SITUATION, YOU COULD REASONABLY ASK WHETHER OUR BRAINS ARE SO USED TO EACH OTHER...

...THAT WE HAVE TRAINED THEM NOT TO THINK OF OURSELVES AS DISTINCT INDIVIDUALS.

BUT IT'S ALSO AN ILLUSTRATIVE EXAMPLE OF A SOCIAL INTERACTION...
(NOTICING THAT A COMPANION'S GLASS NEEDS REFILLING)

...THAT IS AT THE SAME TIME INTERPRETED ON THE BASIS OF A PHYSICAL INTERACTION.
(REFILLING SAID GLASS OR NOT, AS THE CASE MAY BE)

WHAT'S SO DIFFERENT ABOUT THE SOCIAL WORLD AND THE PHYSICAL WORLD ANYWAY?
IS THIS JUST ANOTHER DUALIST ERROR?

I WOULD ARGUE IT'S NOT AN ERROR.
WE WOULD ARGUE IT'S NOT AN ERROR.

BRAINS, AT LEAST, SEEM TO WORK IN THREE DIFFERENT REALMS:
PHYSICAL, SOCIAL, MENTAL.

IDEALLY, ALL THREE SHOULD WORK IN HARMONY WITHIN ANY ONE BRAIN.
BUT ALSO, AND THIS IS THE FUN BIT, IN HARMONY SHARED BETWEEN MULTIPLE BRAINS OWNED BY DIFFERENT PEOPLE.

PERHAPS THE BEST EXAMPLE TO DEMONSTRATE SUCH SOCIAL INTERACTION COMES FROM MUSIC.

PLAYING MUSIC WITH OTHER PEOPLE IS A SKILL THAT REQUIRES A LOT OF PRACTICE.
IT'S HARD ENOUGH WITH JUST TWO PEOPLE, BOTH OF WHOM CAN READ SHEET MUSIC.

OUR OWN DUET REPERTOIRE IS SOMEWHAT LIMITED TO CLASSICAL MUSIC – OR GERMAN CHRISTMAS CAROLS...

...WHICH OUR CHILDREN LOVED TO HEAR US PLAY.
OH DU FRÖH–LICH–E

STOP IT!

ONE OF THE GREAT MYSTERIES OF SOCIAL COGNITION COMES FROM THE WORLD OF JAZZ.

SPECIFICALLY, EXTENDED IMPROVISATIONS, SUCH AS THOSE FOUND IN MILES DAVIS'S KIND OF BLUE.*

*CHRIS'S FAVORITE ALBUM AS A STUDENT AT CAMBRIDGE, 1960–1963

IN THIS KIND OF MUSIC, A SOLOIST IMPROVISES ON A PREDETERMINED SET OF RHYTHMS AND CHORD CHANGES, WHILE THE OTHER PLAYERS IMPROVISE IN SUPPORT OF THE SOLOIST.

ALL OF THEM ARE MUTUALLY ADAPTING TO EACH OTHER.

THIS INVOLVES HIGHLY SKILLED SOCIAL INTERACTIONS, AND PRODUCES DIFFERENT RESULTS EVERY TIME.

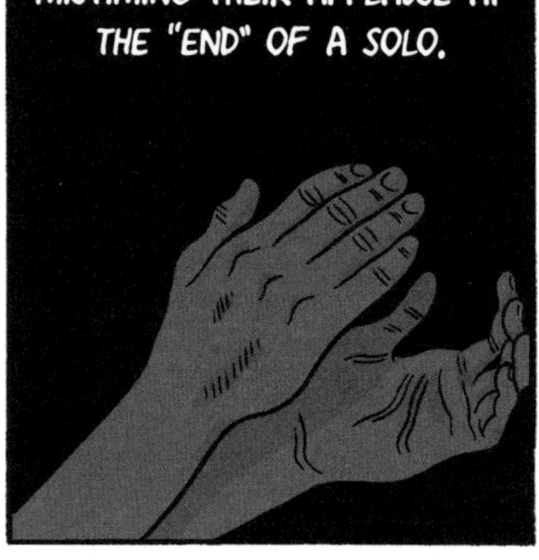
EVEN A SEASONED AUDIENCE CAN BE CAUGHT OUT MISTIMING THEIR APPLAUSE AT THE "END" OF A SOLO.

HARDER STILL IS "FREE IMPROVISATION," IN WHICH THERE ARE NO PREDETERMINED RHYTHMS OR CHORD CHANGES.

THIS IS MY BROTHER FRED FRITH. HE'S A PROFESSOR OF COMPOSITION, DON'T YOU KNOW.

NOW, IN FREE IMPROVISATION, ONE NEEDS VERY SOPHISTICATED SOCIAL INTERACTION SKILLS.
I KNOW ENOUGH TO TEACH IMPROV, BUT MY FRIEND GEORGE HAS GONE A STEP FURTHER...

THAT'S GEORGE LEWIS, PROFESSOR OF AMERICAN MUSIC AT COLUMBIA UNIVERSITY, NEW YORK.
HI.

GEORGE IS ALSO A COMPOSER AND PERFORMER. HE AND FRED HAVE IMPROVISED TOGETHER MANY TIMES.

ALONG WITH MUSIC, I BEGAN TO STUDY PHILOSOPHY, AT A TIME WHEN COMPUTERS WERE STARTING TO TAKE OFF.
I GOT INTERESTED ENOUGH TO BUILD AN ELECTRONIC MUSIC SYSTEM CALLED VOYAGER.

VOYAGER IS AN IMPROVISATION PROGRAM. THE SYSTEM LISTENS TO PLAYERS AS THEY PERFORM...

...AND CAN IMPROVISE WITH THEM, BY USING A SYNTHESIZER OR A PLAYER PIANO.
YOU CAN HEAR VOYAGER ON SOME OF MY ALBUMS.

WHAT VOYAGER CAN'T DO IS WORK OUT WHEN TO STOP. NEGOTIATING WHEN TO STOP IS THE MOST DIFFICULT TASK FACING A GROUP OF FREE IMPROVISERS.
When does the music stop??

THANKS, FRED AND GEORGE.
LEWIS'S WORK WITH VOYAGER SHOWS HOW COMPLEX MUSICAL IMPROVISATION CAN BE...
...BUT IF YOU THINK ABOUT IT, MOST SOCIAL INTERACTIONS INVOLVE A KIND OF IMPROVISATION, AND WE USUALLY FIND THIS EASY!
THE SEEMINGLY INNATE ABILITY TO GO ALONG WITH A GROUP IS THE CENTRAL CONCERN OF SOCIAL COGNITION!

LET'S GET AWAY FROM MUSIC. PLAYING AN INSTRUMENT, LET ALONE IMPROVISING ON ONE, IS TRICKY ENOUGH FOR MANY PEOPLE!

SO HERE'S AN EXAMPLE OF A SOCIAL ABILITY THAT ALMOST EVERYONE HAS: UNDERSTANDING LAUGHTER.

HEE HEE HEE!
HA HA HA!

CAN YOU TELL WE GET ALONG WITH EACH OTHER?
ACTUALLY, NOT BASED ON THIS PAGE, YOU CAN'T.

BECAUSE IT'S A CARTOON. (AND WE MAY BE LYING).*
*WE'RE NOT LYING.

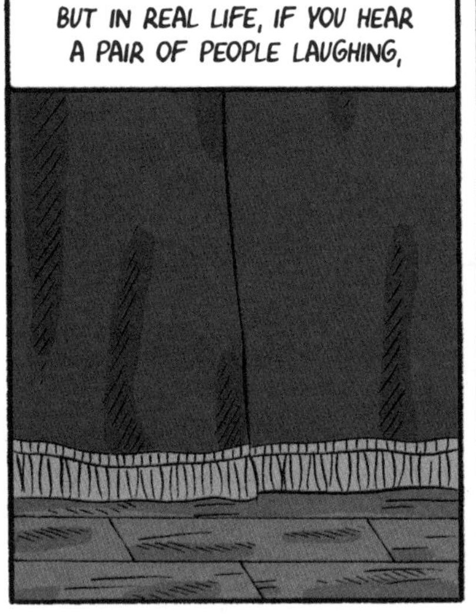
BUT IN REAL LIFE, IF YOU HEAR A PAIR OF PEOPLE LAUGHING,

HARD RESEARCH SAYS YOU CAN TELL IF THEY ARE FRIENDS OR STRANGERS.

THIS FINDING IS BASED ON TWO STUDIES BY A TEAM WORKING BETWEEN LOS ANGELES, USA, AND AARHUS, DENMARK.

FINDING 1:
PEOPLE CAN ACCURATELY ASSESS A RELATIONSHIP JUST BY HEARING A SNIPPET OF TWO PEOPLE LAUGHING TOGETHER.

(YOU DON'T EVEN HAVE TO SEE THE COUPLE IN QUESTION TO MAKE AN ACCURATE GUESS.)

FINDING 2:
THIS IS TRUE EVEN IF YOU'RE LISTENING TO PEOPLE FROM A DIFFERENT CULTURE THAN YOUR OWN, WHO SPEAK A DIFFERENT LANGUAGE.*

*FOR THE MOMENT, WE DON'T KNOW EXACTLY HOW OUR BRAINS CAN DO THIS TRICK, JUST THAT THEY CAN.

THE AUTHORS OF THE STUDY SUGGEST THAT LAUGHTER IS BOTH UNIVERSAL AND A GOOD MEASURE OF HOW COOPERATIVE ANOTHER PERSON – OR GROUP OF PEOPLE – IS GOING TO BE.

HANDY IF YOU'RE A DIPLOMAT, PERHAPS.
WHAT ARE YOU REALLY THINKING?

IT'S A SIMPLE FINDING, BUT BE WARNED! IT'S NOT WITHOUT SOME COMPLEXITY.

THE STATISTICS WITHIN THE SECOND STUDY SHOWED THAT PEOPLE ONLY GUESSED CORRECTLY 53–67% OF THE TIME.

ROUGHLY 2 OUT OF 3 TIMES, IF THAT'S AN EASIER WAY TO THINK ABOUT IT.

BETTER THAN A RANDOM GUESS (50%), BUT IT'S ALSO MORE EVIDENCE THAT NOT ALL HUMANS ARE EQUAL WHEN IT COMES TO SOCIAL COGNITION.

LAUGHTER IS LIKE A SOCIAL GLUE THAT SOME PEOPLE, FOR EXAMPLE PEOPLE WITH AUTISM, FIND DIFFICULT TO FATHOM.

IT'S NOT ABOUT LACKING THE IMPULSE TO LAUGH, OR THE DESIRE TO FIT IN WITH THE GROUP.
HA HA!

IT'S ABOUT NOT EASILY GRASPING WHEN LAUGHTER IS APPROPRIATE.

MORE ON AUTISM IN THE NEXT CHAPTER.

BUT FIRST, A NONSCIENTIFIC OBSERVATION ON LAUGHTER – A PHENOMENON THAT HAS NOT BEEN STUDIED ENOUGH.*

*DON'T WORRY, WE'RE GOING TO GIVE SOME SCIENTIFIC OBSERVATIONS ABOUT LAUGHTER IN CHAPTER 11.

IT SEEMS TO BE AN INSTINCTIVE ACTIVITY...
HA HA!
TICKLE TICKLE

...BUT IT'S TIED UP WITH BOTH COPYING AND LEARNING, TOO. FOR EXAMPLE, CHILDREN LEARN THE RHYTHMS OF JOKES, AND WHEN TO LAUGH...

...LONG BEFORE THEY UNDERSTAND WHY ANY PARTICULAR JOKE IS FUNNY.

LET ME DEMONSTRATE BY TELLING A JOKE.
A JUNIOR DOCTOR WALKS INTO A MENTAL HOSPITAL.*

*THE SETTING ALREADY REVEALS THIS IS AN OLD JOKE.

THE PATIENTS ARE SILENT. BUT EVERY NOW AND THEN...

...ONE OF THEM UTTERS A NUMBER, AND EVERYONE LAUGHS.
15
HA! GOOD ONE.

EVEN THE OTHER DOCTORS JOIN IN AND LAUGH.

WHAT'S GOING ON?
WE ONLY HAVE ONE JOKE BOOK HERE. WE'VE ALL MEMORIZED ALL THE JOKES, SO WE JUST REFER TO THEM BY NUMBER NOW.

HERE'S A FUNNY ONE: 37.

WHY ISN'T ANYONE LAUGHING?
IT'S THE WAY YOU TELL 'EM.
JOKES

HA HA HA!

PHEW, THAT'S AN AWFUL LOT TO LEARN ABOUT LEARNING.
BUT THE KEY LESSON IS A SOCIAL ONE. WE LEARN ABOUT THE WORLD BY WATCHING OTHER PEOPLE...

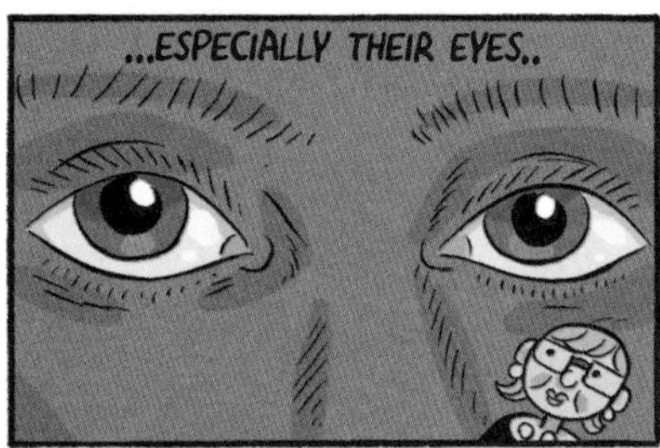
...ESPECIALLY THEIR EYES..

EYES ARE WINDOWS INTO EACH OTHER'S MINDS.

LOOKING AT A PERSON'S EYES CAN INSTANTLY TELL YOU SOME OF THEIR THOUGHTS AND EVEN INTENTIONS.
IS THIS A SKILL WE ARE BORN WITH? OR ONE WE LEARN? WE DON'T KNOW. IT MIGHT EVEN BE A SKILL WE THINK WE HAVE, BUT ACTUALLY DON'T, AS WE'LL SEE IN THE NEXT CHAPTER!

CLOSE OBSERVATION OF OTHERS SETS US ALL ON THE PATH TO COPYING, THE ULTIMATE SHORTCUT TO LEARNING NEW THINGS.

WE SEE WHAT THEY DO, AND WE TRY TO IMAGINE OURSELVES BEING THAT PERSON, DOING THAT THING.

ALTHOUGH SIMPLY COPYING SOMEONE ISN'T ENOUGH FOR MOST OF US TO SUCCEED!

WE HAVE TO PRACTICE, AND IN DOING SO, HELP WIRE UP OUR BRAINS TO BE ABLE TO DO WHATEVER IT IS.
WHETHER THAT'S MUSICAL IMPROV OR CLIMBING TREES.
SOME ABILITIES START SO EARLY THAT WE SUSPECT THE BRAIN COMES WITH BUILT-IN INSTRUCTIONS FOR UNDERSTANDING THEM...
...SUCH AS HOW TO RECOGNIZE A FACE...
...OR THE IMPORTANCE OF PAYING ATTENTION TO THE MOVEMENTS OF LIVING THINGS.
BRAINS PLACE GREAT VALUE ON CURIOSITY. OUR CHILDHOOD INSTINCT FOR CLIMBING AND EXPLORING FUNCTIONS AS A NEAT METAPHOR FOR OUR ADULT WORKING LIVES.
COPYING REQUIRES SOCIAL INTERACTION. HOW DOES THIS PART OF THE PROCESS WORK? AND WHY DO SOME PEOPLE LEARN HOW TO COPY THE "RIGHT" THINGS MORE EASILY THAN OTHERS? LET'S FIND OUT!

Chapter 4

Teaching is a tool, copying is an instinct.
WE KNOW THAT ONE EXCELLENT METHOD OF LEARNING IS BY COPYING.
LOOK, I'M COPYING.

WE ALSO LEARN THROUGH BEING TAUGHT.
WHICH IS SUBTLY DIFFERENT, BUT OFTEN VERY EFFECTIVE.

FIRST, YOU PULL THE LACES TIGHT.

THE DIFFERENCE MATTERS BECAUSE TEACHING APPEARS TO BE ESPECIALLY HUMAN.
DEFINING HOW HUMANS ARE DIFFERENT FROM OTHER ANIMALS IS ONE OF THOSE THINGS NO ONE HAS QUITE MANAGED TO PUT THEIR FINGER ON.
A BIT LIKE DEFINING CONSCIOUSNESS.

WE HUMANS LIKE TO BELIEVE THAT OUR BRAINS AND MENTAL ABILITIES ARE SPECIAL.

BUT THE TRUTH IS, ALMOST EVERYTHING HUMANS CAN DO IS DONE BY OTHER ANIMALS.

COMMENTS
CLICK-TIKKIKI TIK
OMG – DOLPHINS USE NAMES FOR THEMSELVES AND EACH OTHER (EVEN IF WE CAN'T PRONOUNCE THEM).

COMMENTS
MEGALOLS!
A NATURALIST LEFT HIS CAMERA IN THE JUNGLE; AN APE FOUND IT, AND USED IT TO TAKE THIS SELFIE.

COMMENTS
ARE BEES BETTER THAN PEOPLE?
THESE LOVABLE DRONES SACRIFICED THEIR LIVES FOR THE GOOD OF THE HIVE...

COMMENTS

WE'RE ALL GOING TO DIE ONE DAY.

WE KNOW IT, YOU KNOW IT, AND EVEN ELEPHANTS KNOW IT – FOOTAGE OF THEM RESPECTING THEIR DEAD.

COMMENTS
PLANET OF THE APES IS NON-FICTION!
FOOTAGE OF CHIMPS CHOOSING AND USING TOOLS SHOWS OUR DAYS IN CHARGE ARE NUMBERED...

AND HERE'S ONE YOU MIGHT NOT HAVE HEARD BEFORE – ANIMALS CAN LEARN NOT JUST THROUGH COPYING, BUT THROUGH TEACHING.

TAKE MEERKATS...

MEERKAT ADULTS TEACH THEIR PUPS HOW TO CATCH SCORPIONS WITHOUT GETTING STUNG.

MEERKATS HAVE EVOLVED TO PREPARE LESSONS IN ADVANCE, AND IN STAGES.

IF PUP CRIES LIKE BABY, GIVE DEAD SCORPION.
DEAD

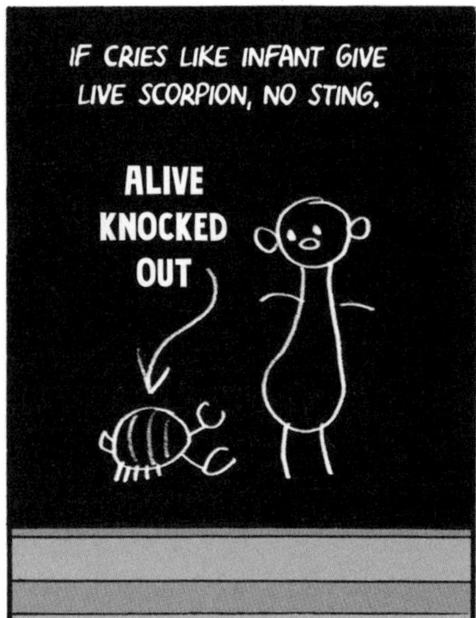
IF CRIES LIKE INFANT GIVE LIVE SCORPION, NO STING.
ALIVE KNOCKED OUT

IF CRIES LIKE SENIOR, LET THEM DO THEIR OWN HUNTING.

ONLY AFTER THIS EXPERIENCE CAN PUPS FORAGE FOR SCORPIONS BY THEMSELVES.

BUT THIS ISN'T LIKE HUMAN TEACHING. HUMANS LEARN HOW TO DO THINGS "THE RIGHT WAY," WHICH PSYCHOLOGISTS CALL OVERIMITATION.

HERE'S OUR COLLEAGUE ANTONIA HAMILTON SHOWING TWO CHILDREN (AROUND FOUR YEARS OLD) HOW TO OPEN A BOX.

WAVE

TAP TAP

TAP TAP

NOW YOU DO IT.

WAVE

TAP
TAP

TAP
TAP

THIS CHILD HAS COPIED THE SEQUENCE EXACTLY. IT'S WHAT THE MAJORITY OF CHILDREN WILL DO.

THE SAME EXPERIMENT HAS BEEN TRIED ON ADULTS, WHO TEND TO OVERIMITATE THE STEPS EVEN MORE STRONGLY THAN YOUNG CHILDREN. IT'S CALLED OVERIMITATION BECAUSE YOU DON'T NEED TO COPY ALL THE STEPS IN ORDER TO OPEN THE BOX.

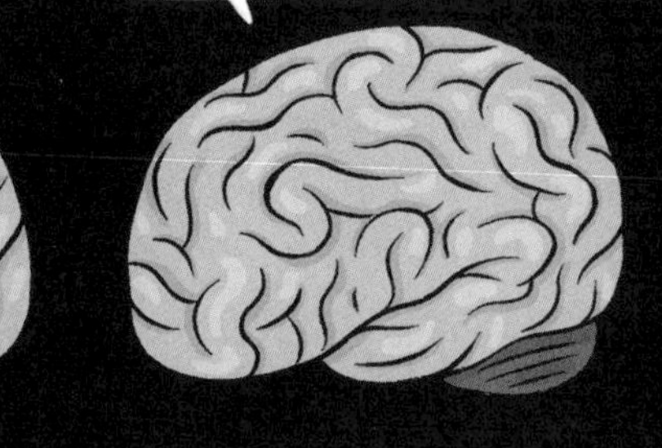
MUST COPY EXACTLY!
MUST COPY EXACTLY!

I DON'T SEE THE POINT OF ALL THAT TAPPING AND WAVING, BUT I'D BETTER DO IT!

OVERIMITATION IS SOCIAL. WE DO IT BECAUSE THIS IS THE WAY OUR GROUP DOES THINGS. AND WE WANT TO FIT IN WITH OUR GROUP.
DELIBERATELY NOT OVERIMITATING CAN BE A WAY TO MARK OURSELVES PART OF ONE GROUP RATHER THAN ANOTHER.
THIS MIGHT BE A HUMAN-ONLY THING. APES, IRONICALLY ENOUGH, DON'T OVERIMITATE.

PSYCHOLOGIST WILLIAM MCDOUGALL WROTE ABOUT HOW AND WHEN HUMANS CHOOSE TO IMITATE EACH OTHER:

MOST ENGLISHMEN WOULD SCORN TO KISS AND EMBRACE ONE ANOTHER FREELY, IF ONLY BECAUSE FRENCHMEN DO THESE THINGS.

THE AUTHORS OF THIS COMIC DO NOT CONDONE XENOPHOBIA, ALTHOUGH THEY DON'T MIND EXPLOITING IT FOR A GAG.
I ♥ PARIS

AND IT DOESN'T HAVE TO BE ABOUT "FOREIGNERS." MY GRANDMOTHERS (BOTH GERMAN) LOOKED DOWN ON EACH OTHER BECAUSE ONE USED LEMON, THE OTHER VINEGAR, IN SALAD DRESSINGS.
CHOOSING TO COPY OR NOT TO IS PART OF OUR SOCIAL TOOLKIT. BUT WHAT HAPPENS WHEN PEOPLE CAN'T USE THIS TOOL?

REMEMBER THE BOX OF SWEETS?

SOME CHILDREN SIMPLY CUT TO THE CHASE AND OPEN THE BOX WITHOUT FOLLOWING THE RITUAL ACTIONS.

THE COMMON FACTOR IN THOSE FEW CHILDREN? A CONDITION KNOWN AS AUTISM.

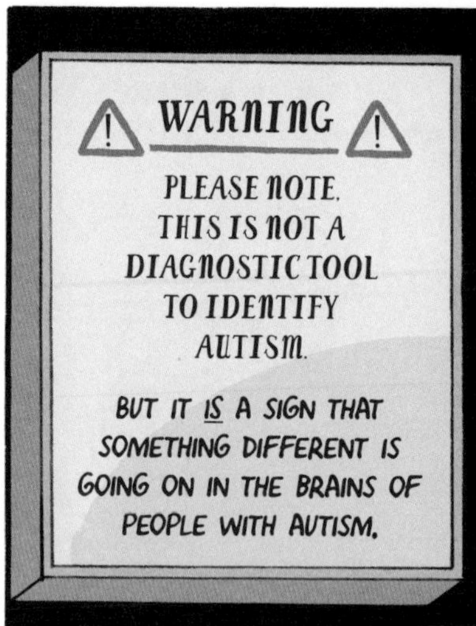
WARNING
PLEASE NOTE. THIS IS NOT A DIAGNOSTIC TOOL TO IDENTIFY AUTISM.
BUT IT IS A SIGN THAT SOMETHING DIFFERENT IS GOING ON IN THE BRAINS OF PEOPLE WITH AUTISM.

AUTISM IS A RELATIVELY RECENT DISCOVERY. ITS STORY BEGINS WITH PIONEERING VIENNESE PEDIATRICIAN HANS ASPERGER.

(THAT'S A HARD "G" AS IN HAMBURGER.)

HE WAS ONE OF THE FIRST PEOPLE TO RECOGNIZE AND DESCRIBE AUTISM. HE PUBLISHED TWO KEY PAPERS IN 1938 AND 1944...
...ONE OF THE DARKEST PERIODS IN GERMANY'S HISTORY.

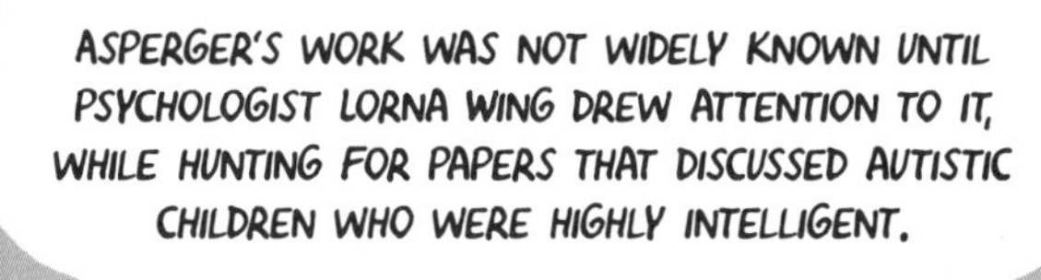
ASPERGER'S WORK WAS NOT WIDELY KNOWN UNTIL PSYCHOLOGIST LORNA WING DREW ATTENTION TO IT, WHILE HUNTING FOR PAPERS THAT DISCUSSED AUTISTIC CHILDREN WHO WERE HIGHLY INTELLIGENT.

(A FEW YEARS LATER, IN 1991, I TRANSLATED ASPERGER'S WORK INTO ENGLISH.)

IT HAS BEEN POINTED OUT THAT ASPERGER WAS NOT IMMUNE TO NAZI IDEOLOGY.* BUT WHERE NAZIS ENCOURAGED THE EXTERMINATION OF SO-CALLED MENTAL DEFECTIVES, ASPERGER WROTE ABOUT:

THE HIGH VALUE TO SOCIETY OF PEOPLE WHO ARE DIFFERENT TO THE POINT OF BEING MISFITS.

*A BIG AND UNRESOLVED QUESTION IS HOW TO SEPARATE A SCIENTIST'S USEFUL DISCOVERIES FROM THEIR IDEOLOGIES.

A VIVID DESCRIPTION OF AUTISM WAS SUPPLIED BY AMERICAN CHILD PSYCHIATRIST LEO KANNER IN 1944. IT TOOK UNTIL THE 1990S TO REALIZE THAT AUTISM IS PRESENT IN MANY FORMS.

THE "CLASSIC" CASE IS A CHILD WHO DOES NOT SPEAK, REPEATS THE SAME ACTIONS OVER AND OVER, AND IS DIFFICULT TO CONTROL. MOST GROW OUT OF THIS.
KANNER USED THE PHRASES "AUTISTIC ALONENESS," "INSISTENCE ON SAMENESS," AND "ISLETS OF ABILITIES."
BUT THIS DOESN'T CAPTURE MANY AUTISTIC CHILDREN, WHO TALK AND INTERACT ON THEIR OWN TERMS. THEY OFTEN FIND COMFORT IN ROUTINES AND DEVELOP SINGULAR OBSESSIONS.

THEN THERE ARE AUTISTIC PEOPLE WHO ARE AS INTELLIGENT AS ANYONE ELSE, BUT STRUGGLE WITH THE SUBTLETIES OF NONVERBAL COMMUNICATION.
FIRST LET ME TELL YOU ABOUT INSECTS STARTING WITH THE LETTER "A," THEN "B," THEN "C"...

LORNA WING IDENTIFIED CERTAIN COMMON FEATURES ACROSS THE
WIDE VARIETY OF CASES, WHICH SHE DESCRIBED AS A SPECTRUM.

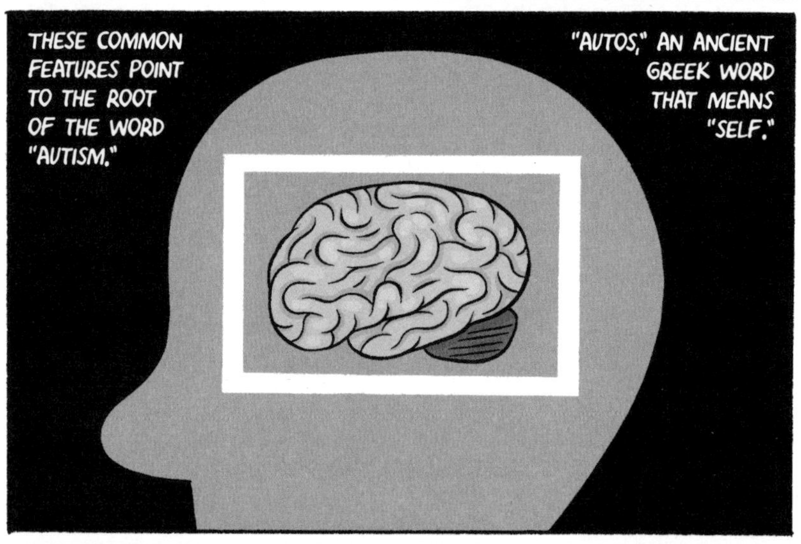
THESE COMMON FEATURES POINT TO THE ROOT OF THE WORD "AUTISM."
"AUTOS," AN ANCIENT GREEK WORD THAT MEANS "SELF."

THE ISM MARKS IT OUT AS A CONDITION, SO IN ITS ESSENCE "AUTISM" MEANS SOMETHING LIKE "BEING WRAPPED UP IN YOURSELF."

PEOPLE DIAGNOSED WITH AUTISM ARE ALL, FOR SOME STILL UNKNOWN REASON, UNABLE TO CONNECT WITH THE PEOPLE AROUND THEM IN THE SPONTANEOUS WAY THAT MOST PEOPLE – NEUROTYPICALS – CAN.

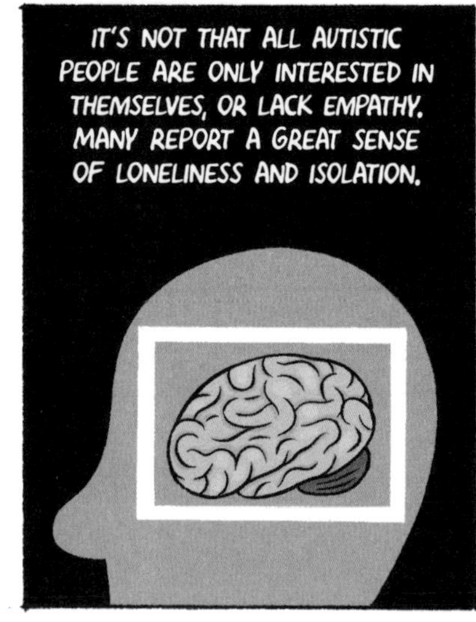
IT'S NOT THAT ALL AUTISTIC PEOPLE ARE ONLY INTERESTED IN THEMSELVES, OR LACK EMPATHY. MANY REPORT A GREAT SENSE OF LONELINESS AND ISOLATION.

INSTEAD, MANY TAKE GREAT PAINS TO LEARN "RULES" OF SOCIAL INTERACTIONS.
THIS IS VERY DIFFICULT – A LITTLE LIKE TRYING TO LEARN A NEW LANGUAGE AS AN ADULT.
1. Look at people's eyes, but don't stare.
2. Say "your dress looks nice," even if it's not true.
3. If man, talk about football.
4. If woman, guess their age and make sure it's younger than they really look.

A COMMON THREAD THAT GOES THROUGH ALL THE SOCIAL PROBLEMS IS THAT THEY CAN'T TRACK OTHER PEOPLE'S THOUGHTS WITHOUT PUTTING IN A LOT OF EFFORT.
RING
RING
NEUROTYPICALS OFTEN JUST KNOW WHAT ANOTHER PERSON IS WANTING OR THINKING...

HELLO.
HELLO.
IS MR. BROWN THERE, PLEASE?

YES.
CLICK
...WHILE PEOPLE WITH AUTISM ARE PRONE TO TAKING THINGS LITERALLY.

ASPERGER SYNDROME* HAS COME TO BE THE TYPICAL LABEL FOR INTELLIGENT PEOPLE WITH AN AUTISM SPECTRUM DISORDER.
*A SYNDROME NO LONGER LISTED BY THE LATEST EDITION OF DSM – THE PSYCHIATRIC DIAGNOSTICS HANDBOOK.

THE STEREOTYPE IS TO BE OBSESSED WITH SOME ESOTERIC PURSUIT.

THAT'S A BIT LIKE TRAINSPOTTERS! I WONDER IF THERE'S SOMETHING SIMILAR GOING ON IN THEIR BRAINS?

SADLY, I MANAGED TO UPSET GREAT SWATHS OF PEOPLE IN BRITAIN WHEN THE PRESS PICKED UP ON THIS COMPARISON...
THE INDEPENDENT
FREE MAGAZINE

WE'RE NOT WEIRDOS! TRAINS HAPPEN TO BE GENUINELY FASCINATING – THAT'S WHY SO MANY PEOPLE LIKE THEM.
THOSE GUYS AT THE END OF THE PLATFORM WHO GET HERE AT 5 A.M., THEY'RE REALLY OBSESSED.

I'M OFTEN ASKED WHAT DREW ME TO STUDY AUTISM.
A LITTLE TOO OFTEN, ACTUALLY.

I'VE COME TO REALIZE THAT PEOPLE VERY MUCH WANT THERE TO BE A "STORY," EXPECTING A PERSONAL CONNECTION OR AT LEAST AN EMOTIONAL EVENT.

THE TRUE ANSWER COMES IN TWO PARTS.
I VISITED THE CHILDREN'S WARD OF A HOSPITAL AS PART OF MY CLINICAL PSYCHOLOGY COURSE.

THERE, I MET SOME AUTISTIC CHILDREN FOR THE FIRST TIME. THEY WERE LIVING ALONGSIDE CHILDREN WITH LEARNING DIFFICULTIES, AND IN MANY RESPECTS TREATED AS THE SAME.

I WAS STRUCK BY HOW WELL THESE CHILDREN DID ON SOME TESTS, WHILE DOING VERY POORLY ON OTHERS.*

*MORE ON THIS ON PAGE 218

I WAS ESPECIALLY INTRIGUED BY THE FACT THAT THEY WERE TOTALLY UNINTERESTED IN ME, VERY UNLIKE THE OTHER CHILDREN.

THIS BRINGS ME TO PART 2: A PAPER I HAPPENED TO READ AROUND THIS TIME THAT DETAILED EXPERIMENTAL WAYS TO UNPACK PRECISELY THIS MIX OF HIGH AND LOW ABILITIES WITH POOR COMMUNICATION.

THE PAPER ITSELF WAS VERY DRY. THE PEOPLE WHO WROTE IT WERE NOT.
BEATE "ATI" HERMELIN
NEIL O'CONNOR

NEIL, AN AUSTRALIAN FROM A SMALL MINING TOWN, DROVE A FLASHY SPORTS CAR AND WAS A STAUNCH COMMUNIST. ATI WAS BORN IN BERLIN, AND CAME TO LONDON VIA PALESTINE. SHE WAS THE EXACT OPPOSITE OF ANY WOMEN SCIENTISTS I HAD MET – SHE WAS STYLISH, AND DIDN'T LET FEAR OF HER PEERS' OPINIONS STAND IN HER WAY.

TO ME, THEY WERE THE AVENGERS* OF THE MAUDSLEY HOSPITAL, WHERE I WAS STUDYING.
I PLUCKED UP THE COURAGE TO APPROACH THEM TO SAY HOW MUCH I ADMIRED THEIR PAPER.
*THAT'S STEED AND PEEL – THE UK'S AVENGERS

IT WENT WELL – THEY TOOK ME ON AS A PHD STUDENT, STUDYING AUTISM.
IF THEY HADN'T, I'D HAVE CONTINUED RESEARCHING OCD.

HERMELIN & O'CONNOR WENT ON TO DO HIGHLY ORIGINAL WORK WITH WHAT WERE THEN CALLED "IDIOT SAVANTS." YES, LIKE RAIN MAN.

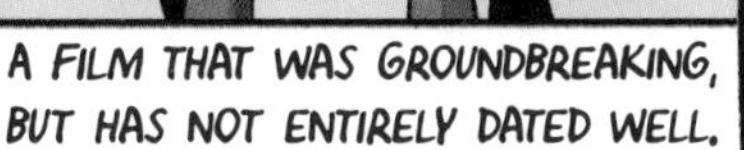
A FILM THAT WAS GROUNDBREAKING, BUT HAS NOT ENTIRELY DATED WELL.

LET ME EMPHASIZE THAT THE VAST, VAST MAJORITY OF PEOPLE WITH AUTISM DO NOT HAVE MENTAL SUPERPOWERS.
BUT BEFORE THIS FILM WAS MADE, MOST PEOPLE HAD NO IDEA AUTISM EXISTED IN ADULTS, OR HAD EVEN HEARD OF IT AT ALL.

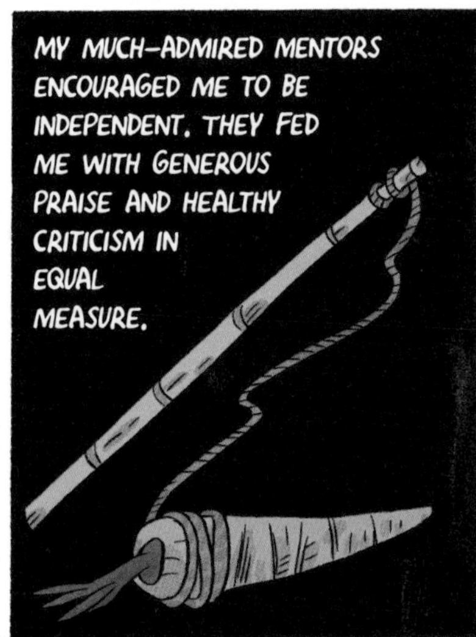
MY MUCH-ADMIRED MENTORS ENCOURAGED ME TO BE INDEPENDENT. THEY FED ME WITH GENEROUS PRAISE AND HEALTHY CRITICISM IN EQUAL MEASURE.

MANY YEARS LATER, IN 1989, I PUBLISHED MY FIRST BOOK ON THE SUBJECT:
AUTISM
Explaining the Enigma*
Uta Frith
*IT DOESN'T EXPLAIN HOW AND WHY AUTISM HAPPENS, IT EXPLAINS WHAT IS ENIGMATIC ABOUT IT. NO, I DIDN'T LIKE THAT SUBTITLE EITHER.

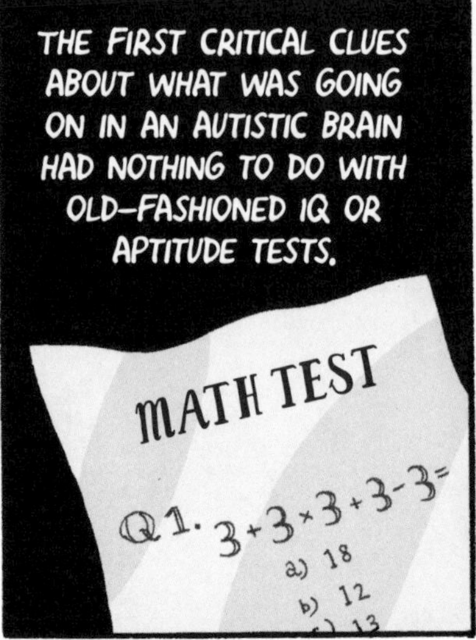
THE FIRST CRITICAL CLUES ABOUT WHAT WAS GOING ON IN AN AUTISTIC BRAIN HAD NOTHING TO DO WITH OLD-FASHIONED IQ OR APTITUDE TESTS.
MATH TEST
Q1. 3+3×3+3-3=
a) 18
b) 12

THEY CAME FROM LORNA WING'S OBSERVATIONS ABOUT THE WAY CHILDREN PLAY.
FLORA
LAURA

FLORA IS TUCKING HER TEDDY IN BED, PRETENDING TO BE ITS MOTHER.

GOOD NIGHT, TEDDY

LAURA IS LINING UP HER TOYS IN A ROW.

FLORA IS PLAYING PRETEND; LAURA IS NOT.

HERE'S ANOTHER OBSERVATION ON THE THEME OF PRETEND, THIS TIME WITH A BABY.

RING! RING!

RING!
RING!

HELLO?

BANANA IS NOT PHONE. MUMMY IS PRETENDING. IS FUNNY. HA!

THE OBSERVATION, AS MADE BY A SCOTTISH COLLEAGUE, ALAN LESLIE:
CHILDREN UNDERSTAND PRETENSE INCREDIBLY EARLY IN LIFE.

FROM THIS EXAMPLE, LESLIE GOT THE IDEA THAT PRETEND PLAY RELIES ON ONE PERSON KNOWING WHAT IS IN ANOTHER PERSON'S HEAD, AND NOT JUST WHAT'S IN THE OUTSIDE WORLD.
WE CALL THIS "THEORY OF MIND."

AUTISTIC CHILDREN LACK PRETEND PLAY...
...SO PERHAPS THEY DON'T HAVE THEORY OF MIND.
(A RARE ACTUAL LIGHTBULB MOMENT, SHARED BY ME AND ALAN)

AT THE SAME TIME, PSYCHOLOGISTS JOSEF PERNER AND HEINZ WIMMER WANTED TO SEE IF THEY COULD OBSERVE THEORY OF MIND IN YOUNG CHILDREN.

THEY WERE REMINDED OF A SCENE FROM AN OLD GERMAN COMIC, MAX UND MORITZ.*

IN THIS SCENE, MAX & MORITZ BELIEVE THE WIDOW BOLTE WILL THINK HER DOG ATE THE CHICKENS.

IN OTHER WORDS, THE SCENE TRADES ON WHAT'S IN HER HEAD, NOT WHAT'S REALLY HAPPENING.

THAT'S "THEORY OF MIND."

Schnupdiwup! da wird nach oben
Schon ein Huhn heraufgehoben.

Max und Moritz
von
Wilhelm Busch

*THIS AND OTHER COMICS BY WILHELM BUSCH ARE KEY TEACHING TOOLS IN THE FRITH HOUSEHOLD.

PERNER AND WIMMER DEVISED AN EXPERIMENT, ADAPTED BY LESLIE, FRITH AND SIMON BARON–COHEN INTO THE "SALLY–ANNE" TEST.
EQUIPMENT REQUIRED: TWO DOLLS AND TWO CONTAINERS.*
*THE ORIGINAL DOLLS WERE PINCHED FROM MARTIN AND ALEX'S TOY BASKET.

LISTEN CAREFULLY. I'M GOING TO TELL YOU A STORY ABOUT SALLY AND ANNE.
SIMON BARON–COHEN

SALLY HAS A MARBLE. SHE'S PUTTING IT IN HER BASKET TO KEEP IT SAFE.

SALLY GOES OUT TO PLAY. BYE–BYE!

WHILE SALLY IS AWAY, ANNE TAKES SALLY'S MARBLE AND PUTS IT IN HER POT.

HERE COMES SALLY AGAIN. NOW, WHERE WILL SHE LOOK FOR HER MARBLE?

IN THE BASKET. THAT'S WHERE SHE PUT IT.
FLORA

IN THE POT. THE MARBLE IS IN THE POT.
LAURA

LAURA'S RESPONSE IS TYPICAL IN CHILDREN WITH AUTISM, AND INDEED ALL CHILDREN UNDER FOUR.
MY THEORY IS THAT THEY DON'T SEPARATE KNOWLEDGE THEY HAVE FROM KNOWLEDGE THAT OTHER PEOPLE HAVE.
IT'S AS IF WE CAN SEE OVER A WALL INTO THEIR HEADS, BUT THEY CAN'T SEE BACK.

LACK OF THEORY OF MIND HAS, SO FAR, EXPLAINED A NUMBER OF PUZZLING FEATURES TO DO WITH AUTISM.

AUTISM SEEMS NOT TO HAMPER UNDERSTANDING OF WORDS OR IDEAS, BUT IT DOES SEEM TO LIMIT PRETEND PLAY, AND EVEN THE TELLING OF LIES.

I THINK IT ALSO EXPLAINS WHY SOME PEOPLE WITH AUTISM TEND TO TAKE WORDS LITERALLY, AND FEEL EXTRA FRUSTRATION WHEN THEY CAN'T "READ BETWEEN THE LINES."

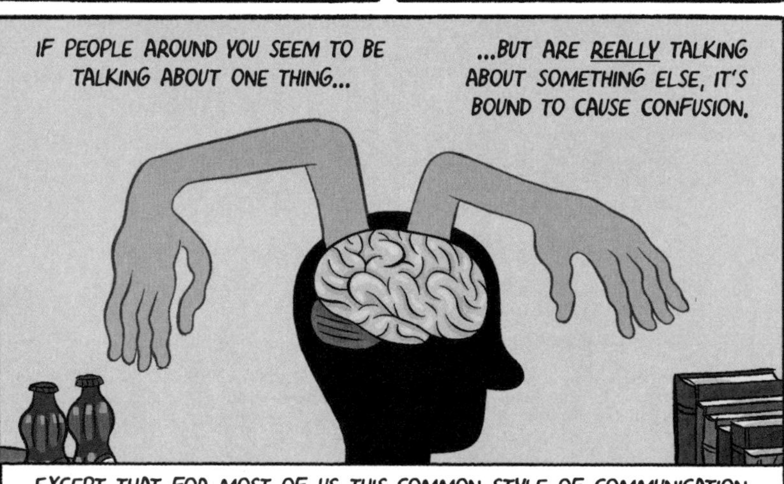
IF PEOPLE AROUND YOU SEEM TO BE TALKING ABOUT ONE THING...
...BUT ARE REALLY TALKING ABOUT SOMETHING ELSE, IT'S BOUND TO CAUSE CONFUSION.
EXCEPT THAT, FOR MOST OF US, THIS COMMON STYLE OF COMMUNICATION STOPS BEING CONFUSING AT QUITE A YOUNG AGE.

HOW YOUNG? WELL, IF THEORY OF MIND IS SO IMPORTANT, I THINK WE'D EXPECT TO SEE SOME FORM OF IT EVEN IN AN INFANT.

REMEMBER OVERIMITATION? TURN BACK A FEW PAGES.

WE KNOW NOT ALL THE TAPS ARE REQUIRED TO OPEN THE BOX, BUT WE "READ BETWEEN THE LINES" THAT THE PERSON WHO DID THE TAPPING IS EXPECTING US TO COPY THEM.

IN OTHER WORDS, WE HAVE A THEORY ABOUT WHAT'S GOING ON IN THEIR MIND.
ALTHOUGH IT'S NOT REALLY PROPER TO CALL IT A "THEORY," SINCE WE'RE NOT ACTIVELY HOLDING IT IN OUR CONSCIOUS THOUGHTS.

PSYCHOLOGIST ÁGNES KOVÁCS MANAGED TO OBSERVE THEORY OF MIND IN BABIES AS YOUNG AS 7 MONTHS, USING ANOTHER BELOVED COMICS CREATION – THE SMURFS.

IN THE TASK, PEOPLE WATCH A SIMPLE ANIMATION THAT SHOWS A SMURF LOOKING FOR A BALL...

...THAT MAY OR MAY NOT BE BEHIND A SCREEN.

IN SOME VERSIONS, THE SMURF HAS SEEN THE BALL ROLL BEHIND THE SCREEN...

...AND NOT ROLL PAST THE OTHER SIDE.

IN OTHER VERSIONS, THE VIEWER SEES THE BALL ROLL AWAY, BUT THE SMURF DOESN'T SEE THIS.

VIEWERS CAN TELL WHETHER OR NOT THE SMURF EXPECTS TO FIND THE BALL BEHIND THE SCREEN...

...BY WATCHING THE SMURF'S EYE GAZE.

IF THE BALL IS MISSING, BUT THE SMURF EXPECTS IT TO BE THERE, MOST VIEWERS – INCLUDING YOUNG BABIES – WILL STARE FOR A LONG TIME, IN SURPRISED SYMPATHY, IF YOU LIKE.

THEORY OF MIND IN SOME FORM AT LEAST EXISTS IN THE VERY YOUNG. I THINK THIS MEANS IT'S AN INNATE ABILITY.

OTHERS DISAGREE WITH ME. BUT EVERYONE AGREES THAT AUTISTIC PEOPLE CAN AND DO LEARN HOW FALSE BELIEFS WORK, EVENTUALLY.

OLDER CHILDREN (AND ADULTS) WITH AUTISM TEND TO "PASS" THE SALLY-ANNE TEST.

THEY EXPLAIN THAT SALLY WILL LOOK FOR THE MARBLE IN HER BASKET, BUT, INTERESTINGLY...

...THEY DON'T LOOK AT THE BASKET.
IF YOU WATCH WHERE PEOPLE LOOK, MOST NEUROTYPICAL* PEOPLE INSTINCTIVELY GLANCE AT THE BASKET BEFORE THEY SPEAK.

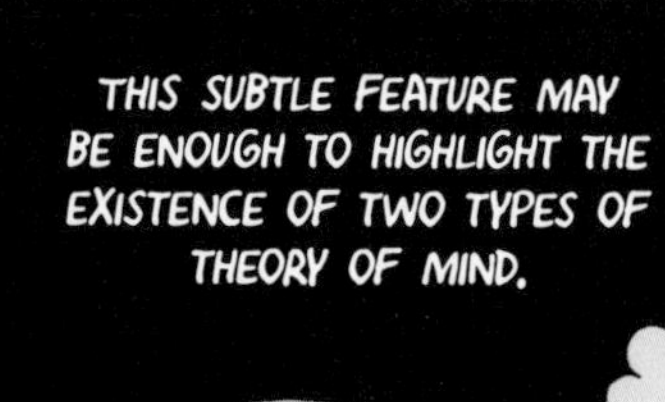
THIS SUBTLE FEATURE MAY BE ENOUGH TO HIGHLIGHT THE EXISTENCE OF TWO TYPES OF THEORY OF MIND.

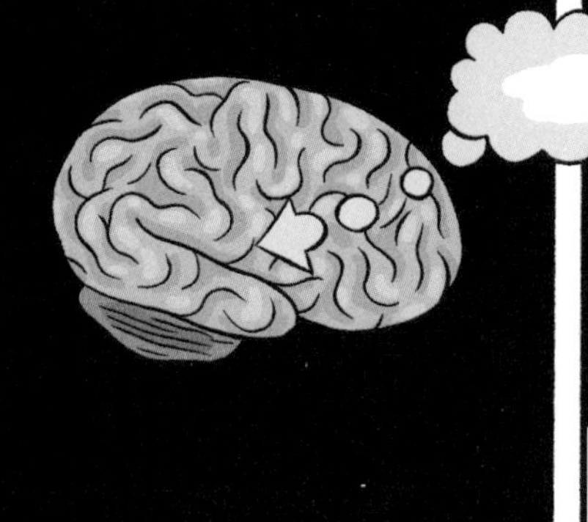

1. AN IMPLICIT FORM, IN WHICH PEOPLE SEEM TO KNOW WHAT ANOTHER PERSON IS THINKING ABOUT WITHOUT HAVING TO THINK ABOUT IT.
(THIS MAY BE THE INNATE FORM THAT'S MISSING IN AUTISM.)

2. AN EXPLICIT FORM, IN WHICH PEOPLE LEARN TO USE LOGICAL REASONING TO DETERMINE A PERSON'S THOUGHTS.

MY TURN!
YES, I COULD SENSE YOUR GROWING IMPATIENCE.

*A WORD USED BY MANY IN THE AUTISTIC COMMUNITY TO DESCRIBE PEOPLE WHO DON'T HAVE THIS DIAGNOSIS. OF COURSE, THERE'S NO SUCH THING AS HAVING A "TYPICAL" BRAIN, IT'S JUST A USEFUL WORD.

THERE IS MORE TO SOCIAL ABILITIES THAN JUST THEORY OF MIND.
SOME OF THE MOST BASIC SOCIAL ABILITIES ARE SHARED BY ANIMALS, TOO.

WE, LIKE MANY ANIMALS, FEEL THE SAME EMOTIONS AS OTHERS – KNOWN AS EMPATHY.
WOOF WOOF

FOR EXAMPLE, IF WE SEE SOMEONE WITH A FEARFUL EXPRESSION, WE INSTINCTIVELY FEEL (AND SHOW) FEAR.
WOOF WOOF

...EVEN WHEN WE ARE NOT AWARE OF SEEING THE EXPRESSION, LET ALONE SEEING WHATEVER IT IS THAT MAKES THE PERSON AFRAID.

IT'S THE SAME FOR DISGUST, BUT ALSO HAPPINESS AND OTHER POSITIVE EMOTIONS, TOO.

WE CAN TELL IF SOMEONE NEARBY HAS SPOTTED SOMETHING EXCITING...
FREE ICE CREAM

...AND OUR BRAINS INSTINCTIVELY PREPARE US TO APPROACH IT.
IN CASE YOU'RE WONDERING, MOST PEOPLE WITH AUTISM EXPERIENCE EMPATHY (AND FEAR LONELINESS) JUST AS MUCH AS EVERYBODY ELSE.

AS FAR AS SOCIAL INSTINCTS GO, THERE IS A LOT TO EXPLORE.
ONE OF MY FAVORITES IS A SHARED PHENOMENON KNOWN AS AFFORDANCE.
LET'S LOOSELY DEFINE IT AS: HOW WE PERCEIVE ENVIRONMENTS AS WAYS TO AFFORD US OUR NEEDS.

TO UNDERSTAND AFFORDANCE, IMAGINE YOU'RE SITTING AT A TABLE WITH TWO FRESH CUPS OF TEA.

IT'S YOUR SENSE OF AFFORDANCE THAT TELLS YOU THE NEARBY CUP IS IN YOUR REACH.

NOW IMAGINE A FRIEND JOINS YOU AT THE TABLE. SHE SITS OPPOSITE.

THE SECOND CUP OF TEA IS WITHIN HER ZONE OF AFFORDANCE.

ALTHOUGH HER CUP OF TEA IS NOT IN YOUR REACH...

...YOUR BRAIN RECOGNIZES HER AS A FRIEND, SO HER ZONE OF AFFORDANCE BECOMES PART OF YOUR OWN. THIS ISN'T JUST A THOUGHT EXPERIMENT, OR A "FACT" BORNE OUT BY EVERYDAY EXPERIENCE. IT HAS BEEN OBSERVED IN LAB CONDITIONS BY MEASURING BRAIN ACTIVITY.

YOUR BRAIN CAN BE DESCRIBED AS HAVING ACTIVATED THE "WE" MODE.
BOTH CUPS ARE NOW "OURS."

AFFORDANCE HAS BEEN EXPLORED IN ANIMALS, TOO.

HERE'S ATSUSHI IRIKI TO EXPLAIN THE STORY OF THE MONKEY AND THE RAKE.
HELLO.

I'VE BEEN ABLE TO TEACH MACAQUES TO USE RAKES TO REACH THINGS. (A SKILL THEY DO NOT HAVE IN THE WILD.)

I'VE ALSO BEEN ABLE TO LOOK INTO THEIR BRAINS AT THE SAME TIME, OBSERVING THE VERY NEURONS THAT LINK TO AFFORDANCE.

WHEN THEY ARE HOLDING THE RAKE...

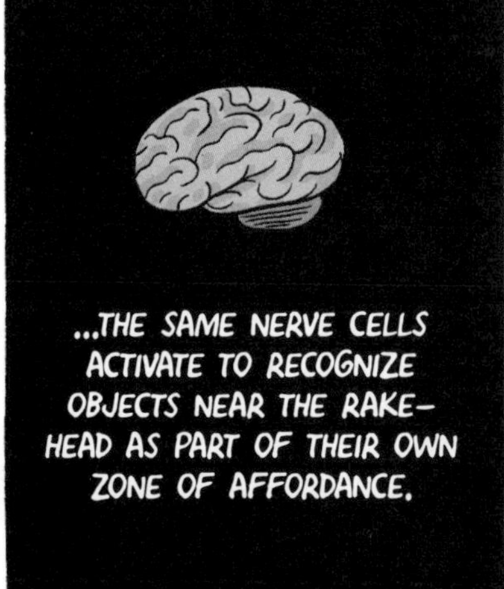
...THE SAME NERVE CELLS ACTIVATE TO RECOGNIZE OBJECTS NEAR THE RAKE-HEAD AS PART OF THEIR OWN ZONE OF AFFORDANCE.

STUDIES REMAIN TO BE DONE TO SHOW IF MONKEYS HAVE, OR CAN LEARN, THE "WE MODE."

HUMANS LOVE THE WE MODE.

IT'S WHAT WE USE WHEN WE ASK SOMEONE AT A TABLE TO PASS THE SALT.

THIS REMINDS ME OF AN ANECDOTE A PARENT ONCE SHARED WITH ME ABOUT HER AUTISTIC SON.

SHE FOUND HIM IN THE KITCHEN, POINTING UP AT A CUPBOARD, OUT OF HIS REACH, LOOKING DEEPLY AGITATED.

HE HAD BEEN THERE FOR SOME MINUTES, SHE DISCOVERED.

SHE OPENED THE CUPBOARD, WHICH CONTAINED A SELECTION OF BISCUITS, AND GAVE ONE TO HER SON.

HER SON WAS NOT GIVEN TO TALKING, SO HIS PARENTS HAD TAUGHT HIM TO POINT, AS A WAY TO TELL PEOPLE WHAT HE WANTED.

BUT, AT THIS STAGE, HE HADN'T WORKED OUT THAT IT WAS A VITAL PART OF THE SYSTEM FOR ANOTHER PERSON TO BE THERE TO SEE HIM POINTING.
CRUNCH

AUTISM, I REMIND YOU, IS NOT A CONDITION THAT MEANS PEOPLE ONLY CARE ABOUT THEMSELVES, OR ARE ANTISOCIAL.
THAT SAID, ONE COMMON FEATURE TENDS TO BE, EVEN AMONG THE MOST INTELLIGENT, A REAL DIFFICULTY PICKING UP ON VARIOUS UNCONSCIOUS SOCIAL CUES.
WHICH CUES?
WELL, THAT'S THE TRICK, ISN'T IT.
THERE ARE CUES THAT TELL WHEN IT IS, IN A SOCIAL SENSE, CORRECT TO IMITATE AND OVER-IMITATE. CUES WHEN IT IS CORRECT TO TAKE ACCOUNT OF WHAT ANOTHER PERSON KNOWS, OR WISHES, OR BELIEVES – OR DOESN'T.
WE CAN'T EVEN SAY WHAT IT IS ABOUT HUMAN SOCIETY THAT MAKES IT DIFFERENT FROM OTHER ANIMAL SOCIETIES.
WE BARELY KNOW HOW HUMAN SOCIAL INTERACTIONS WORK, IN TERMS OF BRAIN FUNCTIONS. AND WE CERTAINLY DON'T HAVE A LIST OF ALL THE SUBTLY DIFFERENT WAYS THESE INTERACTIONS SEEM TO WORK (OR NOT).
ISN'T SCIENCE MARVELOUS! WE'VE DEVOTED YEARS OF OUR LIVES TO TRYING TO UNDERSTAND THINGS LITERALLY NOBODY UNDERSTANDS, AND WE BASICALLY KNOW WE'RE GOING TO FAIL – IN OUR OWN LIFETIMES AT LEAST.
WE WOULDN'T HAVE IT ANY OTHER WAY. CHOOSING ANOTHER CAREER WOULD BE BORING.

Chapter 5

Explaining Empathy
THEY SAY THE MARKER OF A SUCCESSFUL RELATIONSHIP IS GOOD COMMUNICATION.

The latest chapter in the history of neuroscience.
THE TRICK IS GETTING WHATEVER THOUGHTS ARE INSIDE YOUR HEAD INTO YOUR PARTNER'S HEAD.
YOU COULD ARGUE THAT THIS THEME IS AT THE HEART OF MYSTERY STORIES.

WHAT'S SHE SAYING? THEY'RE TALKING TOO FAST.
SHE'S ASKING IF THE MAN HAS SEEN THE SUSPECT. WE KNOW HE HAS, BUT SHE DOESN'T KNOW THAT.
WHY CAN'T HE SEE HOW FRUSTRATING IT IS TRYING TO KEEP UP WITH THE DIALOGUE?
WHY CAN'T SHE LEARN THAT THE DETAILS OF THE DIALOGUE RARELY MATTER VERY MUCH?*
*REGULAR VIEWING INCLUDES DETECTIVE DRAMAS IN GERMAN (A LANGUAGE UTA SPEAKS BUT CHRIS DOESN'T – YET HE STILL HAS TO EXPLAIN WHAT'S HAPPENING...)

ANOTHER OFT-DEMANDED PERSONALITY TRAIT IN A PARTNER IS EMPATHY.
SIMPLY PUT, EMPATHY MEANS "HAVING THE SAME FEELINGS AS SOMEONE ELSE."
PEOPLE ARE PRAISED WHEN THEY SHOW EMPATHY – IT'S A SIGN OF COMPASSION AND GENERALLY THE MARKER OF A NICE PERSON.
BUT EMPATHY MAY BE LITTLE MORE THAN AN INVOLUNTARY RESPONSE – NOT SOMETHING WE CAN REALLY TAKE CREDIT FOR.

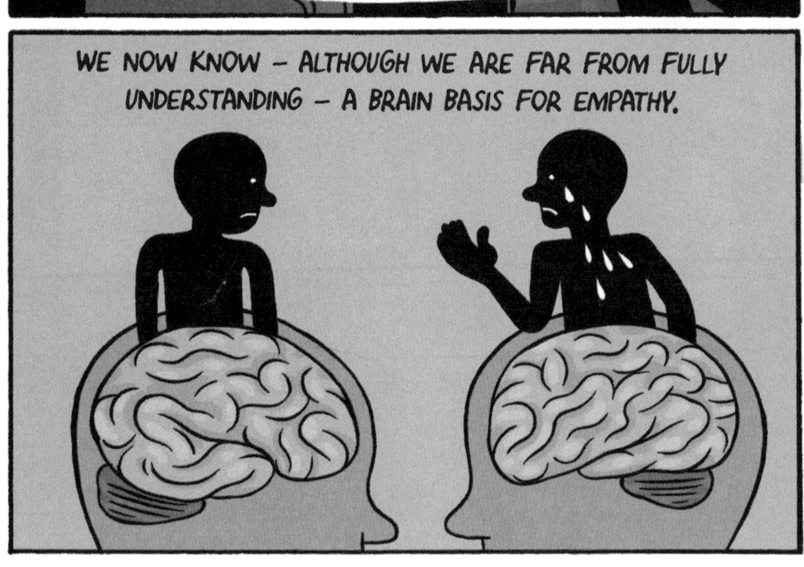
WE NOW KNOW – ALTHOUGH WE ARE FAR FROM FULLY UNDERSTANDING – A BRAIN BASIS FOR EMPATHY.

Psychology is a non-science
HOLD ON JUST A MINUTE!

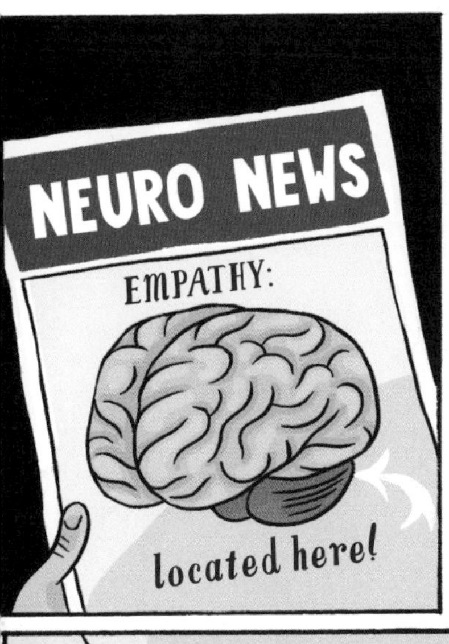
NEURO NEWS
EMPATHY:
located here!

PEOPLE TEND TO BE UNIMPRESSED WITH THESE TYPES OF FINDINGS.

TELL ME HOW TO MAKE MY PARTNER EMPATHIZE WITH ME, THEN I'LL BE IMPRESSED!

WE'RE DETERMINED TO PROVE TO YOU THAT THIS SORT OF WORK IS BOTH USEFUL AND IMPRESSIVE! AND WE HAVE THE TECHNOLOGY TO HELP US DO IT.
VERY BASICALLY, WE NOW KNOW THAT EMPATHY HAS A LINK TO THE BRAIN'S MIRROR SYSTEM.
ALTHOUGH WE ALL THINK WE KNOW WHAT EMPATHY IS, SCIENCE CAN HELP US UNDERSTAND HOW IT REALLY WORKS.

HOW DO PEOPLE FIND OUT ANYTHING ABOUT EMPATHY? WELL, LET'S TALK YOU THROUGH A SEQUENCE OF EVENTS...

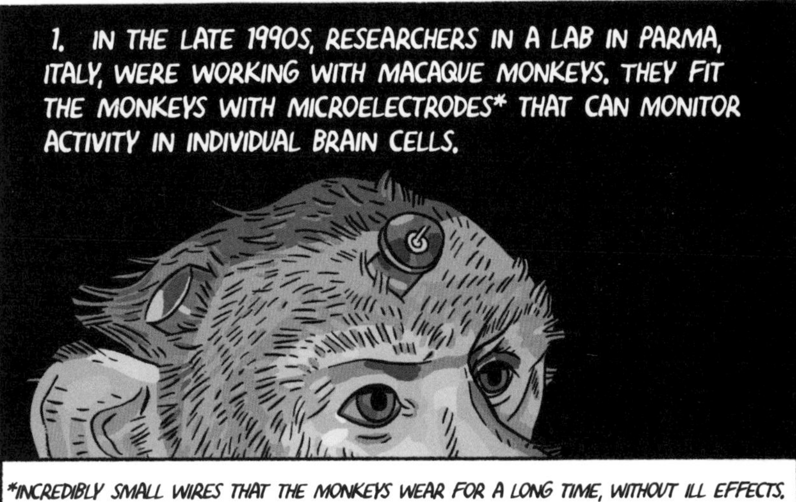
1. IN THE LATE 1990S, RESEARCHERS IN A LAB IN PARMA, ITALY, WERE WORKING WITH MACAQUE MONKEYS. THEY FIT THE MONKEYS WITH MICROELECTRODES* THAT CAN MONITOR ACTIVITY IN INDIVIDUAL BRAIN CELLS.

*INCREDIBLY SMALL WIRES THAT THE MONKEYS WEAR FOR A LONG TIME, WITHOUT ILL EFFECTS.

WHEN ONE OF THE TEAM PICKS UP A PEANUT, SOMETHING NOTICEABLE HAPPENS IN THE MONKEY'S BRAIN.
A SET OF NEURONS ACTIVATE IN A REGION ASSOCIATED WITH MOTOR CONTROL.
PIONEERING NEUROPHYSIOLOGIST GIACOMO RIZZOLATTI

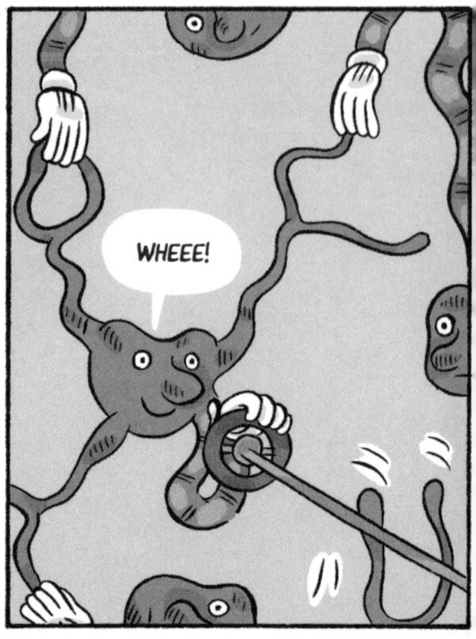
WHEEE!

IT'S PEANUT EATIN' TIME!

IN FACT, THE SAME NEURONS ACTIVATE WHEN THE MONKEY SEES A PERSON CRACKING OPEN A PEANUT...

CRACK

...AS IF THE MONKEY WERE ITSELF OPENING A PEANUT.

2. A FEW YEARS LATER IN LONDON, A PSYCHOLOGY PROFESSOR (JAMIE WARD) GAVE A LECTURE.
HE DESCRIBED THE PHENOMENON OF SYNESTHESIA – THE EXPERIENCE OF, FOR EXAMPLE...

BOOM
...SEEING A COLOR (YELLOW) WHEN YOU HEAR A PARTICULAR SOUND (BOOM).

I THOUGHT THAT HAPPENED TO EVERYONE.
USUALLY AT LEAST ONE PERSON PER LECTURE SAYS THAT!*
I FEEL A TOUCH ON MY OWN FACE WHEN I SEE SOMEONE ELSE BEING TOUCHED.
THAT'S A NEW ONE ON ME!
*NO ONE KNOWS HOW MANY PEOPLE ARE SYNESTHETIC – IT IS LIKELY NO MORE THAN 1 IN EVERY 300, AND NO FEWER THAN 1 IN EVERY 2,000. PRETTY RARE, IN OTHER WORDS.

HOW WOULD YOU LIKE TO SEE THE INSIDE OF A BRAIN SCANNER?
IN THE SCANNER, IT TURNS OUT THAT SYNESTHETES SHOW THE SAME BRAIN ACTIVATION TO WATCHING SOMEONE ELSE BEING TOUCHED, AS THEY DO WHEN THEY ARE TOUCHED THEMSELVES.

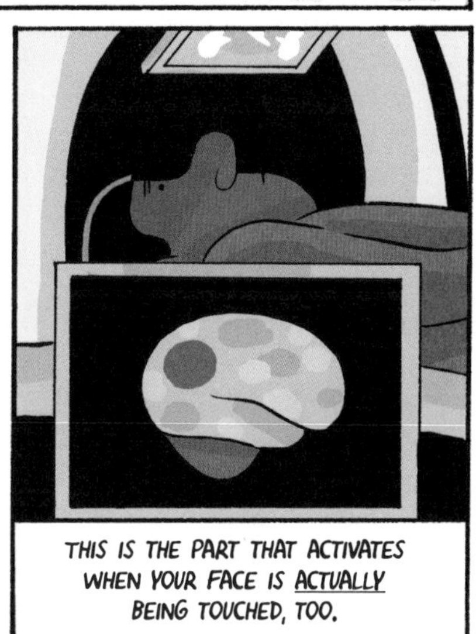
THIS IS THE PART THAT ACTIVATES WHEN YOUR FACE IS ACTUALLY BEING TOUCHED, TOO.

AS ALWAYS, THE EXPERIMENTERS PERFORM SOME CONTROL TRIALS, TOO. THEY SCAN NON-SYNESTHETES.

IT TURNS OUT THAT THEIR BRAINS SHOW THE SAME PATTERN OF NEURON ACTIVATION.

THE DIFFERENCE IS, THEY DO NOT FEEL IT HAPPENING.

AND SO, THE DISCOVERY: IN OUR BRAINS, THE SAME NEURONS DEAL WITH BEING TOUCHED AS SEEING SOMEONE ELSE BEING TOUCHED.

WE CALL THEM "MIRROR NEURONS." MIRROR NEURONS PLAY A HUGE ROLE IN EMPATHY: WHEN WE SEE SOMEONE ELSE EXPERIENCING AN EMOTION, WE EXPERIENCE IT TOO.

DOES THIS DISCOVERY SOUND MORE IMPRESSIVE, NOW THAT YOU KNOW THE DETAILS OF WHAT LED UP TO IT?

I ADMIT, IT'S NOT MUCH OF AN OBSERVATION TO SAY "HUMANS ARE GOOD AT COPYING EACH OTHER."
BUT CONSIDER THE PHENOMENON OF UNCONSCIOUS COPYING.

YOU MAY WELL HAVE SEEN ONE PERSON COPYING ANOTHER WITHOUT ACTIVELY REALIZING WHAT THEY ARE DOING.

HOW DOES THIS COME ABOUT? WELL, SINCE OUR MIRROR NEURONS ACTIVATE WHENEVER WE SEE SOMEONE DO SOMETHING...

...IT'S A SMALL STEP FOR THE SAME PATTERN OF NEURONS TO ACTIVATE AGAIN, MAKING US PHYSICALLY DO THE SAME THING.
HOW IS IT THAT WE DON'T COPY PEOPLE ALL THE TIME, THEN?

ANOTHER PART OF THE BRAIN CAN TELL THE MIRROR NEURONS WHEN TO SWTICH ON, AND WHEN TO SWITCH OFF. THAT'S ENOUGH ABOUT MIRROR NEURONS FOR NOW. I WANT TO TALK ABOUT A MORE "USELESS" EXAMPLE OF PSYCHOLOGY RESEARCH: TICKLING.

AMPUTEES ARE VERY FAMILIAR WITH "PHANTOM LIMB" SYNDROME – SENSATIONS IN THE MISSING LIMB.

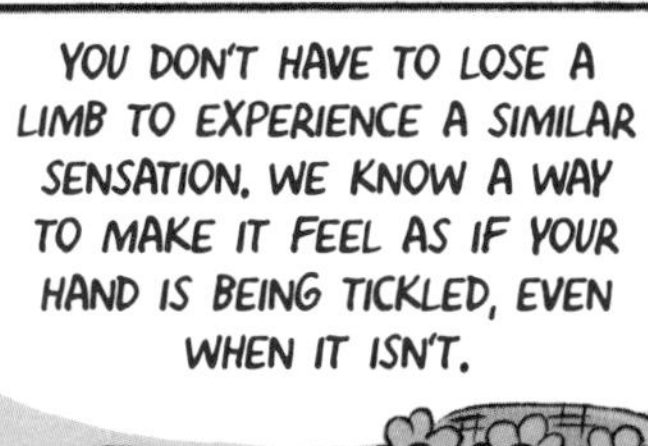
YOU DON'T HAVE TO LOSE A LIMB TO EXPERIENCE A SIMILAR SENSATION. WE KNOW A WAY TO MAKE IT FEEL AS IF YOUR HAND IS BEING TICKLED, EVEN WHEN IT ISN'T.

FIRST, YOU NEED A RUBBER HAND. LIKE THIS ONE.

NEXT, ARRANGE YOURSELF CAREFULLY SO THAT YOUR RIGHT HAND IS HIDING BEHIND A SCREEN...
...AND THE RUBBER HAND IS IN FRONT OF THE SCREEN, WHERE YOU CAN SEE IT.
STICK A TUBE UP THE RUBBER HAND TO MAKE IT LOOK LIKE AN ARM.

AT THIS POINT YOU NEED A FRIEND TO HELP.

ASK YOUR FRIEND TO TICKLE YOUR REAL RIGHT HAND AND THE RUBBER RIGHT HAND AT THE SAME TIME, AND WITH THE SAME STROKE.
AFTER ABOUT ONE MINUTE, YOUR BRAIN STARTS TO REGISTER THE RUBBER HAND AS YOUR OWN HAND.
THE TICKLING SENSATION YOU FEEL IN YOUR REAL HAND NOW SEEMS TO BE COMING FROM THE RUBBER HAND.

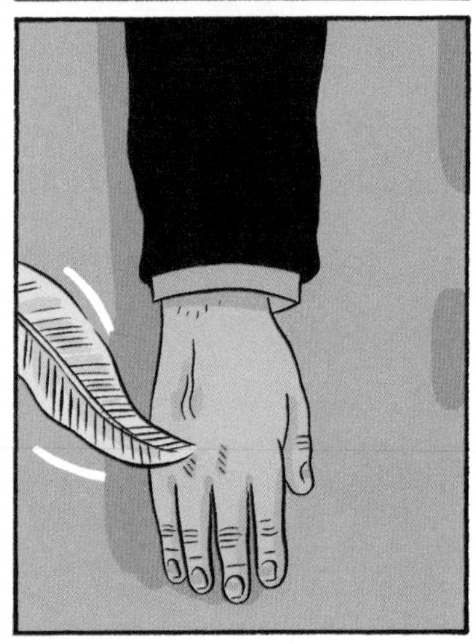

FUN, BUT IS THIS KNOWLEDGE USEFUL?
WELL, IT'S NOT OBVIOUSLY USEFUL IN ITSELF. BUT IT ABSOLUTELY DOES FURTHER HUMAN UNDERSTANDING OF HOW BRAINS WORK.

NO IT DOESN'T!
Psychology is a non-science!
PEOPLE ALREADY KNOW ABOUT TICKLING!
WASTE OF MONEY! GO AND DO SOME PROPER RESEARCH!

PERHAPS IN ORDER TO REALLY IMPRESS THE READER WITH THE ACHIEVEMENTS OF NEUROSCIENTISTS, WE SHOULD DELVE INTO THE PAST.
THERE WAS A LONG CHAIN OF DISCOVERIES BEFORE WE REACHED THE POINT THAT WE CAN SEE NEURONS IN ACTION.
THE HISTORY OF NEUROSCIENCE

LET'S START WITH JOHANNES MÜLLER (1801–1858).

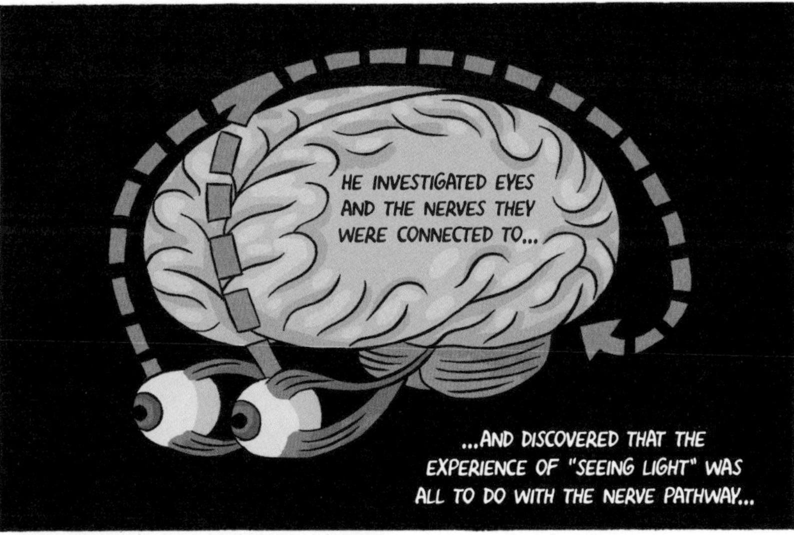
HE INVESTIGATED EYES AND THE NERVES THEY WERE CONNECTED TO...
...AND DISCOVERED THAT THE EXPERIENCE OF "SEEING LIGHT" WAS ALL TO DO WITH THE NERVE PATHWAY...

...NOT THE NERVE ENDINGS IN THE EYEBALL ITSELF.
THE EYE
THE RETINA DETECTS LIGHT AND COLOR, BUT CANNOT INTERPRET THEM.

MÜLLER PROVED THE POINT BY DEMONSTRATING THAT YOU CAN MAKE YOUR BRAIN "SEE" LIGHTS BY APPLYING PRESSURE TO THE EYES, EVEN WHEN THEY'RE CLOSED.

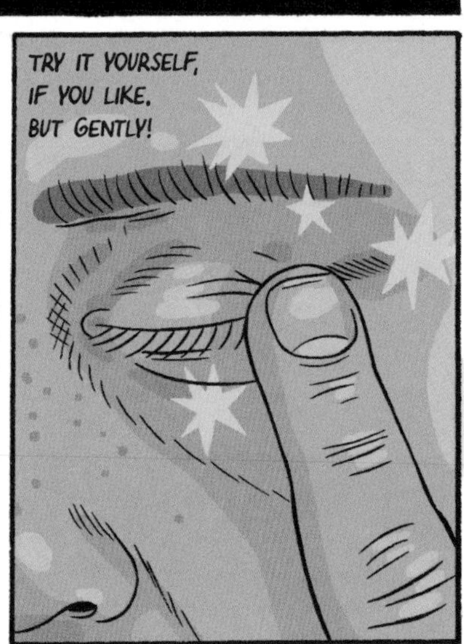
TRY IT YOURSELF, IF YOU LIKE. BUT GENTLY!

HERMANN VON HELMHOLTZ (1821–1894) IS MOST REMEMBERED AS A PHYSICIST, BUT HE WAS, IN THE MAIN, A PROFESSOR OF PHYSIOLOGY.

HELMHOLTZ HAD THE IDEA TO TRY TO MEASURE THE SPEED OF A NERVE IMPULSE. HIS SUPERVISOR, ONE JOHANNES MÜLLER, TOLD HIM NOT TO BOTHER.
IT'S AN ELECTRICAL IMPULSE, YOU DOLT! THAT MEANS IT MOVES AT THE SPEED OF LIGHT.

NONETHELESS, HELMHOLTZ PERSEVERED – AND PROVED MÜLLER WRONG. IT TRAVELS FAR SLOWER! (THERE'S RESISTANCE IN NERVE CELLS)

TICK TICK

NERVE IMPULSES TRAVEL BETWEEN 24.6 M/S AND 38.4 M/S (ROUGHLY 55–70 MPH).

HELMHOLTZ WAS, IN TURN, MENTOR TO WILHELM WUNDT...

...WHO EARNS A PASSING MENTION AS THE PERSON WHO COINED THE CONCEPT OF "PSYCHOLOGY" AS A UNIQUE AREA OF STUDY.*

*ALTHOUGH, IN DOING SO, HE SHOULDERS SOME OF THE BLAME FOR THAT ONGOING PROBLEM OF PEOPLE BELIEVING THE MIND TO BE SEPARATE FROM THE BODY. TSK.

HERE'S A NAME YOU'LL PROBABLY RECOGNIZE: IVAN PAVLOV (1849–1936).

YES, THE RUSSIAN DOCTOR WTH THE DOGS.

PAT PAT

YOU KNOW THE STORY: PAVLOV RINGS A BELL...

?

...AND VERY SOON AFTERWARDS, HE GIVES THE DOG ITS DINNER.

(IN FACT, PAVLOV USED A VARIETY OF NOISE MAKERS, INCLUDING BUZZERS, METRONOMES, AND TUNING FORKS. HE MAY OR MAY NOT HAVE USED A BELL.)

AFTER REPEATING A FEW TIMES, THE SOUND OF THE BELL HAS BECOME A PRIOR* IN THE DOG'S BRAIN. IT STARTS TO SALIVATE ON HEARING THE SOUND.

PAVLOV DESCRIBED THIS AS "CONDITIONING" – ONE OF THE MOST INFLUENTIAL FINDINGS IN ALL OF PSYCHOLOGY. IT'S PROOF THAT WHATEVER MAY BE HARDWIRED INTO A BRAIN, BRAINS CAN BE REWIRED.

*HAD YOU FORGOTTEN ABOUT BAYESIAN PRIORS FROM CHAPTER 2? YOU'LL REMEMBER THEM NOW!

ITALIAN BRAIN ENTHUSIAST CAMILLO GOLGI WAS FASCINATED BY CELLS. CRUCIALLY, HE DISCOVERED A WAY TO STAIN NERVE CELLS SO THAT THEY COULD BE SEEN UNDER A MICROSCOPE.
GOLGI BELIEVED THE CELLS HE FOUND FUNCTIONED AS A CONTINUOUS ENTITY...
...A BELIEF CHALLENGED BY SPANISH NEUROSCIENTIST SANTIAGO RAMÓN Y CAJAL*...
...WHO THOUGHT THAT THERE WERE LOTS OF CELLS, BUT ALL CONNECTED.
C. GOLGI (1843–1926)
S. RAMÓN Y CAJAL (1852–1934)
CAMILLO GOLGI, C. 1890
*PLEASE NOTE THE FULL-LENGTH SURNAME "RAMÓN Y CAJAL."
(AND, WHILE YOU'RE AT IT, YOU MIGHT BE INTERESTED TO NOTE THAT WHAT IS OFTEN SIMPLY CALLED "TOURETTE'S SYNDROME" SHOULD ACTUALLY BE CALLED "GILLES DE LA TOURETTE'S SYNDROME.")

RAMÓN Y CAJAL STUDIED THE STRUCTURE AND FUNCTION OF NEURONS IN MUCH GREATER DETAIL, AND HIS WORK PROVIDES THE MAJOR FOUNDATION FOR MODERN NEUROSCIENCE.
BOTH MEN ALSO DESERVE CREDIT FOR BEING THE FIRST NEUROSCIENCE ARTISTS, MAKING BEAUTIFUL AND DETAILED DRAWINGS OF NEURONS AND OTHER STRUCTURES.
S. RAMÓN Y CAJAL, C. 1912
THE TWO EVENTUALLY SHARED A NOBEL PRIZE FOR THEIR WORK, BUT THEY CONTINUED TO GIVE LECTURES ATTACKING EACH OTHER'S BELIEFS ABOUT NEURONS.
IT'S VERY CLEAR THAT RAMÓN Y CAJAL WAS RIGHT – NEURONS ARE DISTINCT AND DISCRETE ENTITIES THAT LINK TOGETHER, ALTHOUGH SOME NEURONS CAN INDEED BE VERY LONG.

WHILE SOME PEOPLE WERE STUDYING THE FINE DETAIL OF THE BRAIN'S WORKINGS, OTHER NEUROSCIENCE PIONEERS WERE LOOKING AT THE BIGGER PICTURE.

WHAT HAPPENS IN BRAINS THAT DON'T WORK PROPERLY?...

...ASKED PAUL BROCA (1824–1880).

HE HELPED MOVE THE WORLD AWAY FROM THE DELIGHTFUL BUT NONSENSICAL STUDY OF PHRENOLOGY.
III
V

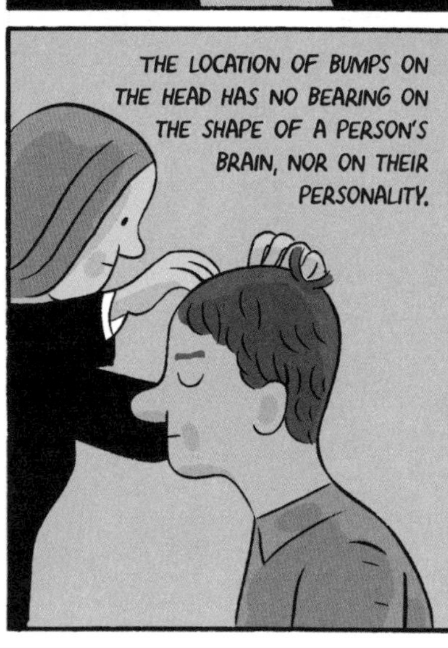
THE LOCATION OF BUMPS ON THE HEAD HAS NO BEARING ON THE SHAPE OF A PERSON'S BRAIN, NOR ON THEIR PERSONALITY.

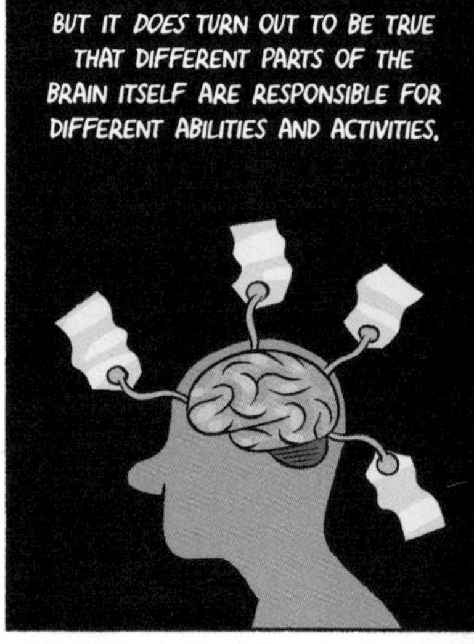
BUT IT DOES TURN OUT TO BE TRUE THAT DIFFERENT PARTS OF THE BRAIN ITSELF ARE RESPONSIBLE FOR DIFFERENT ABILITIES AND ACTIVITIES.

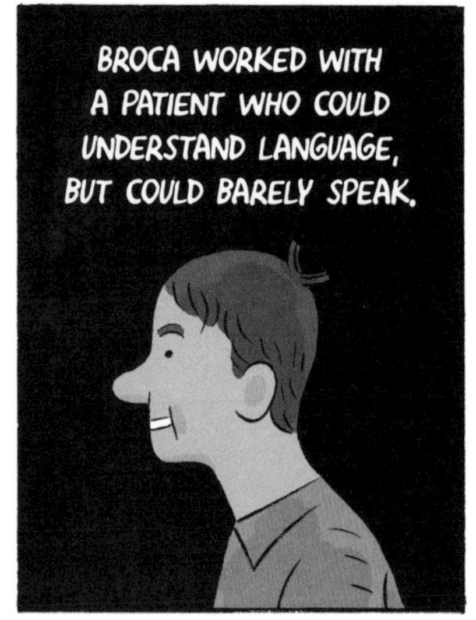
BROCA WORKED WITH A PATIENT WHO COULD UNDERSTAND LANGUAGE, BUT COULD BARELY SPEAK.

HE DISCOVERED A BRAIN REGION THAT WAS RESPONSIBLE FOR SPEECH <u>PRODUCTION</u> – STILL KNOWN AS "BROCA'S AREA" – THAT IS NOT ALSO INVOLVED IN SPEECH <u>COMPREHENSION</u>.

THE TASK OF MAPPING OUT THE BRAIN ACCORDING TO FUNCTION IS STILL ONGOING (SEE PAGE 19).

A LOT OF EARLY BRAIN MAPPING WAS THE RESULT OF STUDYING PEOPLE WITH LESIONS...

...A FANCY TERM THAT SIMPLY MEANS HAVING HOLES IN THE HEAD.

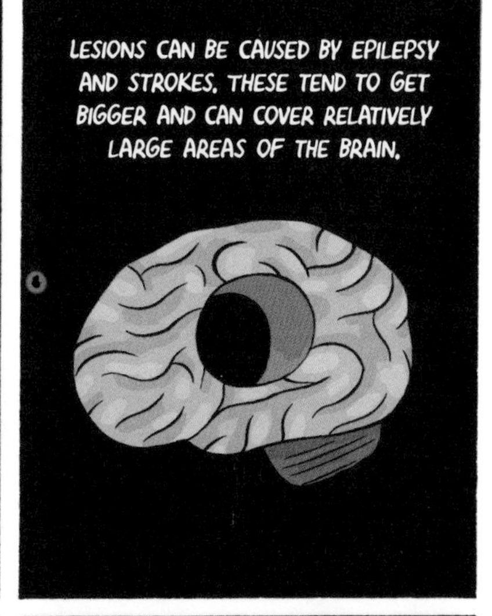
LESIONS CAN BE CAUSED BY EPILEPSY AND STROKES. THESE TEND TO GET BIGGER AND CAN COVER RELATIVELY LARGE AREAS OF THE BRAIN.

SO FAR IN THIS POTTED HISTORY OF NEUROSCIENCE, I COUNT 6 WHITE EUROPEAN MALES, MOST WITH BEARDS...

WELL, WHILE WE'RE TALKING ABOUT BRAIN MAPPING, LET'S INTRODUCE TATSUJI INOUYE (1881–1976), FROM JAPAN.

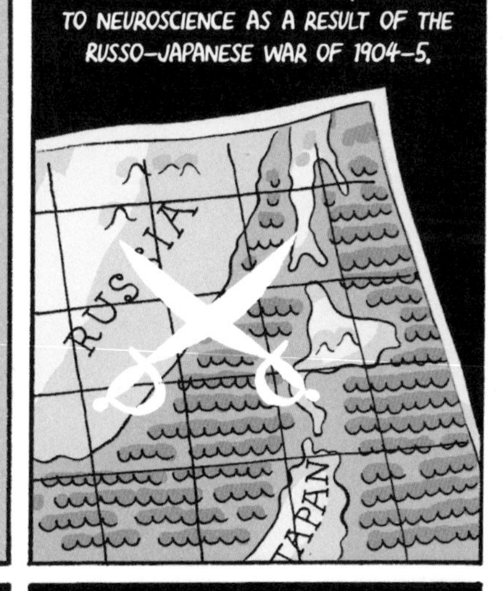
A TRAINED OPHTHALMOLOGIST, HE CAME TO NEUROSCIENCE AS A RESULT OF THE RUSSO-JAPANESE WAR OF 1904–5.
RUSSIA
RUSSIA
JAPAN

HE DIDN'T FIGHT, BUT HE DID SEE A LOT OF SOLDIERS WHO HAD SURVIVED BEING SHOT IN THE HEAD.

THE THING IS, BULLETS CAN CREATE VERY NEAT LESIONS IN THE BRAIN.

INOUYE WAS ABLE TO MAP OUT THE VISUAL CORTEX IN SOME DETAIL. IT REMAINS ONE OF THE BEST-MAPPED BRAIN REGIONS.
VISUAL CORTEX

SEEMS A LITTLE TOKENISTIC. AND WHAT ABOUT WOMEN?

WELL, LET'S MOVE ON TO IDA HYDE (1857–1945).

IN 1921, HYDE INVENTED A SET OF MICROELECTRODES THAT SHE USED TO LOCATE AND MEASURE NERVE IMPULSES IN SEA CREATURES.

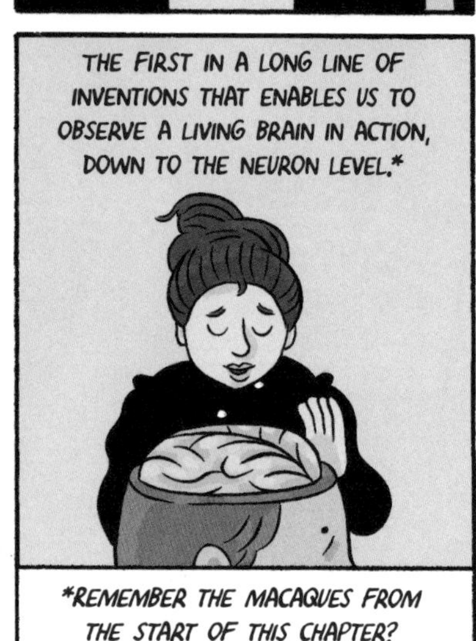
THE FIRST IN A LONG LINE OF INVENTIONS THAT ENABLES US TO OBSERVE A LIVING BRAIN IN ACTION, DOWN TO THE NEURON LEVEL.*

*REMEMBER THE MACAQUES FROM THE START OF THIS CHAPTER?

SHE WAS ALSO A VOCAL ADVOCATE FOR GETTING WOMEN INTO THE SCIENCES, AND FOR GETTING EQUAL PAY.

IS THAT STRICTLY RELEVANT HERE?
IT'S ALWAYS RELEVANT.

THIS BRINGS US TO OUR FIRST STILL-LIVING SCIENTIST: ELIZABETH WARRINGTON (BORN 1931).

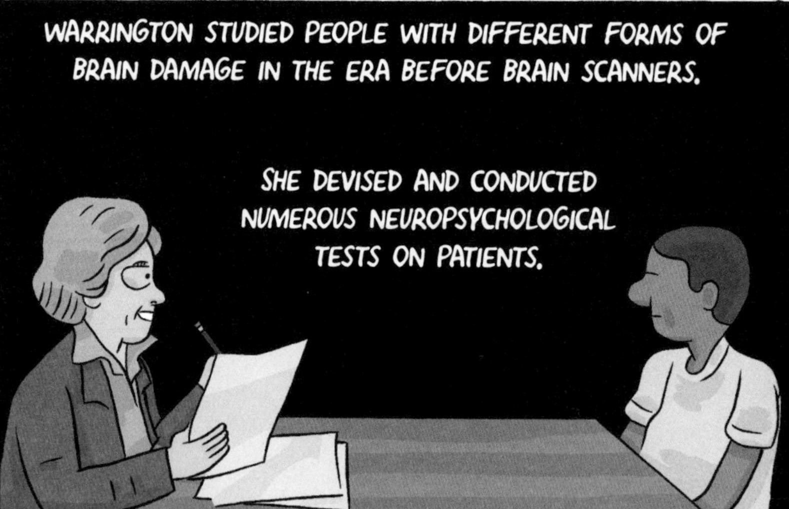
WARRINGTON STUDIED PEOPLE WITH DIFFERENT FORMS OF BRAIN DAMAGE IN THE ERA BEFORE BRAIN SCANNERS.
SHE DEVISED AND CONDUCTED NUMEROUS NEUROPSYCHOLOGICAL TESTS ON PATIENTS.

THE WAY PATIENTS RESPONDED TO THE TESTS ALLOWED WARRINGTON TO DETERMINE WHERE LESIONS WERE LOCATED IN THEIR BRAINS.

TEST PAPER.
Q1.
Q2.
Q3.
Q4.

SORT OF LIKE REVERSE BRAIN-MAPPING. INSTEAD OF SEEING A BULLET HOLE AND DETERMINING THE EFFECTS ON THE PATIENT...

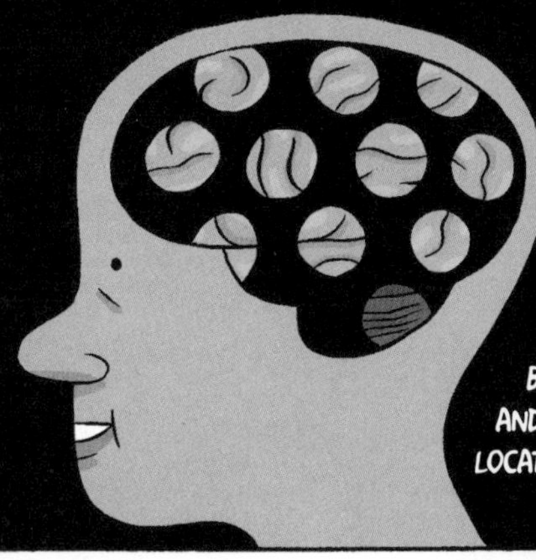

...WARRINGTON WAS ABLE TO EXAMINE A PATIENT BY TALKING TO THEM, AND THEN DISCOVER THE LOCATION (IF ANY) OF AN INTERNAL HOLE.

BRENDA MILNER (BORN 1918)...

...INVESTIGATED MEMORY, IN PARTICULAR ITS LINK TO THE HIPPOCAMPUS.

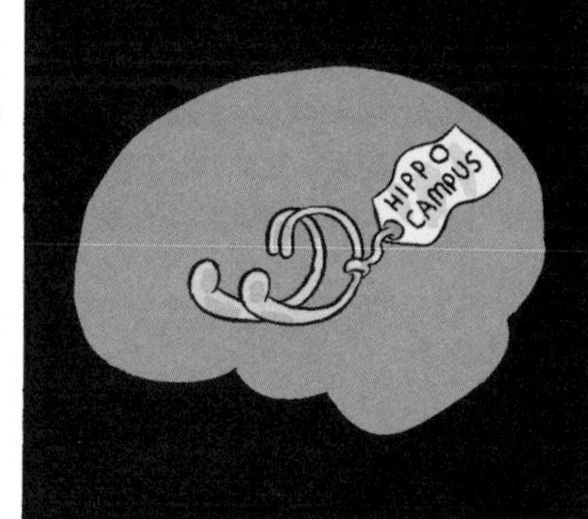

HIPPO CAMPUS

SHE WORKED WITH A FAMOUS AMNESIC PATIENT NAMED "H. M.," WHO COULD NOT REMEMBER NEW EVENTS...

...BUT, AS MILNER FOUND, COULD STILL PICK UP NEW SKILLS.

MILNER WAS THE FIRST TO IDENTIFY A DISTINCTION BETWEEN "EPISODIC MEMORY" AND "PROCEDURAL MEMORY," TO USE THE TECHNICAL TERMS.

EPISODIC: EVENTS I HAVE EXPERIENCED
PROCEDURAL: THINGS I HAVE LEARNED HOW TO DO

SHE ALSO LOCATED A VERY SPECIFIC SUBSET OF MEMORY: THINGS THAT YOU KNOW YOU REMEMBER, STORED IN THE HIPPOCAMPUS. WE'LL GET ONTO THIS IN A COUPLE OF CHAPTERS.

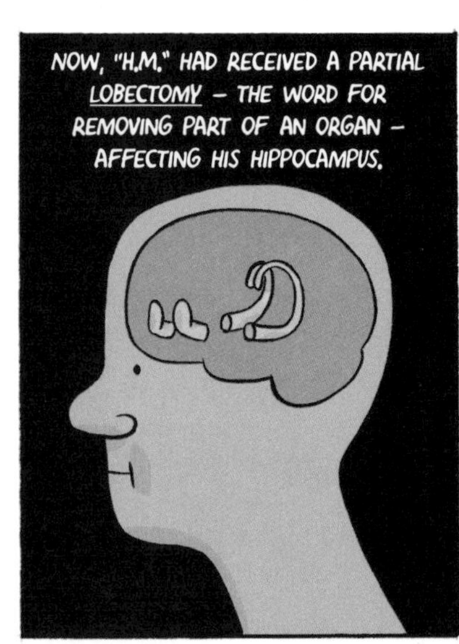
NOW, "H.M." HAD RECEIVED A PARTIAL LOBECTOMY – THE WORD FOR REMOVING PART OF AN ORGAN – AFFECTING HIS HIPPOCAMPUS.

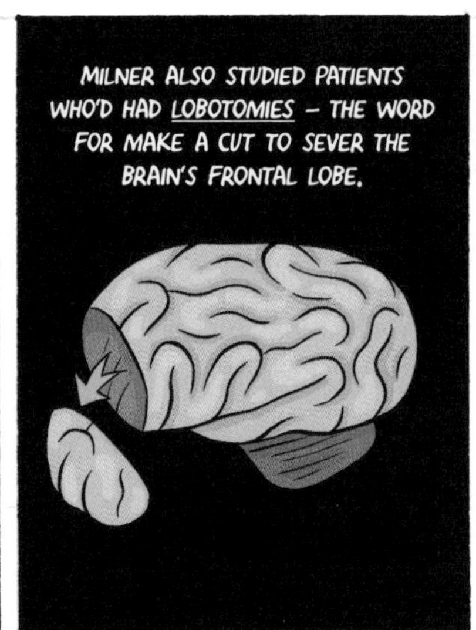
MILNER ALSO STUDIED PATIENTS WHO'D HAD LOBOTOMIES – THE WORD FOR MAKE A CUT TO SEVER THE BRAIN'S FRONTAL LOBE.

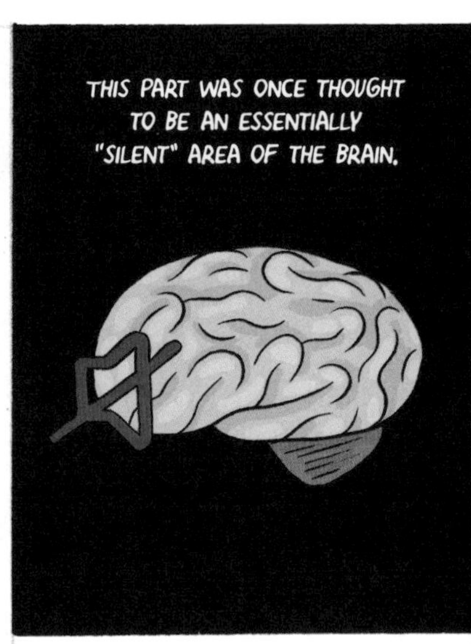
THIS PART WAS ONCE THOUGHT TO BE AN ESSENTIALLY "SILENT" AREA OF THE BRAIN.

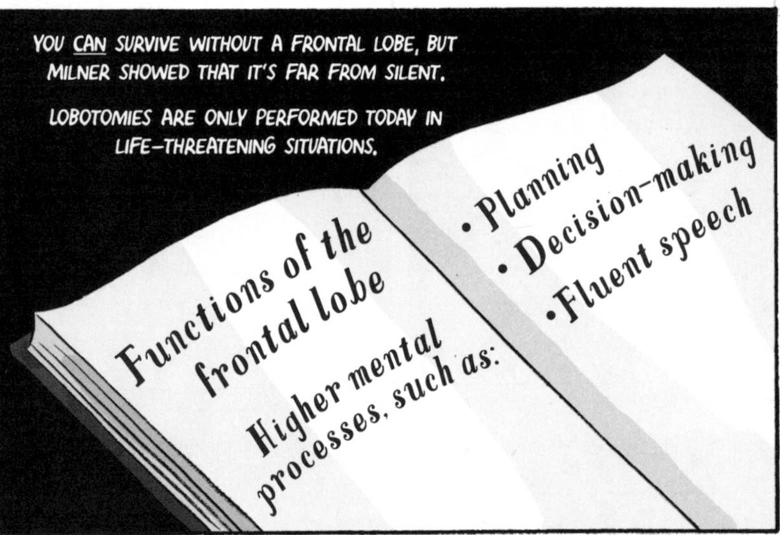
YOU CAN SURVIVE WITHOUT A FRONTAL LOBE, BUT MILNER SHOWED THAT IT'S FAR FROM SILENT.
LOBOTOMIES ARE ONLY PERFORMED TODAY IN LIFE-THREATENING SITUATIONS.
Functions of the frontal lobe
Higher mental processes, such as:
• Planning
• Decision-making
• Fluent speech

THE NEXT BIG STEP IN NEUROSCIENCE CAME FROM WELL BEYOND THE STUDY OF PSYCHOLOGY OR MEDICINE.

ENTER NORBERT WIENER (1894–1964), FOUNDER OF THE FIELD OF CYBERNETICS.
AND ALSO CLAUDE SHANNON (1916–2001), FOUNDER OF INFORMATION THEORY.

CYBERNETICS IS STRONGLY ASSOCIATED WITH COMPUTING, BUT IT'S MORE GENERALLY THE STUDY OF INFORMATION AND CONTROL.
FOR INSTANCE, HOW CAN THE RIGHT INFORMATION BE PROCESSED AND COMMUNICATED IN ORDER TO STEER THIS SHIP?

BOTH PROMOTED THE CONCEPT THAT INFORMATION (FOR EXAMPLE, LANGUAGE)...
hello.

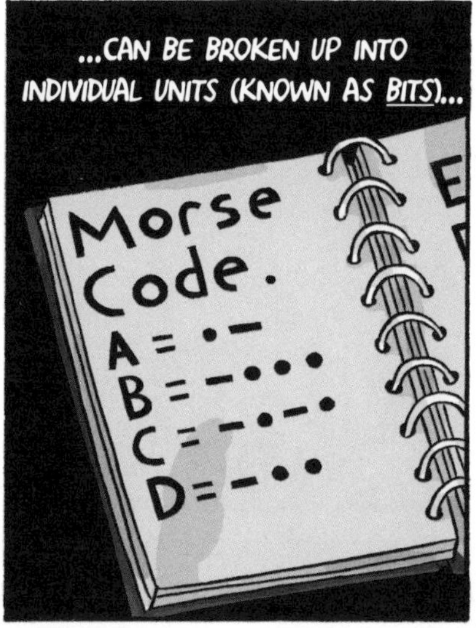
...CAN BE BROKEN UP INTO INDIVIDUAL UNITS (KNOWN AS BITS)...
Morse Code.
A = •—
B = —•••
C = —•—•
D = —••

...THAT CAN BE COMMUNICATED FROM ONE PERSON TO ANOTHER, OR INDEED FROM ONE MACHINE TO ANOTHER.
NOT TO MENTION, ONE NEURON TO ANOTHER.

WIENER DREW AN EXPLICIT PARALLEL BETWEEN BITS OF INFORMATION AS USED TO CONTROL COMPUTERS...

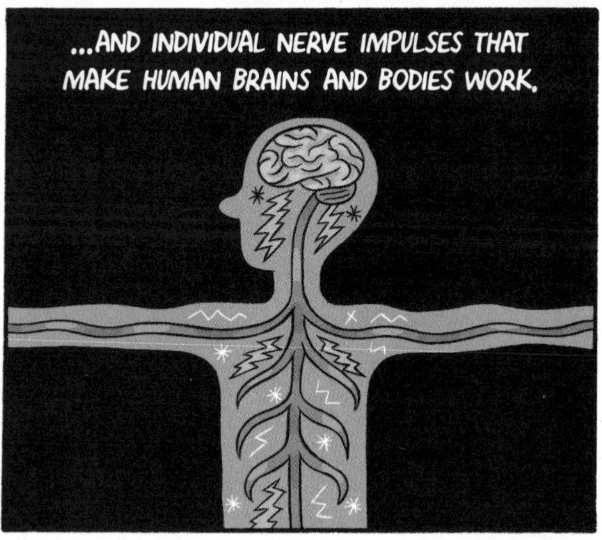
...AND INDIVIDUAL NERVE IMPULSES THAT MAKE HUMAN BRAINS AND BODIES WORK.

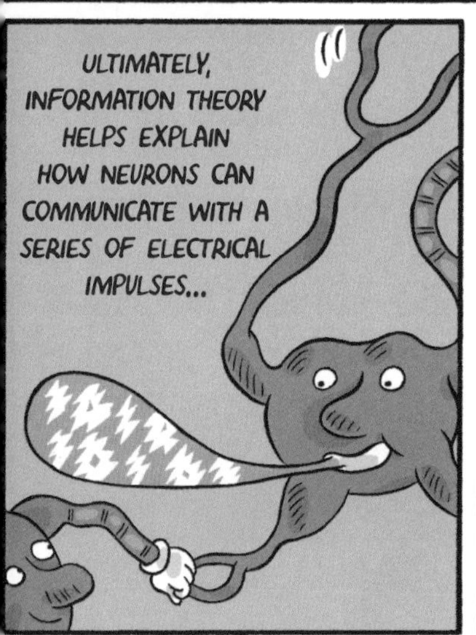
ULTIMATELY, INFORMATION THEORY HELPS EXPLAIN HOW NEURONS CAN COMMUNICATE WITH A SERIES OF ELECTRICAL IMPULSES...

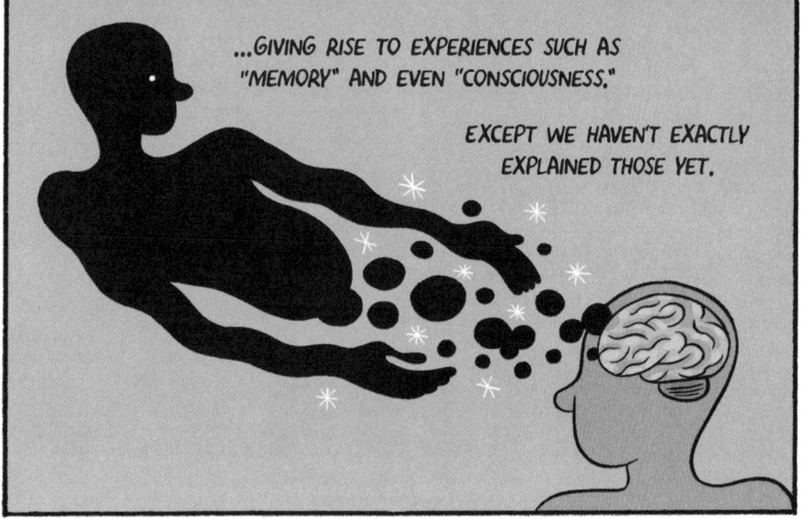
...GIVING RISE TO EXPERIENCES SUCH AS "MEMORY" AND EVEN "CONSCIOUSNESS."
EXCEPT WE HAVEN'T EXACTLY EXPLAINED THOSE YET.

ADD INTO THE MIX ALAN TURING (1912–1954) AND JOHN VON NEUMANN (1903–1957).

BETWEEN THEM (AND VARIOUS OTHERS), THEY INVENTED THE BASIC ARCHITECTURE FOR THE MODERN COMPUTER.

Turing, in particular, pursued the idea that a computer program could describe virtually any process - including, perhaps, human conscious thought.

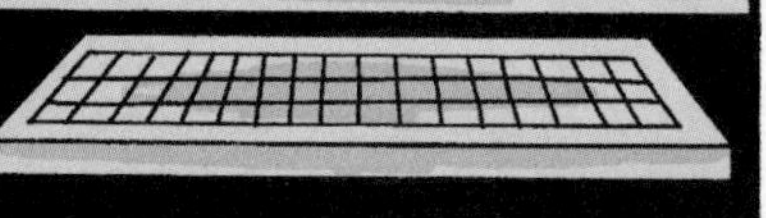

A COMPUTER CONVERSATION PROGRAM THAT PASSES FOR HUMAN IS SAID TO HAVE PASSED THE "TURING TEST."

PASSING THE TEST WOULD APPEAR TO RELY ON A MACHINE'S ABILITY TO MIMIC HUMAN CONVERSATION STYLES.

SOME COMPUTER PROGRAMMERS CLAIM TO HAVE WRITTEN PROGRAMS THAT HAVE PASSED THE TURING TEST ALREADY.

IT SEEMS LIKELY THAT THE TEST WILL BE PASSED EASILY, WITH FULLY REPEATABLE RESULTS, WITHIN THE COMING DECADE.

CHRIS'S OFFICIAL FAVORITE FILM (ACCORDING TO AT LEAST ONE BIO)* IS BLADE RUNNER, WHICH FEATURES A VARIATION ON THE TURING TEST DUBBED THE VOIGHT-KAMPFF TEST.

TAKING ON A THEME COMMON IN PHILIP K. DICK'S WORK, THE STORY IS ABOUT THE CONFUSION OF REAL EXPERIENCES AND ARTIFICIAL MEMORIES.

IT'S INCREASINGLY CLEAR THAT, WITHIN THE BRAIN ITSELF, THERE'S NO WAY TO TELL THE DIFFERENCE.

*HIS ACTUAL FAVORITE FILM IS OF COURSE HIGHLY VARIABLE.
TWO STRONG CONTENDERS ARE: ZAZIE DANS LE MÉTRO AND MY NEIGHBOUR TOTORO.

THIS BRIEF HISTORY SHOWS THE FOUNDATIONS THAT LEAD TO MODERN NEUROSCIENCE.
WHAT WE ATTEMPT THESE DAYS IS A COMBINATION OF TWO CHALLENGES.

1) DEVISING CLEVER TASKS THAT TRY TO ISOLATE MENTAL FUNCTIONS INTO SMALL, OBSERVABLE UNITS.

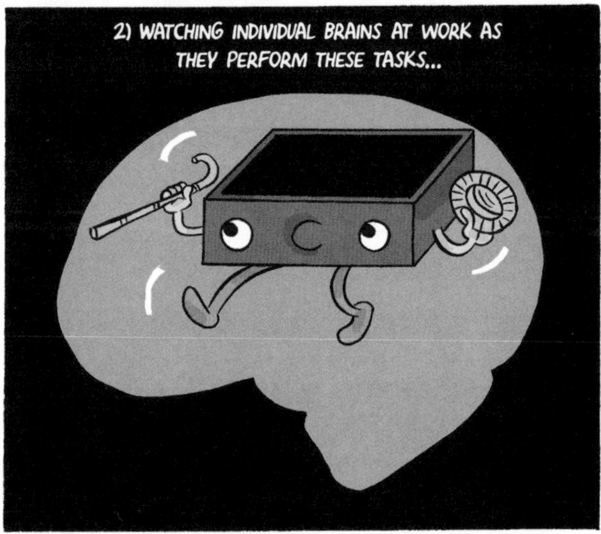
2) WATCHING INDIVIDUAL BRAINS AT WORK AS THEY PERFORM THESE TASKS...

...AND THEN TRYING TO DISCERN WHAT IS COMMON TO ALL BRAINS...

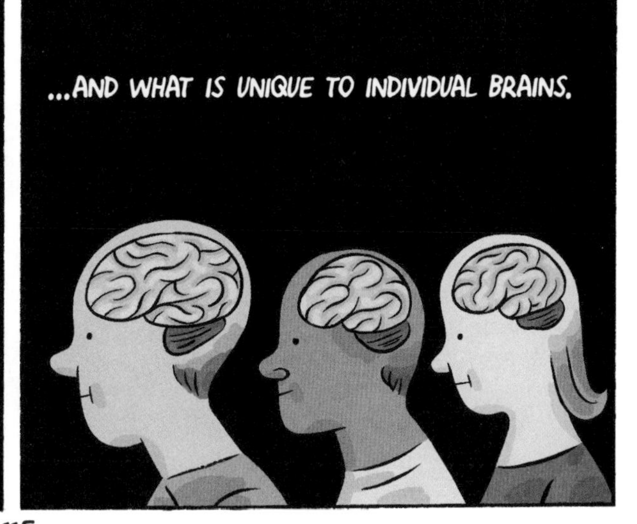
...AND WHAT IS UNIQUE TO INDIVIDUAL BRAINS.

ONE FEATURE THAT SEEMS TO BE COMMON TO ALL (HUMAN) BRAINS, IS THE PRESENCE OF MIRROR NEURONS. BUT THERE'S SOME VARIATION IN HOW STRONGLY THEY AFFECT US.

THESE NEURONS FORCE US TO SHARE IN THE SENSATIONS THAT WE CAN SEE OTHER PEOPLE EXPERIENCING.

FROM OUR OWN DAILY EXPERIENCE, IT IS NOT SURPRISING TO LEARN THAT WE ARE ABLE TO RECOGNIZE WHEN SOMEONE ELSE IS BEING TOUCHED.

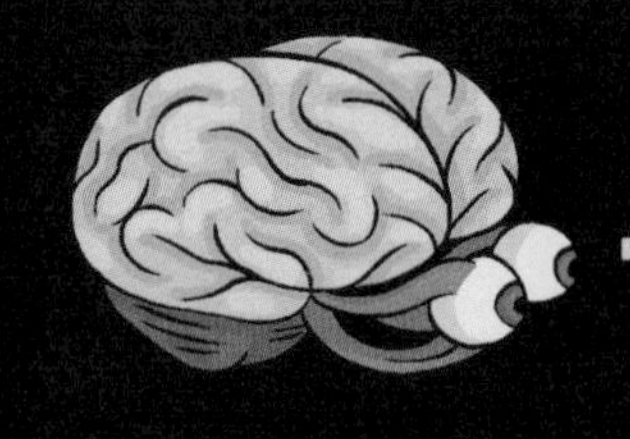

BUT IT IS A TRIUMPH OF SCIENCE AND RESEARCH THAT WE HAVE LEARNED THIS ABILITY IS IN FACT WIRED INTO OUR BRAINS.

WHICH IS TO SAY, IT HAPPENS <u>WHETHER WE WISH IT TO OR NOT.</u>

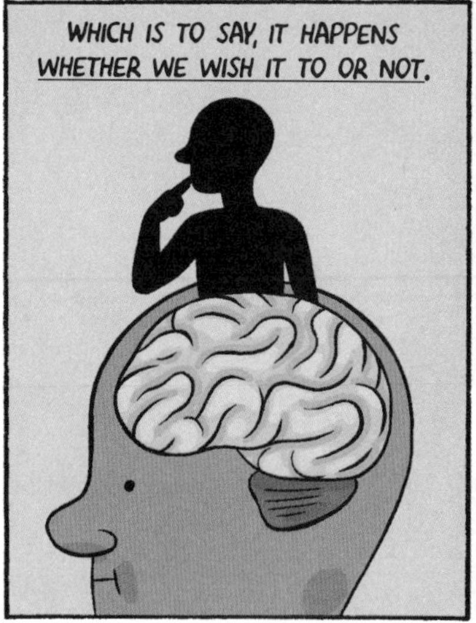

MORE THAN THAT, IN SOME PEOPLE IT CAUSES THE ADDED SENSATION OF FEELING TOUCHED THEMSELVES, IN EXACTLY THE SAME PLACE.
IT'S IMPORTANT TO BE CLEAR THAT WE DON'T YET KNOW IF MIRROR NEURONS ARE SOMETHING INNATE TO HUMAN BRAINS.
IT MAY WELL BE THAT THEY ARE FORMED AFTER A PERIOD OF LEARNING.

WHAT IS INNATE IS THE BASIC MACHINERY, IN THE BRAIN'S BODY-CONTROL SYSTEM, THAT LETS ONE PERSON COPY ANOTHER.

BUT MIRROR NEURONS RAISE THE OLD PROBLEM OF SELF-AWARENESS:

IF THE SAME NEURONS ARE ACTIVE WHEN I DO SOMETHING AS WHEN I SEE SOMEONE ELSE DOING SOMETHING, HOW DO I KNOW WHO IS DOING THE DOING?
READ ON TO FIND OUT MORE!

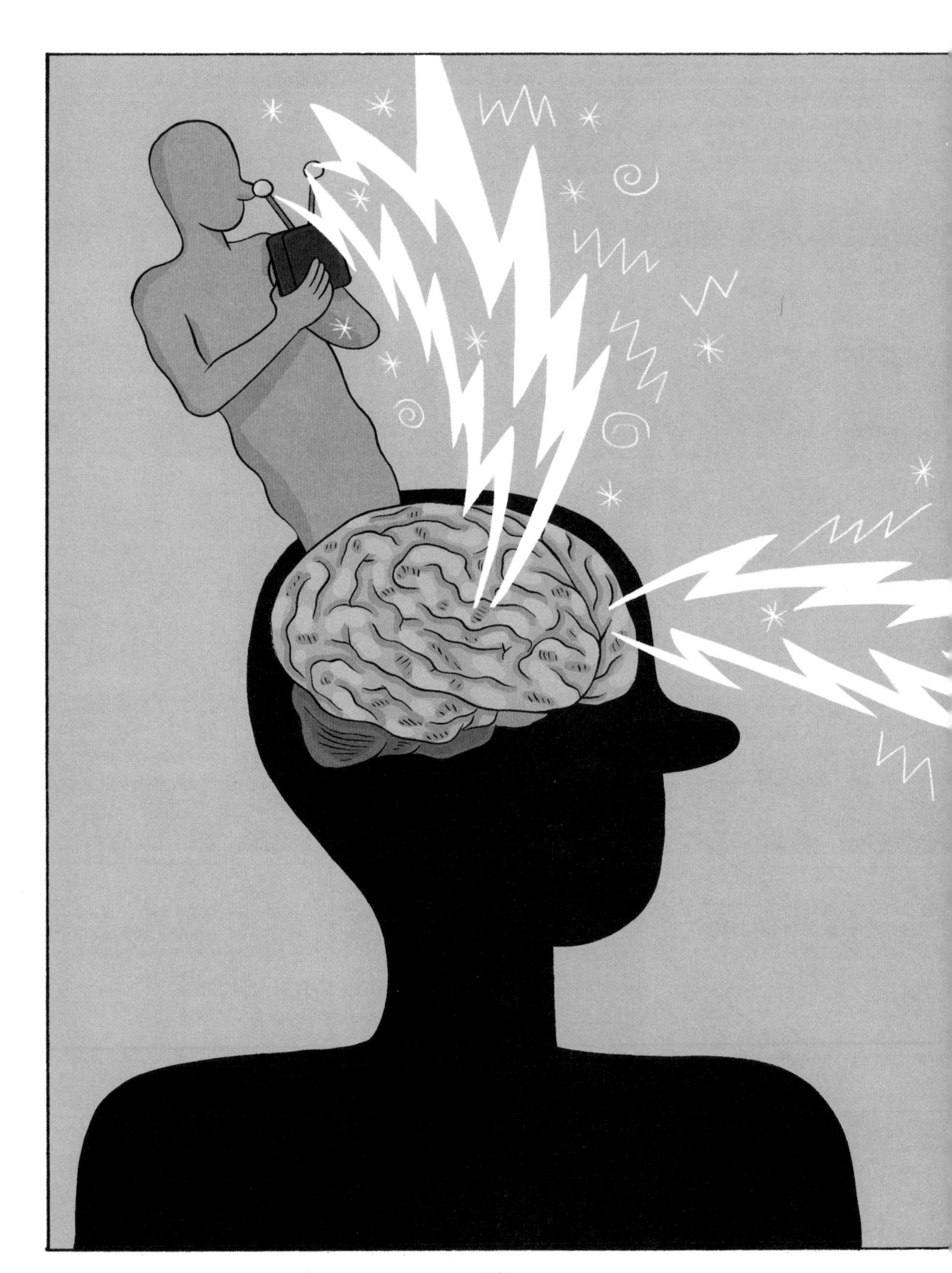

Chapter 6
How the brain knows about itself.

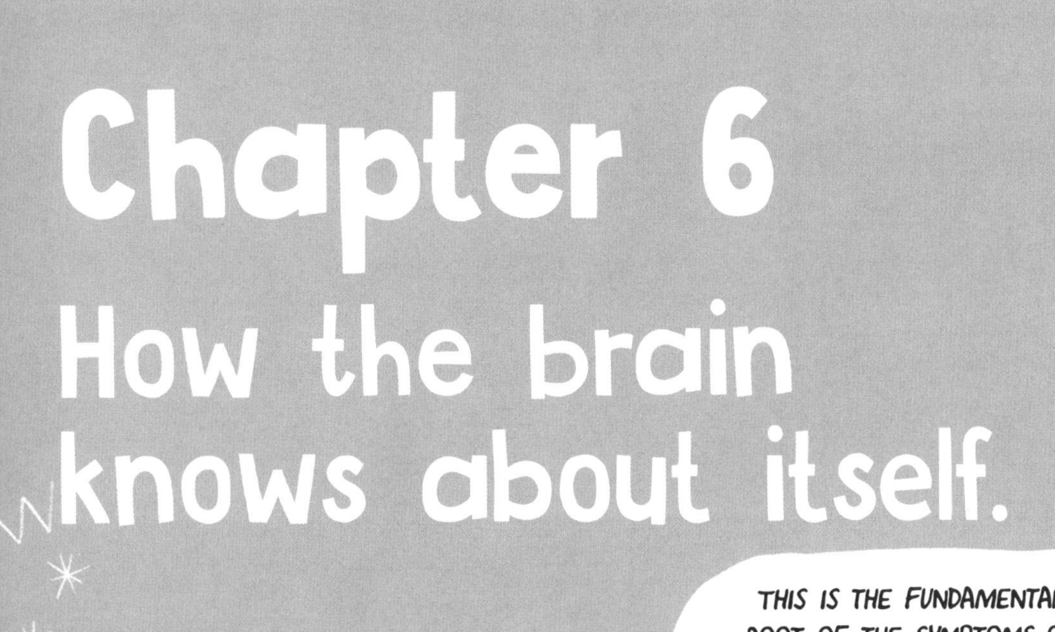

OUR KNOWLEDGE OF OUR OWN BODIES IS FRAGILE AND OUR BRAINS CAN EASILY BECOME CONFUSED.
THIS IS THE FUNDAMENTAL ROOT OF THE SYMPTOMS OF THE DISORDER KNOWN AS SCHIZOPHRENIA.
BEFORE I GET INTO THAT, BACK TO TICKLING.

ONE OF MY OLD PHD STUDENTS, SARAH-JAYNE BLAKEMORE, HAS DEVISED A WAY TO ALLOW PEOPLE TO TICKLE THEMSELVES.

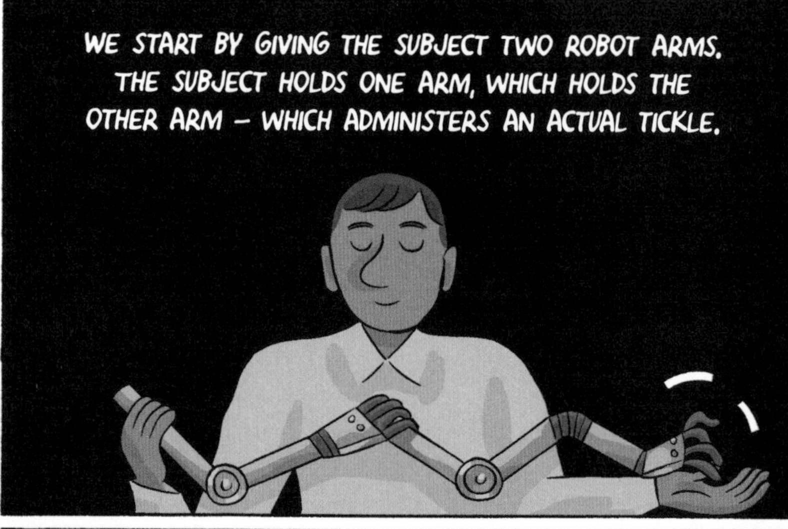
WE START BY GIVING THE SUBJECT TWO ROBOT ARMS. THE SUBJECT HOLDS ONE ARM, WHICH HOLDS THE OTHER ARM – WHICH ADMINISTERS AN ACTUAL TICKLE.

THE SENSATION IS NOT REALLY TICKLISH. MORE LIKE SCRATCHING YOURSELF GENTLY WITH A PENCIL.

I INTRODUCED A KEY FEATURE – A TIME DELAY. THE SUBJECT MOVES THE FIRST ARM AS NORMAL...
...BUT THE SECOND ARM ONLY RESPONDS A FRACTION OF A SECOND LATER.

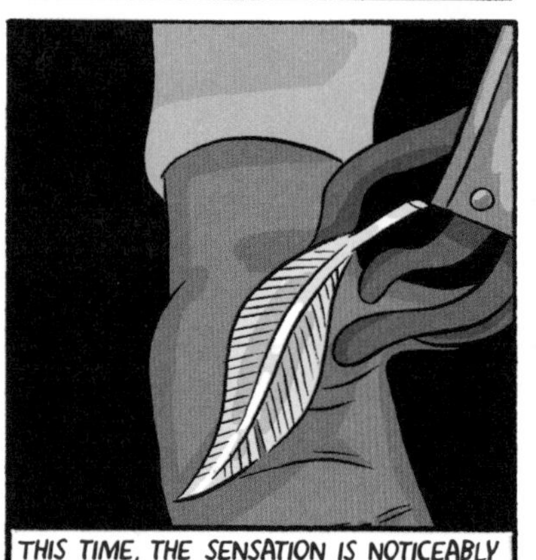
THIS TIME, THE SENSATION IS NOTICEABLY TICKLISH. ALSO, INTERESTINGLY, THE TICKLER DOES NOT NOTICE THAT THERE HAS BEEN A TIME DELAY – ONLY THAT IT FEELS TICKLISH THIS TIME.

IF YOU TICKLE YOURSELF (WITHOUT USING ROBOT ARMS), YOUR BRAIN KNOWS:
1. THERE IS AN INTENTION TO ACT – TO TICKLE YOURSELF.
2. EXACTLY WHICH MUSCLES YOU HAVE COMMANDED TO MOVE.

3. WHAT SENSATIONS TO EXPECT

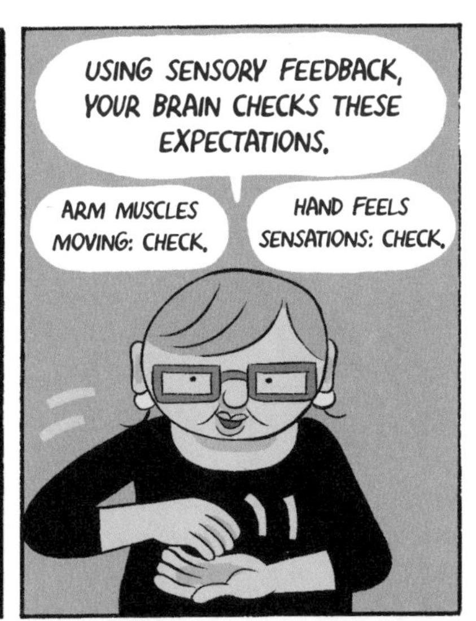
USING SENSORY FEEDBACK, YOUR BRAIN CHECKS THESE EXPECTATIONS.
ARM MUSCLES MOVING: CHECK.
HAND FEELS SENSATIONS: CHECK.

WHEN SOMEONE ELSE TICKLES YOU, YOUR BRAIN DOESN'T HAVE THIS FULL SET OF PREDICTIONS. PART OF THE FRISSON OF "REAL" TICKLING IS THAT YOU HAVEN'T EXACTLY PREDICTED THE SEQUENCE OF SENSATIONS.

IT TURNS OUT THAT AN ERROR IN THE PREDICTION LOOP MAY HELP EXPLAIN THE SYMPTOMS OF SCHIZOPHRENIA...

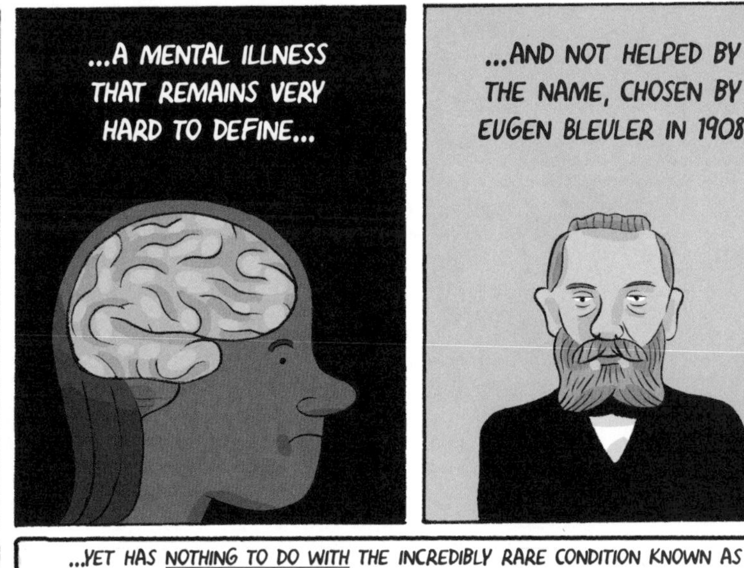
...A MENTAL ILLNESS THAT REMAINS VERY HARD TO DEFINE...

...AND NOT HELPED BY THE NAME, CHOSEN BY EUGEN BLEULER IN 1908.

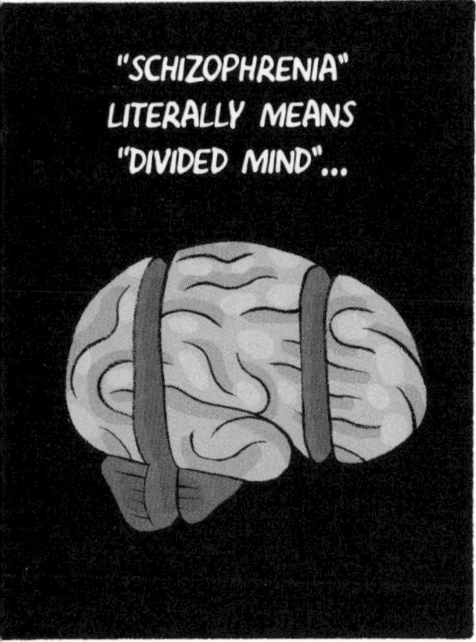
"SCHIZOPHRENIA" LITERALLY MEANS "DIVIDED MIND"...

...YET HAS NOTHING TO DO WITH THE INCREDIBLY RARE CONDITION KNOWN AS "SPLIT PERSONALITY" OR, MORE CORRECTLY, "DISSOCIATIVE IDENTITY DISORDER."

SCHIZOPHRENIA WAS FIRST PROPERLY EXPLORED ONLY IN THE LAST 120 YEARS OR SO. ATTEMPTS TO FIND REFERENCES TO A SIMILAR DISEASE IN THE PAST REMAIN CONTROVERSIAL, AS DO POSTHUMOUS DIAGNOSES OF CREATIVE AND INFLUENTIAL PEOPLE.

ENGLISH POET
JOHN CLARE

DUTCH PAINTER
VINCENT VAN GOGH

CATHOLIC REFORMER
TERESA OF AVILA

THE DISEASE IS CHARACTERIZED ESPECIALLY BY HEARING VOICES, A TYPE OF HALLUCINATION...

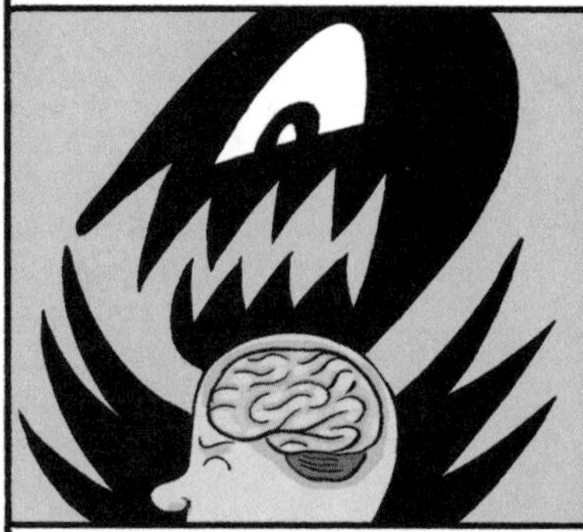

...AND HAVING FALSE BELIEFS ABOUT THE WAY THINGS ARE – DELUSIONS.

BUT PEOPLE EXPERIENCING THEM DON'T REALIZE THEY ARE DELUSIONAL. RECOGNIZING SCHIZOPHRENIA IS HARD.

THE ROOT OF THE DIFFICULTY LIES WITH THE DEFINITION OF A DELUSION. IF ENOUGH PEOPLE WITHIN A PEER GROUP BELIEVE SOMETHING, IT'S SIMPLY "THE TRUTH."
I'M NOT HALLUCINATING. I'M JUST SEEING THE WORLD HOW IT REALLY IS.

BUT IF ONLY ONE PERSON BELIEVES IT, IT'S A DELUSION.

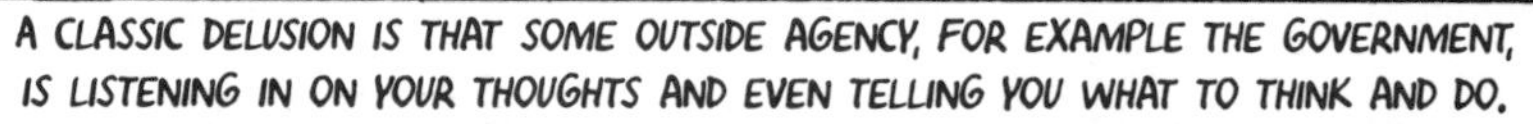
A CLASSIC DELUSION IS THAT SOME OUTSIDE AGENCY, FOR EXAMPLE THE GOVERNMENT, IS LISTENING IN ON YOUR THOUGHTS AND EVEN TELLING YOU WHAT TO THINK AND DO.

IT DOESN'T TAKE A VAST LEAP OF IMAGINATION TO THINK THAT THIS SORT OF OCCURRENCE COULD, ONE DAY, LITERALLY BE TRUE.
THOUGHT POLICE

BUT IT'S FAR MORE LIKELY NOT TO BE TRUE.
BIG BRO Co.

MOST SCHIZOPHRENIA MODELS ARE TESTED USING RATS AND MICE. THEIR BRAINS AND BODIES DO SHOW VARIOUS PHYSIOLOGICAL CHANGES ASSOCIATED WITH SCHIZOPHRENIA BUT IT'S IMPOSSIBLE TO KNOW IF THEY EXPERIENCE DELUSIONS. WE DON'T KNOW IF SCHIZOPHRENIA EVEN EXISTS OUTSIDE OF HUMAN CULTURE.

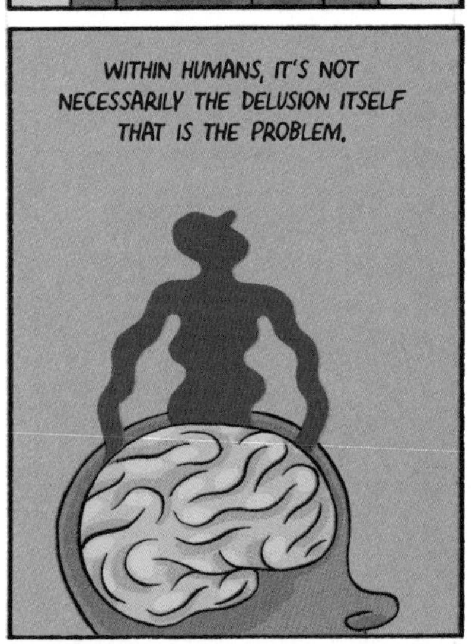
WITHIN HUMANS, IT'S NOT NECESSARILY THE DELUSION ITSELF THAT IS THE PROBLEM.

A SMALL NUMBER OF PEOPLE EXPERIENCE DELUSIONS OR HALLUCINATIONS THAT ARE ESSENTIALLY BENIGN. THEY MIGHT HAVE A FORM OF SCHIZOPHRENIA – BUT THEY WOULDN'T BE DIAGNOSED.

ANGELS ARE WATCHING OVER YOU...

EVEN FOR THE MAJORITY, WHO EXPERIENCE UNPLEASANT DELUSIONS, THE ILLNESS DOESN'T CAUSE ANY PHYSICAL HARM – HOWEVER, IT MAY CAUSE STRANGE BEHAVIOR, IF PEOPLE BEGIN ACTING ON THEIR DELUSIONS.

THE QUEEN IS TALKING TO YOU PERSONALLY...
...YOU'D BETTER GO SEE HER.

SCHIZOPHRENIA IS OFTEN FIRST RECOGNIZED THROUGH A MIX OF:

- SOCIAL WITHDRAWAL
- DISORGANIZED SPEECH
- LACK OF EMOTIONAL RESPONSES
- EVEN AN INABILITY TO FEEL PLEASURE

...MOST OF WHICH MAKE IT HARDER FOR A PERSON TO RELATE TO OTHER PEOPLE.

IN OTHER WORDS, SCHIZOPHRENIA'S MAJOR CONSEQUENCE IS HOW IT REDEFINES A PERSON'S RELATION TO OTHER PEOPLE.

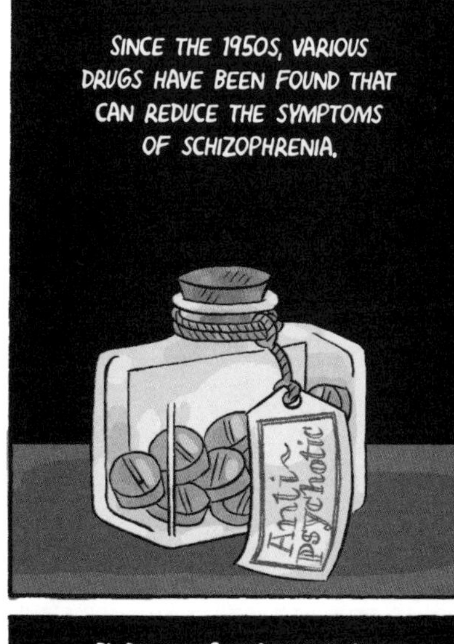
SINCE THE 1950S, VARIOUS DRUGS HAVE BEEN FOUND THAT CAN REDUCE THE SYMPTOMS OF SCHIZOPHRENIA.
Anti-Psychotic

A TREMENDOUS ADVANCE, BUT FAR FROM PERFECT. ALMOST ALL THE DRUGS CAUSE UNPLEASANT SIDE EFFECTS...

DROWSINESS
WEIGHT GAIN
NAUSEA
MEMORY LOSS

...AND TAKING PILLS RELIES ON A PERSON A) REMEMBERING TO DO IT...
M T W Th F S Su

...B) BELIEVING THAT THEY ARE UNWELL, AND THAT PILLS WILL MAKE THEM BETTER, EVEN IF THE PILLS ALSO MAKE THEM PHYSICALLY ILL.

WORSE, THE ONLY REWARD IS THAT THE PILLS HELP PEOPLE "FIT IN" BETTER.

I NEVER LIKED PEOPLE MUCH ANYWAY.

THERE IS A DESPERATE NEED FOR NEW DRUGS – AND OTHER TREATMENTS – BUT THEY'RE ALL PREFERABLE TO BEING LOCKED UP IN A CUCKOO'S NEST, AREN'T THEY?

DIAGNOSIS AND MANAGEMENT OF SCHIZOPHRENIA HAS LONG BEEN A POLITICAL QUESTION...

CENTURIES AGO, PEOPLE WITH ALMOST ANY SORT OF MENTAL DISORDER – ESPECIALLY SCHIZOPHRENIA – WERE CHAINED UP AND RIDICULED.

MORE RECENTLY, THEY WERE SENT TO ASYLUMS...

...WHERE MANY WERE STILL RESTRAINED, IF NOT WITH METAL CHAINS ANYMORE.

NOVELS AND THRILLERS EVEN SUGGESTED THAT "THE POWERS THAT BE" COULD USE ASYLUMS TO DISPOSE OF POLITICAL ENEMIES, BY LABELING THEM AS DELUSIONAL SCHIZOPHRENICS.

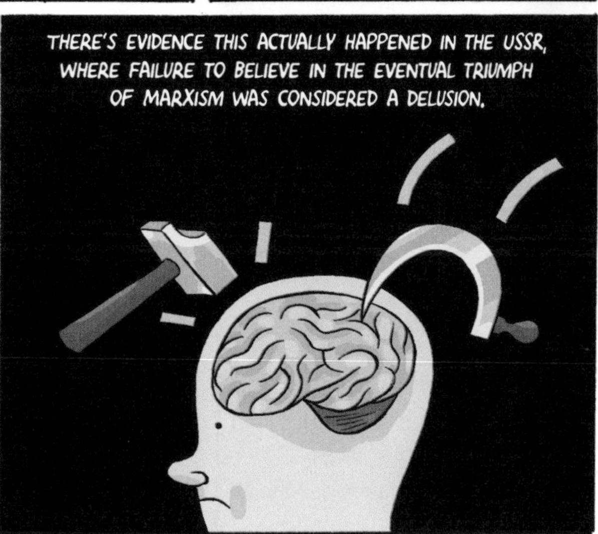
THERE'S EVIDENCE THIS ACTUALLY HAPPENED IN THE USSR, WHERE FAILURE TO BELIEVE IN THE EVENTUAL TRIUMPH OF MARXISM WAS CONSIDERED A DELUSION.

IN THEORY, ASYLUMS ARE SAFE PLACES FOR PEOPLE TO LIVE, AND INMATES GENUINELY BENEFITTED FROM BEING THERE.*
*ALTHOUGH, SADLY, IT'S VERY POSSIBLE THE ABUSE SCANDALS OF RECENT YEARS ARE NOT A NEW PHENOMENON.

OR AT LEAST, THAT'S HOW IT SEEMED TO ME, BASED ON MY REGULAR VISITS TO SHENLEY HOSPITAL (JUST NORTH OF LONDON) TO INTERVIEW AND TEST PATIENTS, MOSTLY THOSE DIAGNOSED WITH SCHIZOPHRENIA.

IT'S A BASIC PART OF RESEARCH THAT YOU NEED TO TEST OUT THEORIES.

SO YES, IT WAS A SPECIAL BENEFIT TO PEOPLE LIKE ME TO HAVE ACCESS TO ASYLUMS WHERE I COULD MEET, SPEAK TO, AND RUN TESTS ON PEOPLE WITH ALL SORTS OF DIAGNOSES.
DO YOU SEE A FACE HERE?

BUT PERHAPS THE BEST EXPERIENCES I HAD AS A RESEARCHER WERE THE YEARS I SPENT WORKING IN A "REGULAR" HOSPITAL.
NORTHWICK PARK

(NOW WE'RE IN NORTH WEST LONDON)

I WAS PART OF A RESEARCH GROUP LED BY TIM CROW, A PSYCHIATRIST.
OUR OFFICES AND LABS WERE IN THE MIDDLE OF A PSYCHIATRIC WARD, COMPLETE WITH PATIENTS, RUN BY EVE JOHNSTONE, ALSO A PSYCHIATRIST.

IT WAS A PRODUCTIVE SETUP: THE INPATIENTS WERE BORED, AND KEEN TO DO OUR EXPERIMENTS. THE DOCTORS, TOO, WERE KEEN TO GET INVOLVED IN RESEARCH.

DR. JOHNSTONE HAD ME MEASURING AREAS IN PHOTOS OF POSTMORTEM BRAINS,* AND, IN TIME, LOOKING AT VERY EARLY SCANS OF BRAIN STRUCTURE.

IT'S ALSO HERE WE LEARNED THAT DRUGS INFLUENCE HALLUCINATIONS AND DELUSIONS...
...BY BLOCKING RECEPTORS FOR A NEUROTRANSMITTER CALLED DOPAMINE.

AND I GOT TO USE AN ULTRA MODERN PDP-11 COMPUTER, TO RUN THE FIRST EXPERIMENTS ENABLING ME TO FIND OUT THAT PATIENTS HAD DIFFICULTY KEEPING TRACK OF THEIR OWN ACTIONS.

*I DIDN'T REMOVE THE BRAINS MYSELF! A TRAINED PATHOLOGIST REMOVED THEM AND TOOK THE NECESSARY PHOTOS, NOT ME.

WE DIDN'T GET ANY NEARER TO UNDERSTANDING THE ROOT CAUSES OF SCHIZOPHRENIA.

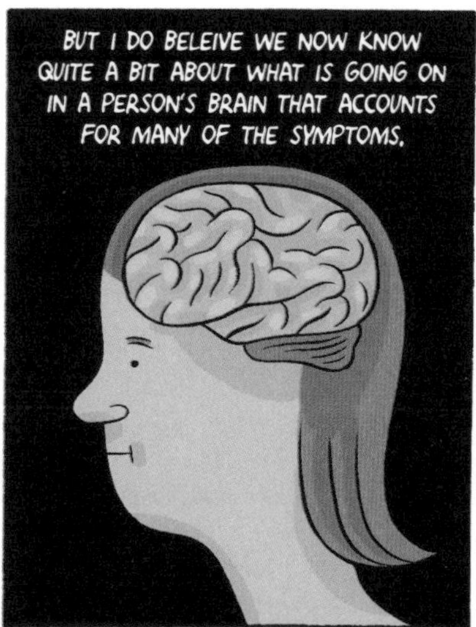
BUT I DO BELEIVE WE NOW KNOW QUITE A BIT ABOUT WHAT IS GOING ON IN A PERSON'S BRAIN THAT ACCOUNTS FOR MANY OF THE SYMPTOMS.

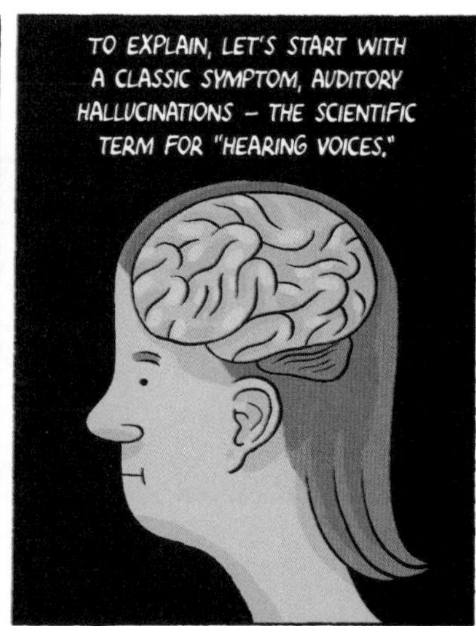
TO EXPLAIN, LET'S START WITH A CLASSIC SYMPTOM, AUDITORY HALLUCINATIONS – THE SCIENTIFIC TERM FOR "HEARING VOICES."

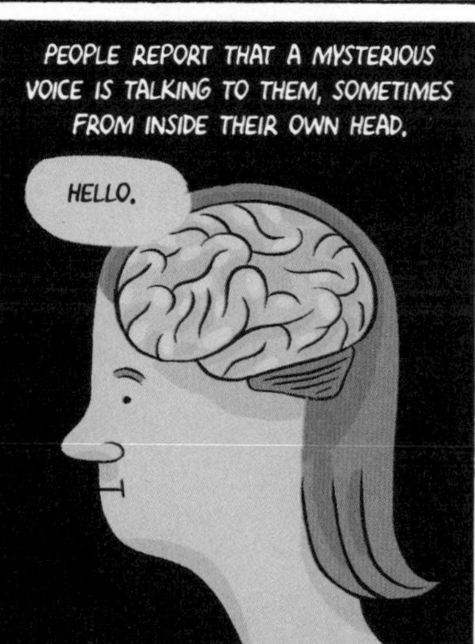
PEOPLE REPORT THAT A MYSTERIOUS VOICE IS TALKING TO THEM, SOMETIMES FROM INSIDE THEIR OWN HEAD.
HELLO.

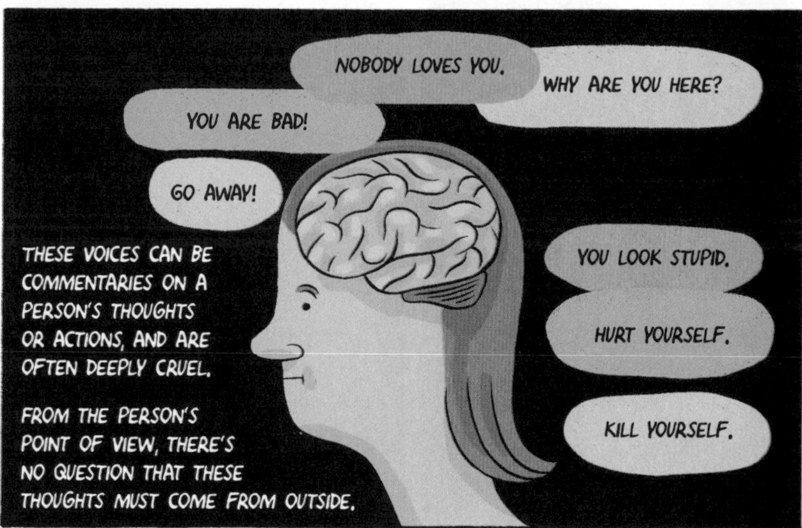
NOBODY LOVES YOU.
WHY ARE YOU HERE?
YOU ARE BAD!
GO AWAY!
THESE VOICES CAN BE COMMENTARIES ON A PERSON'S THOUGHTS OR ACTIONS, AND ARE OFTEN DEEPLY CRUEL.
FROM THE PERSON'S POINT OF VIEW, THERE'S NO QUESTION THAT THESE THOUGHTS MUST COME FROM OUTSIDE.
YOU LOOK STUPID.
HURT YOURSELF.
KILL YOURSELF.

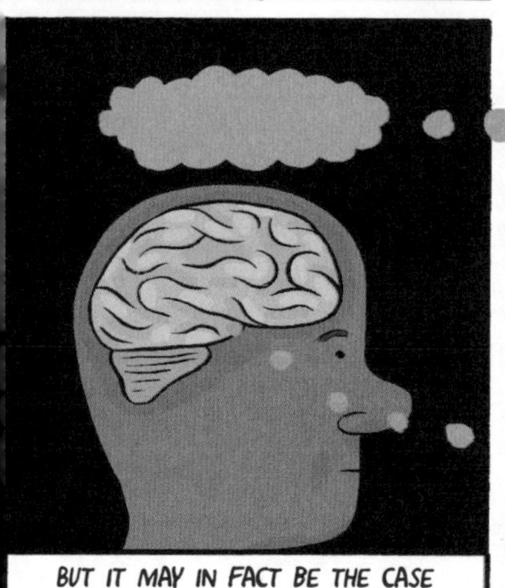
BUT IT MAY IN FACT BE THE CASE THAT THESE "VOICES" ARE FROM THE PERSON'S OWN BRAIN...

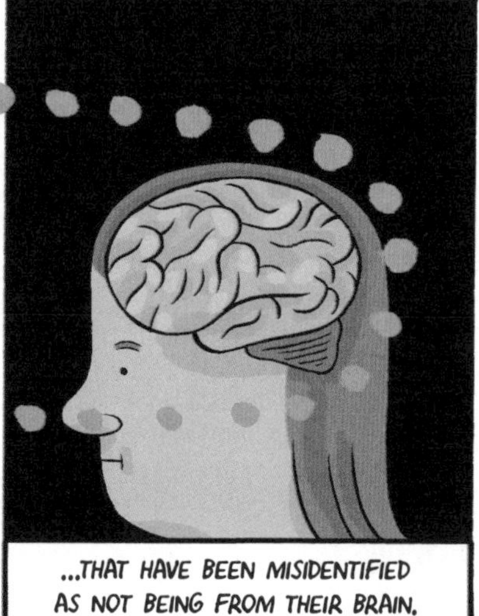
...THAT HAVE BEEN MISIDENTIFIED AS NOT BEING FROM THEIR BRAIN.

LET ME ATTEMPT TO EXPLAIN, BRIEFLY, HOW THIS HAPPENS.*
*JUST IN CASE YOU HAVEN'T READ MY BOOK, THE COGNITIVE NEUROPSYCHOLOGY OF SCHIZOPHRENIA, DON'T WORRY, NO ONE ELSE HAS, EITHER.

EVERY TIME YOU DO OR THINK ANYTHING, ACTIVITY HAPPENS IN YOUR BRAIN. IN BIOLOGICAL TERMS, PATHWAYS OF NEURONS ARE ACTIVATED. DIFFERENT PARTS OF EACH PATHWAY PERFORM DIFFERENT FUNCTIONS. YOU COULD PICTURE THIS AS A WEB OF CONNECTIONS THAT GROWS BIGGER AND BIGGER.

WE DON'T KNOW THE FULL EXTENT OF THESE PATHWAYS OR THEIR FUNCTIONS, BUT HERE ARE SOME THINGS THAT DEFINITELY TAKE PLACE, AS A RESULT OF A COMBINATION OF NEURON ACTIVATION:

1. SOME SORT OF ACTION HAPPENS, FOR EXAMPLE MOVING A MUSCLE, OR THINKING A THOUGHT.

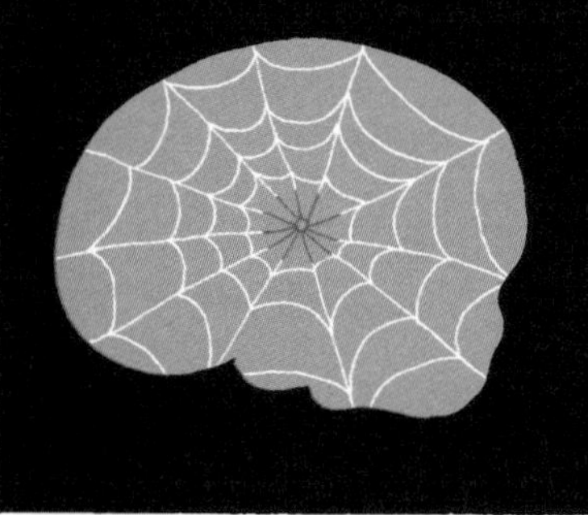

2. YOUR BRAIN IS AWARE THAT THIS ACTION IS TAKING PLACE.

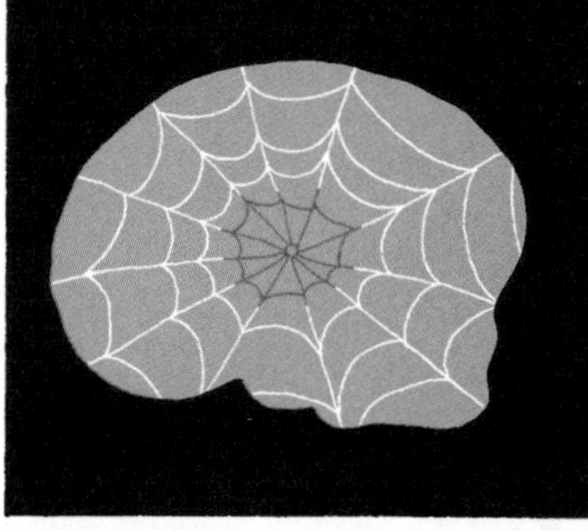

3. YOUR BRAIN STORES A MEMORY OF PERFORMING THIS ACTION.

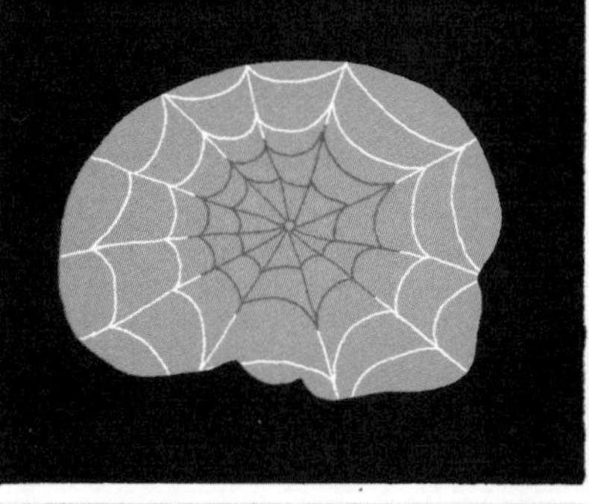

4. SEPARATELY, YOUR BRAIN STORES A MEMORY OF DECIDING TO PERFORM THE ACTION.

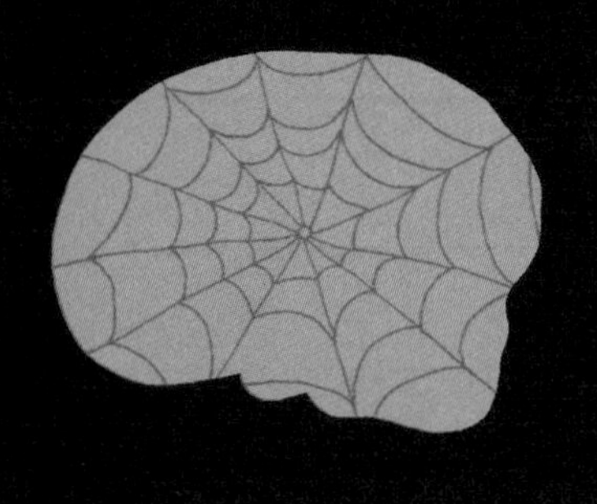

AS THE NAME IS MEANT TO SUGGEST, A SCHIZOPHRENIC EPISODE IN THE BRAIN MEANS THAT THESE FOUR ACTIVITIES (AND WHO KNOWS HOW MANY OTHERS) DO NOT WORK TOGETHER IN HARMONY.

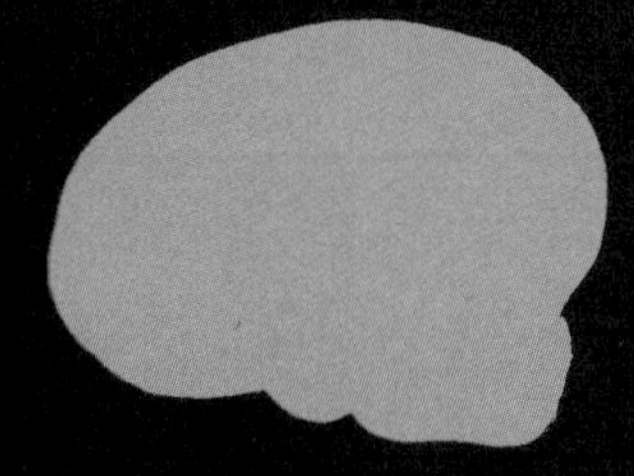

SOMEWHERE IN THE WEB, CONNECTIONS ARE BROKEN – OR PERHAPS NEVER FORM AT ALL.

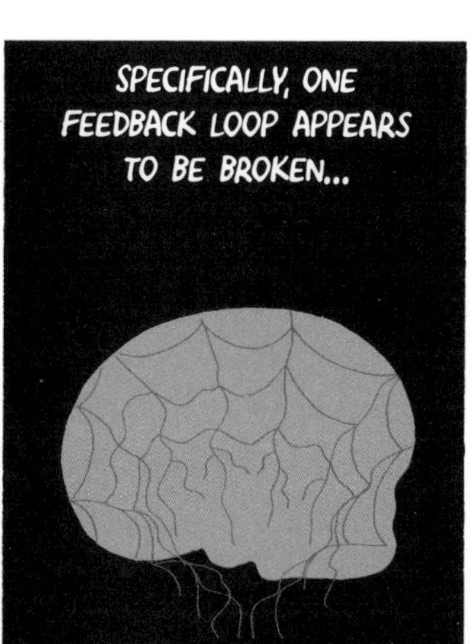
SPECIFICALLY, ONE FEEDBACK LOOP APPEARS TO BE BROKEN...

...THE LOOP THAT ALLOWS YOU TO KNOW THAT IT IS YOU WHO IS THINKING A THOUGHT.
I'M HAVING A THOUGHT.
IT'S MY THOUGHT, AND NO ONE ELSE'S.

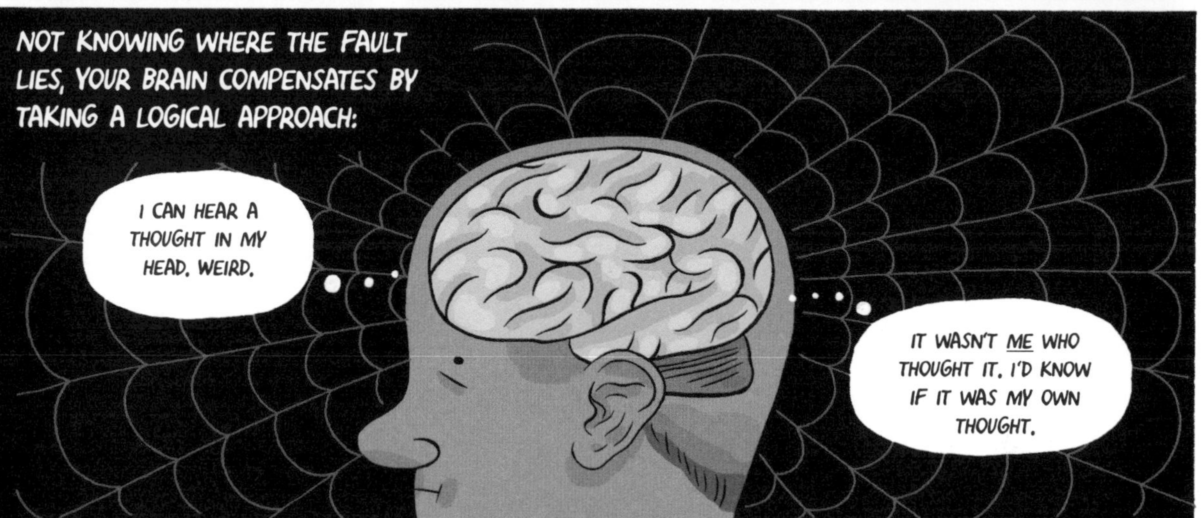
NOT KNOWING WHERE THE FAULT LIES, YOUR BRAIN COMPENSATES BY TAKING A LOGICAL APPROACH:
I CAN HEAR A THOUGHT IN MY HEAD. WEIRD.
IT WASN'T ME WHO THOUGHT IT. I'D KNOW IF IT WAS MY OWN THOUGHT.

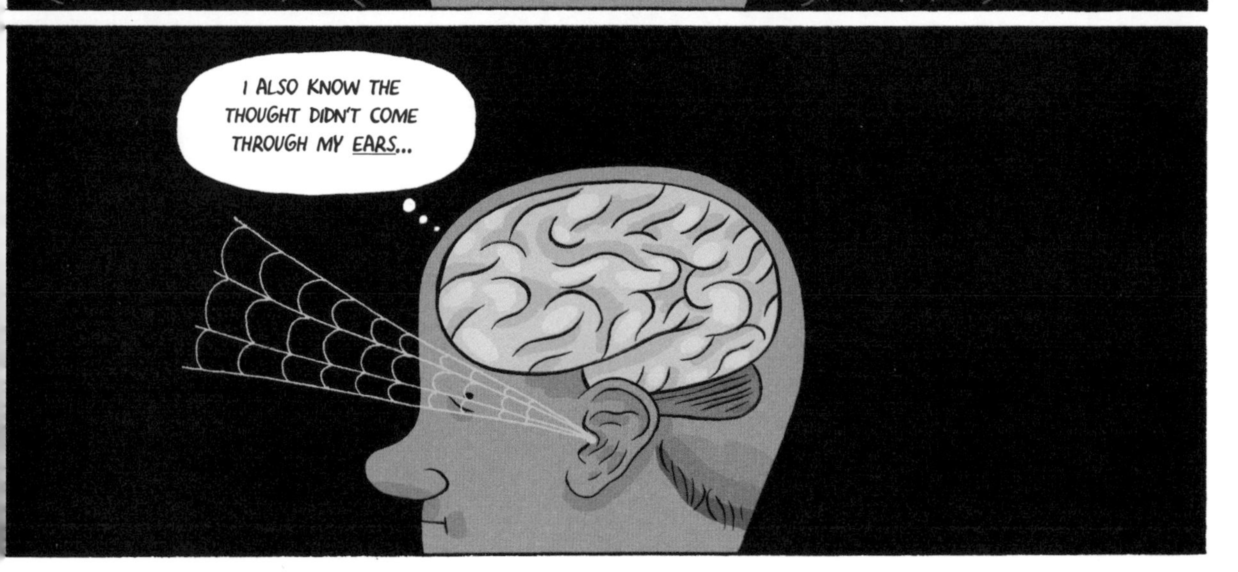
I ALSO KNOW THE THOUGHT DIDN'T COME THROUGH MY EARS...

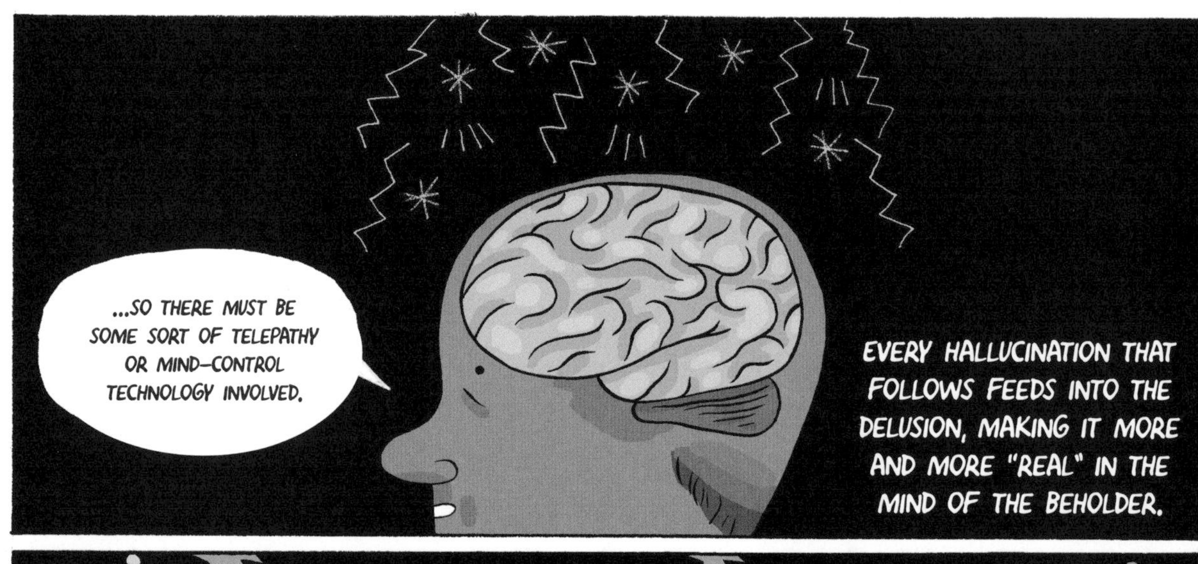
...SO THERE MUST BE SOME SORT OF TELEPATHY OR MIND-CONTROL TECHNOLOGY INVOLVED.
EVERY HALLUCINATION THAT FOLLOWS FEEDS INTO THE DELUSION, MAKING IT MORE AND MORE "REAL" IN THE MIND OF THE BEHOLDER.

SOME OUTSIDE AGENCY MUST BE COMMUNICATING WITH ME. MAYBE IT'S GOVERNMENT SPIES... ...OR ALIENS!
EVERYONE WHO DENIES IT MUST BE PART OF THE CONSPIRACY!
IT'S ALMOST IMPOSSIBLE TO IMAGINE THAT A ROGUE THOUGHT COULD COME FROM YOUR OWN BRAIN – WE HAVE NO PRIOR EXPERIENCE OF THAT SENSATION.
AND WITH EVERY ATTEMPT TO UNDERSTAND AND EXPLAIN WHAT IS GOING ON, A DELUSION BECOMES MORE COMPLEX.

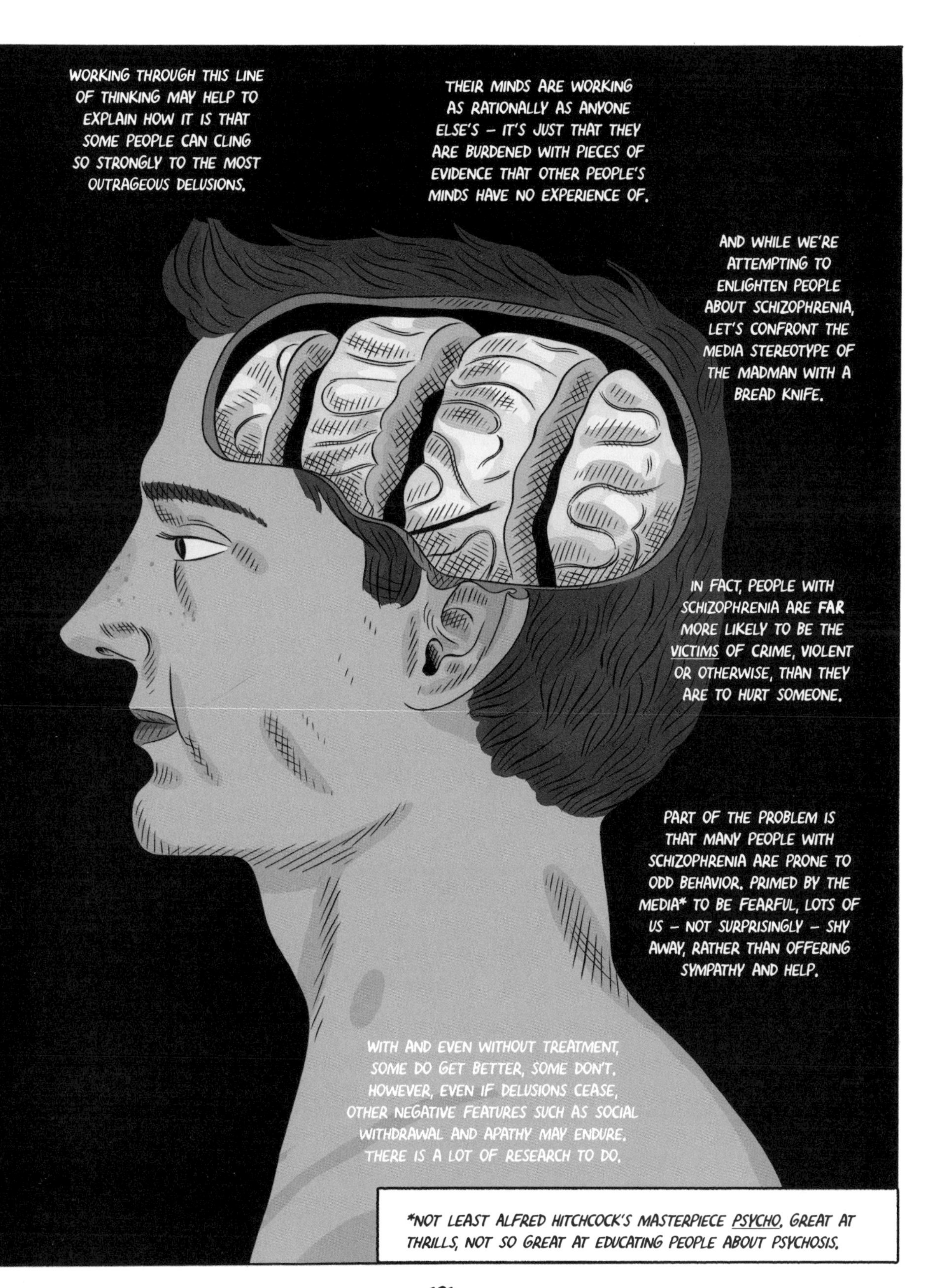
WORKING THROUGH THIS LINE OF THINKING MAY HELP TO EXPLAIN HOW IT IS THAT SOME PEOPLE CAN CLING SO STRONGLY TO THE MOST OUTRAGEOUS DELUSIONS.
THEIR MINDS ARE WORKING AS RATIONALLY AS ANYONE ELSE'S – IT'S JUST THAT THEY ARE BURDENED WITH PIECES OF EVIDENCE THAT OTHER PEOPLE'S MINDS HAVE NO EXPERIENCE OF.
AND WHILE WE'RE ATTEMPTING TO ENLIGHTEN PEOPLE ABOUT SCHIZOPHRENIA, LET'S CONFRONT THE MEDIA STEREOTYPE OF THE MADMAN WITH A BREAD KNIFE.
IN FACT, PEOPLE WITH SCHIZOPHRENIA ARE FAR MORE LIKELY TO BE THE VICTIMS OF CRIME, VIOLENT OR OTHERWISE, THAN THEY ARE TO HURT SOMEONE.
PART OF THE PROBLEM IS THAT MANY PEOPLE WITH SCHIZOPHRENIA ARE PRONE TO ODD BEHAVIOR. PRIMED BY THE MEDIA* TO BE FEARFUL, LOTS OF US – NOT SURPRISINGLY – SHY AWAY, RATHER THAN OFFERING SYMPATHY AND HELP.
WITH AND EVEN WITHOUT TREATMENT, SOME DO GET BETTER, SOME DON'T. HOWEVER, EVEN IF DELUSIONS CEASE, OTHER NEGATIVE FEATURES SUCH AS SOCIAL WITHDRAWAL AND APATHY MAY ENDURE. THERE IS A LOT OF RESEARCH TO DO.
*NOT LEAST ALFRED HITCHCOCK'S MASTERPIECE PSYCHO, GREAT AT THRILLS, NOT SO GREAT AT EDUCATING PEOPLE ABOUT PSYCHOSIS.

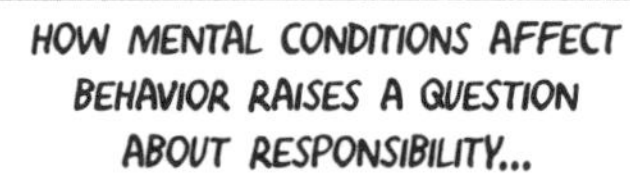
HOW MENTAL CONDITIONS AFFECT BEHAVIOR RAISES A QUESTION ABOUT RESPONSIBILITY...

...AND, ABOVE ALL, FREE WILL.

IF YOU DECIDE TO DO SOMETHING, GOOD OR BAD, AND THEN DO IT, THERE IS NO QUESTION YOU ARE RESPONSIBLE FOR DOING IT. THAT'S THE VERY DEFINITION OF BEING RESPONSIBLE.

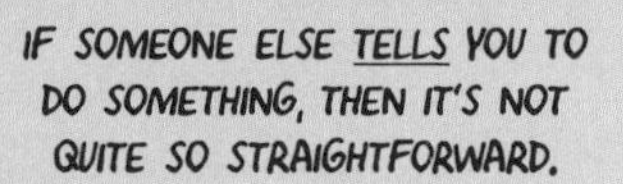
IF SOMEONE ELSE TELLS YOU TO DO SOMETHING, THEN IT'S NOT QUITE SO STRAIGHTFORWARD.

OTHERS MAY JUDGE HOW EASY OR HARD IT IS FOR YOU TO COMPLY OR RESIST. ESPECIALLY IF, FOR EXAMPLE, YOU ARE "FOLLOWING ORDERS."

SO WHAT HAPPENS IF YOU BELIEVE SOMEONE ELSE IS TELLING YOU TO DO SOMETHING, EVEN THOUGH IN FACT THIS IS NOT THE CASE?

IF YOUR BRAIN IS SUFFERING FROM A DISCONNECT OF SELF-AWARENESS, IS YOUR FREE WILL DIMINISHED?

WE'D LIKE TO CLARIFY THAT WE BELIEVE THAT PEOPLE DO HAVE FREE WILL.*
EVEN THOUGH WE ALSO BELIEVE THAT HUMAN ACTIONS CAN, ULTIMATELY, BE EXPLAINED BY OUR BODIES' BIOLOGY, CHEMISTRY, AND PHYSICS.
*MORE TO COME ON FREE WILL IN CHAPTER 11

IMAGINE IF YOU, AND MOST PEOPLE AROUND YOU, DIDN'T HAVE FREE WILL.

THE ONLY WAY TO EXPLAIN OUR THOUGHTS AND MOVEMENTS IS IF THEY ARE...

...PRECISELY CONTROLLED AND DETERMINED BY SOMETHING BEYOND OURSELVES.
FUN, BUT TOTALLY CRAZY!

NOW, THE PHILOSOPHERS AMONG YOU MIGHT WELL ASK, "HOW CAN YOU PROVE THAT THIS ISN'T WHAT IS HAPPENING?"

OR, AS I LIKE TO PUT IT, "HOW DO YOU KNOW THAT IT IS YOU WHO IS IN CONTROL OF YOUR ACTIONS?"

I THINK NEUROSCIENCE HAS SOLVED THIS PROBLEM. FIRST, WE KNOW ABOUT NEURONS THAT GIVE US THE IMPULSE TO COPY WHAT WE SEE IN OTHERS: MIRROR NEURONS.

THESE CAN ONLY WORK IF THE BRAIN HAS A CONCEPT OF "SELF" AND "OTHER."

SECOND, WE KNOW THAT BRAINS WORK THROUGH PREDICTIONS. I KNOW IT'S ME MOVING BECAUSE I CAN PREDICT WHAT IS GONG TO HAPPEN WHEN I MOVE...

...AND THEREFORE I CAN IGNORE THE SENSATIONS THAT WILL OCCUR. THAT'S WHY I CAN'T TICKLE MYSELF.

(UNLESS I USE TWO SETS OF ROBOT ARMS WITH ENOUGH TIME DELAY TO FOOL MY OWN BRAIN.)

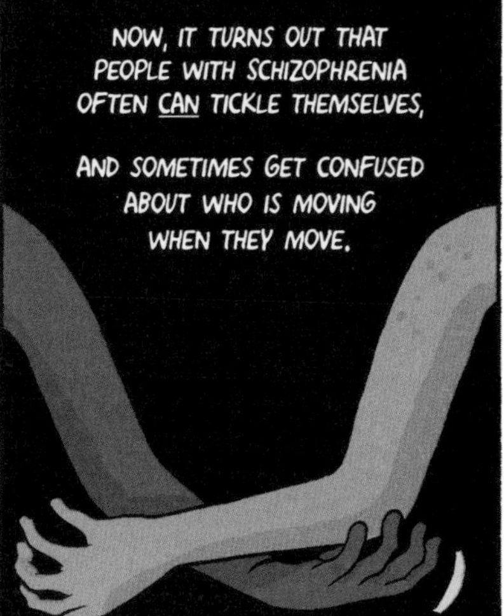
NOW, IT TURNS OUT THAT PEOPLE WITH SCHIZOPHRENIA OFTEN CAN TICKLE THEMSELVES,
AND SOMETIMES GET CONFUSED ABOUT WHO IS MOVING WHEN THEY MOVE.

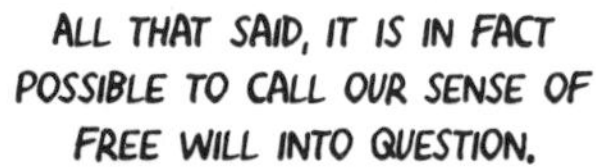
ALL THAT SAID, IT IS IN FACT POSSIBLE TO CALL OUR SENSE OF FREE WILL INTO QUESTION.

AND WE'RE NOT TALKING ABOUT HYPNOTISM, OR EVEN SIMPLE PERSUASION.

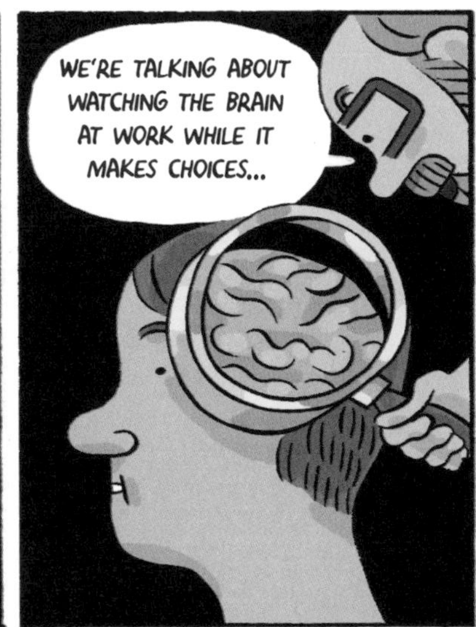
WE'RE TALKING ABOUT WATCHING THE BRAIN AT WORK WHILE IT MAKES CHOICES...

...WHICH, AS IT TURNS OUT, DOESN'T HAPPEN IN THE WAY THAT MOST PEOPLE THINK IT MIGHT.
THE CLASSICAL EXPERIMENT INVOLVES MEASURING A PERSON'S BRAINWAVES (HERE USING AN EEG MACHINE)...

...THEN ASKING THEM TO WIGGLE THEIR FINGERS "WHENEVER THEY FEEL LIKE IT."

THE EEG MACHINE MEASURES ELECTRICAL ACTIVITY IN THE BRAIN, USUALLY CALLED BRAINWAVES.
THESE BRAINWAVES APPEAR AS LINES SCROLLING ACROSS A SCREEN.

CRUCIALLY, THE LINES MOVE SLOWLY ENOUGH THAT CHANGES IN BRAINWAVES CAN BE OBSERVED IN REAL TIME.

THIS TYPE OF EXPERIMENT WAS FIRST DEVISED BY AMERICAN PHYSIOLOGIST BENJAMIN LIBET IN THE 1970s.

A CURIOUS FINDING IS THAT THE RESEARCHER IS ABLE TO "SEE" WHEN THE PERSON IS ABOUT TO MOVE THEIR FINGERS...

...HERE IT COMES...

...UP TO 300 MILLISECONDS BEFORE THE PERSON HIMSELF REPORTS HAVING DECIDED TO MOVE HIS FINGERS.

HERE'S A GOOD OLD-FASHIONED LINE GRAPH TO UNPICK LIBET'S FINDINGS. IT SHOWS A SNAPSHOT OF BRAIN ACTIVITY, REPRESENTED BY THE LONG WAVY LINE.

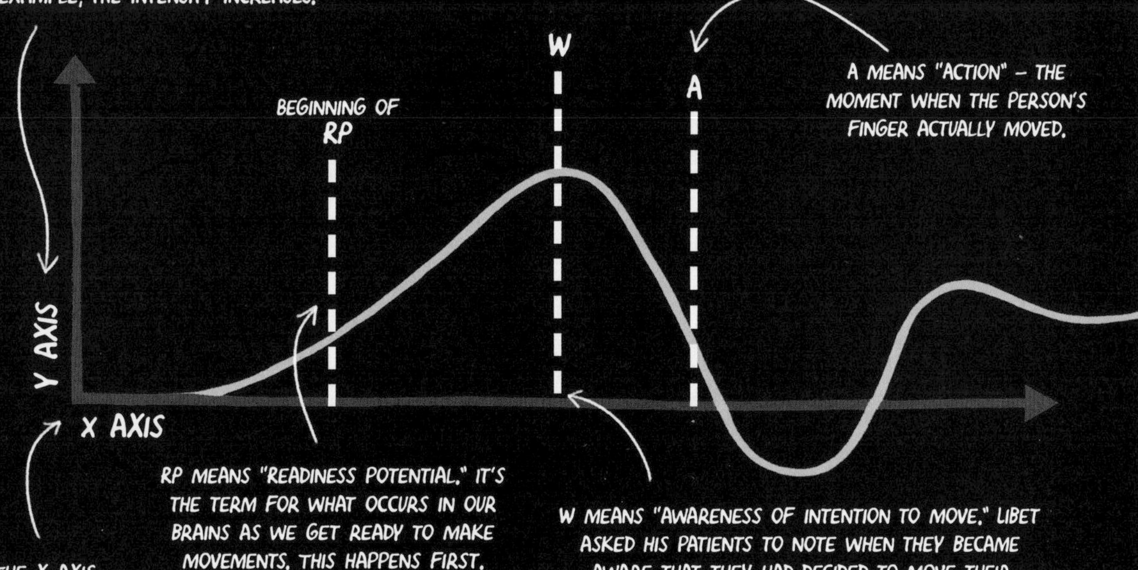
THE Y AXIS RECORDS THE INTENSITY OF BRAINWAVES. WHEN MAKING A VOLUNTARY DECISION TO MOVE A FINGER, FOR EXAMPLE, THE INTENSITY INCREASES.
W
A
A MEANS "ACTION" – THE MOMENT WHEN THE PERSON'S FINGER ACTUALLY MOVED.
BEGINNING OF RP
Y AXIS
X AXIS
RP MEANS "READINESS POTENTIAL." IT'S THE TERM FOR WHAT OCCURS IN OUR BRAINS AS WE GET READY TO MAKE MOVEMENTS. THIS HAPPENS FIRST.
W MEANS "AWARENESS OF INTENTION TO MOVE." LIBET ASKED HIS PATIENTS TO NOTE WHEN THEY BECAME AWARE THAT THEY HAD DECIDED TO MOVE THEIR FINGERS (BY LOOKING AT A LITTLE CIRCULAR "CLOCK").
THE X AXIS SHOWS WHAT HAPPENS OVER TIME.

THE POINT OF ALL THIS IS TO SHOW THAT RP BEGINS BEFORE W. WHICH MEANS, SOMEONE WATCHING THE EEG RECORDING AS IT HAPPENS CAN SEE THAT A PERSON IS GETTING READY TO MOVE THEIR FINGER A FRACTION OF A SECOND BEFORE THEY ARE AWARE OF DECIDING TO DO IT.

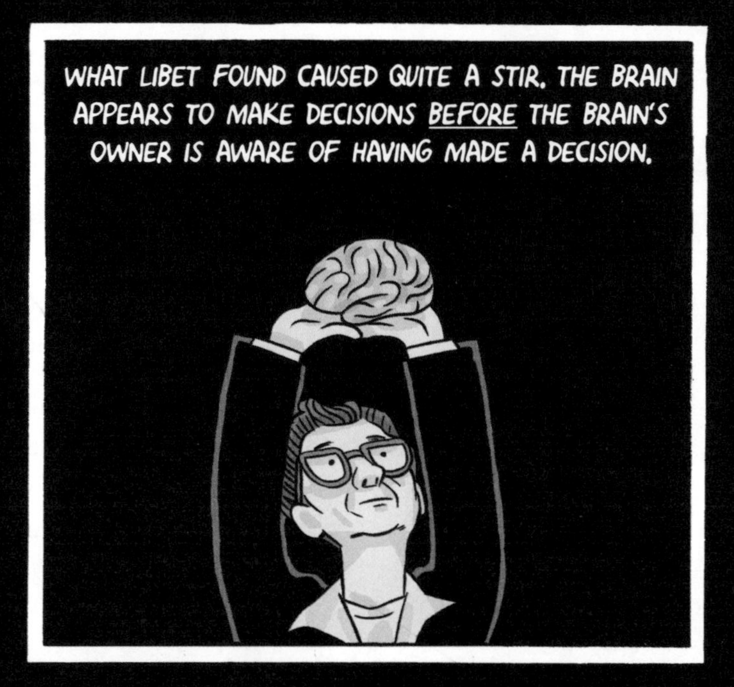
WHAT LIBET FOUND CAUSED QUITE A STIR. THE BRAIN APPEARS TO MAKE DECISIONS BEFORE THE BRAIN'S OWNER IS AWARE OF HAVING MADE A DECISION.

SOME PEOPLE, LIBET INCLUDED, IMMEDIATELY WONDERED IF THIS MEANS THAT PEOPLE DO NOT, IN FACT, HAVE FREE WILL.
IF OUR BRAINS MAKE DECISIONS BEFORE OUR "SELVES" DO, HOW CAN WE BE FREE TO CHOOSE THINGS?
LIBET REFUSED TO BELIEVE THAT FREE WILL IS AN ILLUSION. INSTEAD, HE THEORIZED THAT OUR SELVES ARE ABLE TO OVERRIDE BRAIN-BASED DECISIONS.
AN ABILITY HE CALLED "FREE WON'T."
BUT THERE'S NO EVIDENCE FOR SUCH AN ABILITY.

IS LIBET'S EXPERIMENT PROOF...

...THAT WE ARE NOT IN CONTROL OF OUR DECISIONS?

AND WHAT DOES THIS ALL HAVE TO DO WITH SCHIZOPHRENIA?

I'M GETTING TO THAT! BUT FIRST A REMINDER THAT THE EEG CAN'T READ THOUGHTS – IT JUST SHOWS WHERE AND WHEN PATTERNS OF NEURAL ACTIVITY OCCUR.

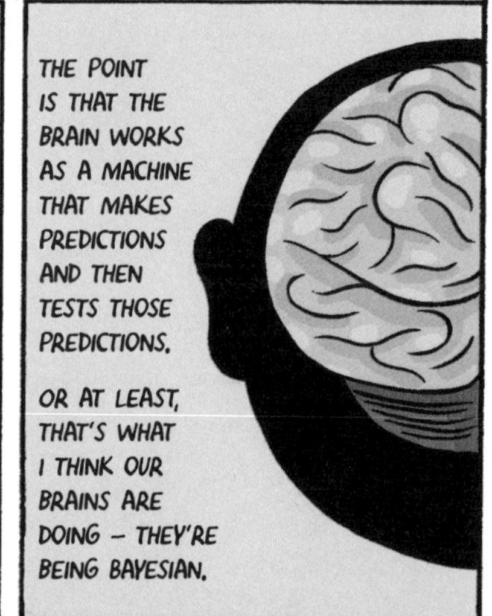
THE POINT IS THAT THE BRAIN WORKS AS A MACHINE THAT MAKES PREDICTIONS AND THEN TESTS THOSE PREDICTIONS.
OR AT LEAST, THAT'S WHAT I THINK OUR BRAINS ARE DOING – THEY'RE BEING BAYESIAN.

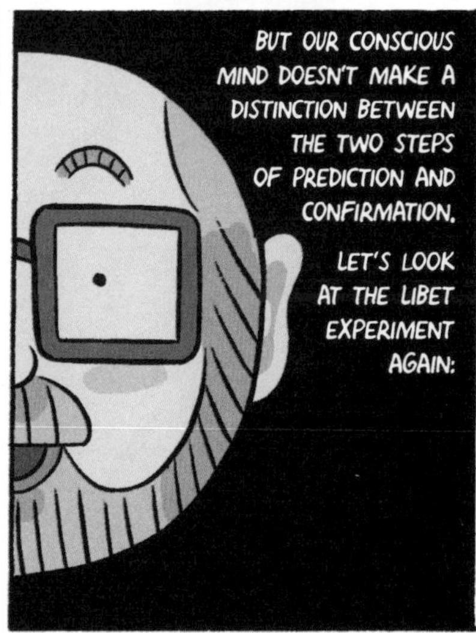
BUT OUR CONSCIOUS MIND DOESN'T MAKE A DISTINCTION BETWEEN THE TWO STEPS OF PREDICTION AND CONFIRMATION.
LET'S LOOK AT THE LIBET EXPERIMENT AGAIN:

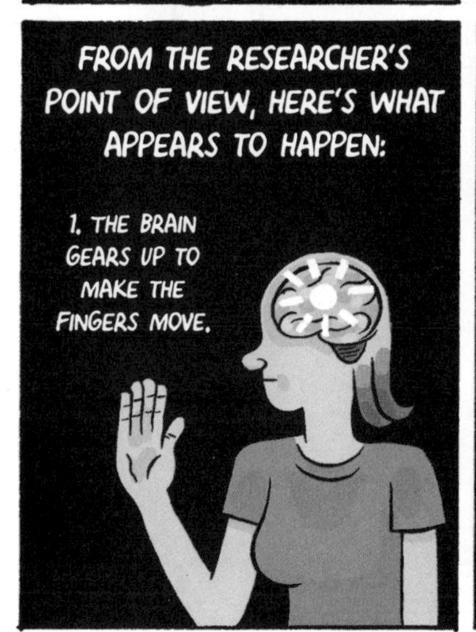
FROM THE RESEARCHER'S POINT OF VIEW, HERE'S WHAT APPEARS TO HAPPEN:
1. THE BRAIN GEARS UP TO MAKE THE FINGERS MOVE.

2. SOME MILLISECONDS LATER, THE BRAIN RECORDS MAKING A DECISION TO MOVE AND WE PERCEIVE THAT WE DECIDED TO MOVE.

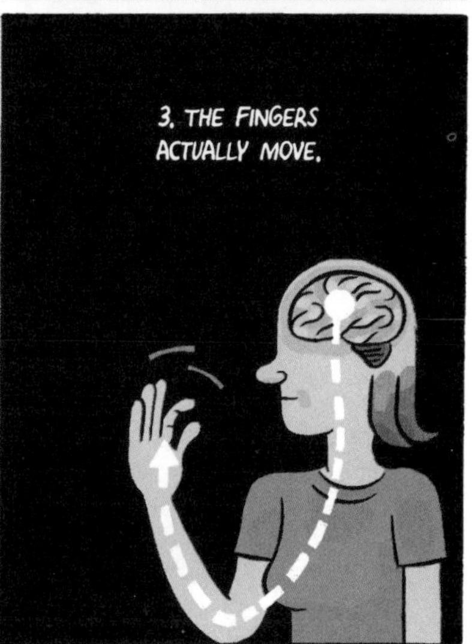
3. THE FINGERS ACTUALLY MOVE.

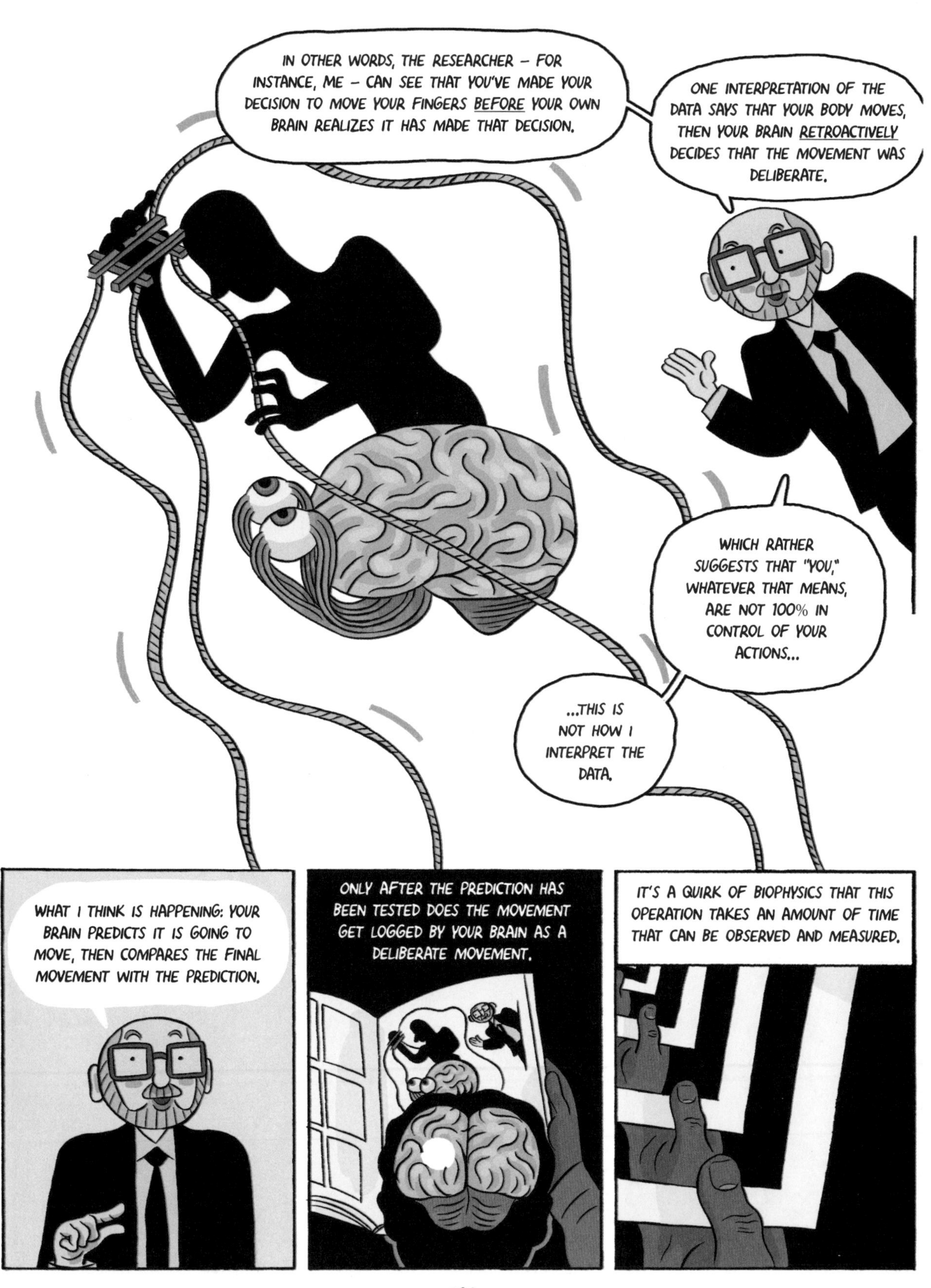
IN OTHER WORDS, THE RESEARCHER – FOR INSTANCE, ME – CAN SEE THAT YOU'VE MADE YOUR DECISION TO MOVE YOUR FINGERS BEFORE YOUR OWN BRAIN REALIZES IT HAS MADE THAT DECISION.
ONE INTERPRETATION OF THE DATA SAYS THAT YOUR BODY MOVES, THEN YOUR BRAIN RETROACTIVELY DECIDES THAT THE MOVEMENT WAS DELIBERATE.
WHICH RATHER SUGGESTS THAT "YOU," WHATEVER THAT MEANS, ARE NOT 100% IN CONTROL OF YOUR ACTIONS...
...THIS IS NOT HOW I INTERPRET THE DATA.
WHAT I THINK IS HAPPENING: YOUR BRAIN PREDICTS IT IS GOING TO MOVE, THEN COMPARES THE FINAL MOVEMENT WITH THE PREDICTION.
ONLY AFTER THE PREDICTION HAS BEEN TESTED DOES THE MOVEMENT GET LOGGED BY YOUR BRAIN AS A DELIBERATE MOVEMENT.
IT'S A QUIRK OF BIOPHYSICS THAT THIS OPERATION TAKES AN AMOUNT OF TIME THAT CAN BE OBSERVED AND MEASURED.

IT'S A FURTHER QUIRK OF BIOPHYSICS THAT THIS OPERATION DOESN'T ALWAYS RUN SMOOTHLY.
SPECIFICALLY, WHEN A PERSON IS AFFECTED BY SCHIZOPHRENIA, I THINK THEIR BRAIN IS PRONE TO EXPERIENCE A FAILURE OF FEEDBACK.
SOMETHING IS BROKEN ON THE PREDICTION-TESTING-CONFIRMATION LOOP, MEANING THAT ACTIONS THEY PERFORM DO NOT GET LOGGED BY THE BRAIN AS "THEIR" ACTIONS.
IF YOUR PREDICTIONS ABOUT THE CONSEQUENCES OF YOUR OWN ACTIONS KEEP FAILING, A PLAUSIBLE CONCLUSION IS THAT SOMEONE ELSE IS INTERFERING WITH YOUR BRAIN.

TO MY MIND, THERE'S NO QUESTION THAT PEOPLE DO HAVE CONTROL OVER THEIR ACTIONS (AT LEAST, MOST OF THE TIME).
BUT THE SCARY THOUGHT REMAINS THAT IF I COULD SEE INSIDE YOUR BRAIN RIGHT NOW...

...I'D SEE YOUR FINGERS PREPARING THEMSELVES TO TURN THE PAGE...

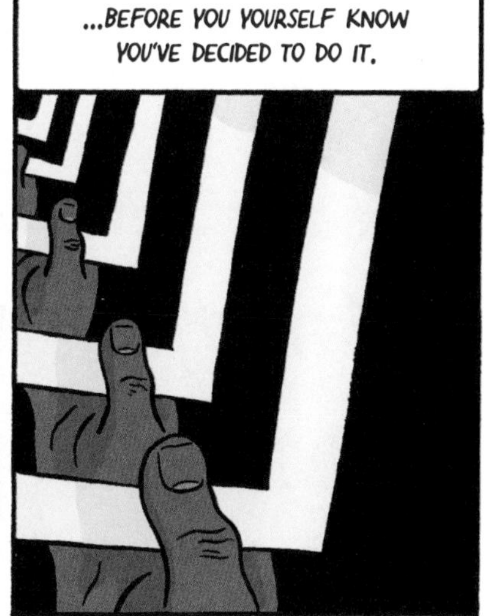
...BEFORE YOU YOURSELF KNOW YOU'VE DECIDED TO DO IT.

Interlude:

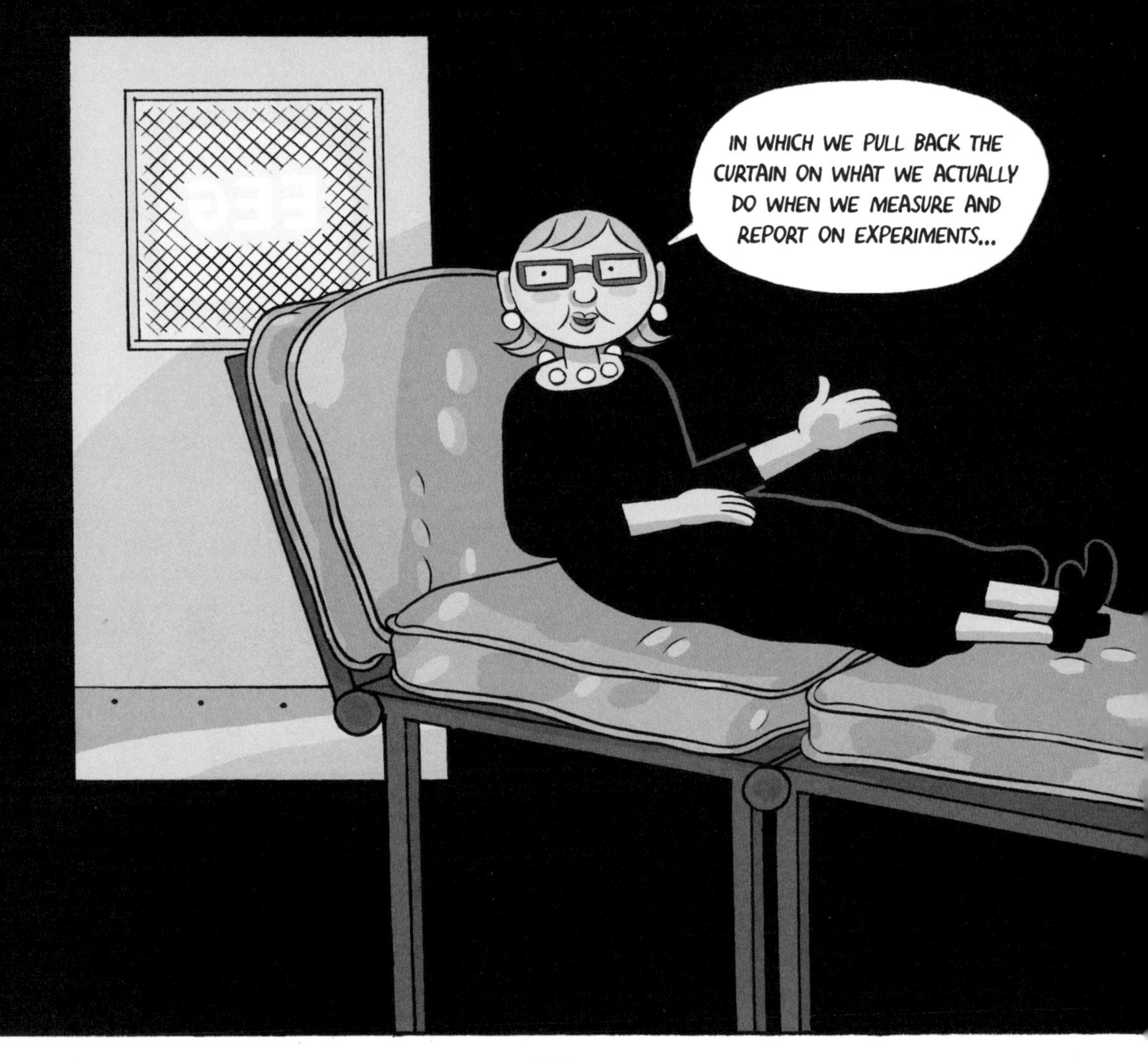
IN WHICH WE PULL BACK THE CURTAIN ON WHAT WE ACTUALLY DO WHEN WE MEASURE AND REPORT ON EXPERIMENTS...

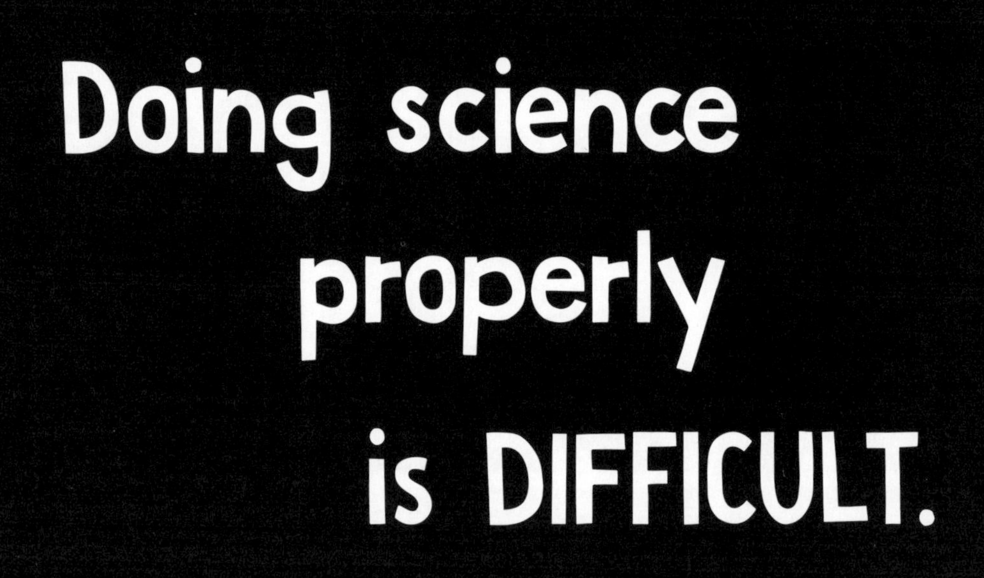
Doing science properly is DIFFICULT.

...NOT BECAUSE WE'VE BEEN CHEATING ALL ALONG, BUT BECAUSE IT'LL HELP SHOW THE SCALE OF THE CHALLENGE OF NEUROSCIENCE AS A WHOLE PROJECT.

WE'LL START WITH ANOTHER BRIEF LOOK AT THE LIBET EXPERIMENT FROM THE END OF THE LAST CHAPTER.

AS IT HAPPENS, IT HAS PROVED RATHER EASY FOR OTHER PEOPLE TO REPEAT THE EXPERIMENT,
AND FIND THE SAME RESULT.

WE'VE EVEN FOUND THE SAME SORT OF RESULT WHEN USING DIFFERENT METHODS OF OBSERVATION...

...FOR EXAMPLE, USING MRI SCANS, INSTEAD OF EEG.
MRI USES POWERFUL MAGNETS TO SHOW WHICH PARTS OF A BRAIN ARE MOST ACTIVE AT ANY GIVEN MOMENT.

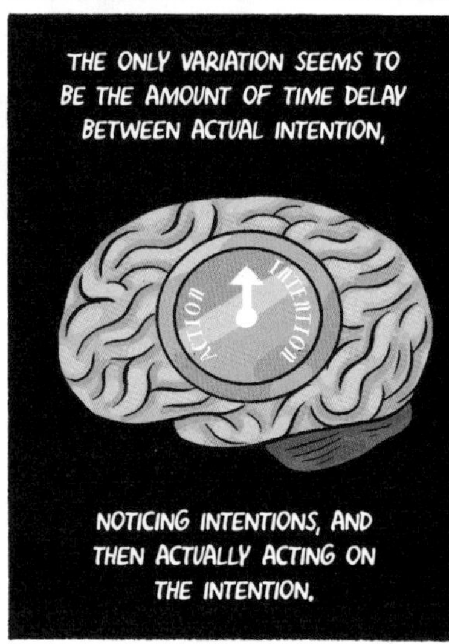
THE ONLY VARIATION SEEMS TO BE THE AMOUNT OF TIME DELAY BETWEEN ACTUAL INTENTION,
ACTION
INTENTION
NOTICING INTENTIONS, AND THEN ACTUALLY ACTING ON THE INTENTION.

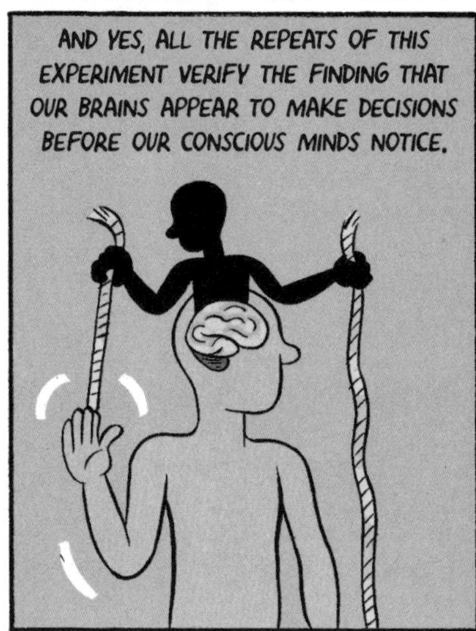
AND YES, ALL THE REPEATS OF THIS EXPERIMENT VERIFY THE FINDING THAT OUR BRAINS APPEAR TO MAKE DECISIONS BEFORE OUR CONSCIOUS MINDS NOTICE.

IGNORING THE CONTROVERSY OVER HOW TO INTERPRET HIS RESULTS, LIBET'S EXPERIMENT IS A ROBUST EXAMPLE OF ONE IN WHICH THE RAW DATA HAS BEEN VERIFIED BY OTHER SCIENTISTS.

THIS IS DELIGHTFUL TO US NEUROSCIENTISTS – IT'S AS CLOSE AS WE KNOW HOW TO GET TO TRUTH.

(UNTIL, OF COURSE, SOME OTHER SCIENTIST DEMONSTRATES WHY THOSE RESULTS WEREN'T ACTUALLY TRUE.)
!

LET'S TALK ABOUT TRUTH FOR A BIT.
WE BELIEVE THERE ARE OBJECTIVE TRUTHS IN THIS WORLD.

FURTHERMORE, WE BELIEVE THAT SCIENCE IS THE BEST WAY TO FIND THOSE TRUTHS.

AS SCIENTISTS, IT'S OUR JOB TO INVESTIGATE THE WORLD, TO FIND OUT MORE ABOUT HOW IT WORKS.

THAT MEANS FORMING AN IDEA ABOUT HOW SOMETHING WORKS – FOR INSTANCE, THE BRAIN – THEN TESTING IT OUT. AGAIN AND AGAIN.

AND YES, IT OFTEN MEANS THAT WE'LL DISCOVER OUR IDEAS WERE WRONG – SO, WE RETHINK THEM, PUT FORWARD A NEW IDEA, AND TEST THIS ONE OUT.

AND, IF YOU THINK ABOUT WHERE OUR KNOWLEDGE OF THE BRAIN WAS IN, SAY, 1850, YOU CAN SEE BOTH HOW WELL AND HOW QUICKLY THIS METHOD CAN WORK.

AS FAR AS SCIENTIFIC KNOWLEDGE GOES, WE GET OUR TRUTH FROM OUR RESEARCH, OUR COLLEAGUES, AND OF COURSE SCIENTIFIC JOURNALS.
WHICH MEANS WE HAVE TO BELIEVE THAT JOURNALS ARE RELIABLE SOURCES OF TRUTH.
AND WE DO!
BUT LET'S TALK ABOUT A FEW STUMBLING BLOCKS IN THE WORLD OF SCIENCE REPORTING.

AS A RULE, PAPERS PUBLISHED IN SCIENTIFIC JOURNALS COME WITH AN AURA OF TRUTH. ESPECIALLY THE PICTURES – CHARTS, BRAIN IMAGES, AND SUCH.

nature

USING PICTURES TO COMMUNICATE IDEAS IS ALWAYS MORE EFFECTIVE THAN WORDS ALONE.
WINK
WINK
Science

YOU KNOW THOSE PICTURES YOU OFTEN SEE IN ARTICLES SHOWING A BRAIN WITH A NICE RED SPLODGE ON IT, INDICATING THE LOCATION OF SOME BIT OF THINKING?
Science.
The location of love*
*NOT A REAL ARTICLE...

THOSE ARE SOMEWHAT BOGUS – NO LIVE BRAIN SCAN WILL SHOW SUCH AN IMAGE.

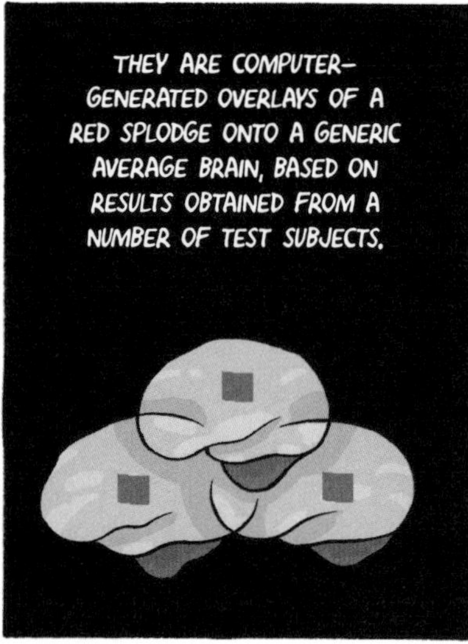
THEY ARE COMPUTER-GENERATED OVERLAYS OF A RED SPLODGE ONTO A GENERIC AVERAGE BRAIN, BASED ON RESULTS OBTAINED FROM A NUMBER OF TEST SUBJECTS.

TWO THINGS TO UNSPOOL HERE:
1. HOW TO INTERPRET RESULTS FROM A MIX OF TEST SUBJECTS ACROSS MULTIPLE TRIALS OF A SINGLE TEST.
2. THE CONCEPT OF THE "AVERAGE BRAIN."

THERE'S NO SUCH THING AS AN AVERAGE BRAIN.
BEAR OF LITTLE BRAIN?
SMARTER THAN THE AVERAGE BEAR?
LIKE EVERY OTHER BODY PART, BRAINS VARY IN SHAPE AND SIZE, AND CERTAINLY VARY IN THE NUMBER AND EXACT DENSITY OF NEURONS LOCATED IN ANY GIVEN BRAIN REGION.

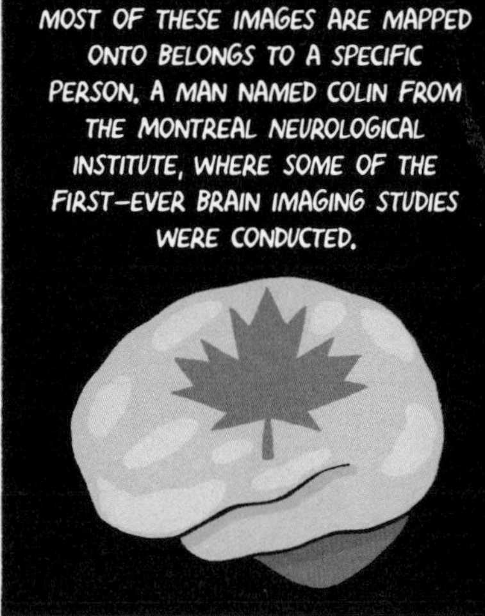
IN FACT, THE "AVERAGE BRAIN" THAT MOST OF THESE IMAGES ARE MAPPED ONTO BELONGS TO A SPECIFIC PERSON, A MAN NAMED COLIN FROM THE MONTREAL NEUROLOGICAL INSTITUTE, WHERE SOME OF THE FIRST-EVER BRAIN IMAGING STUDIES WERE CONDUCTED.

IN VIRTUALLY ALL SUCH REPORTS, USING A GENERIC BRAIN ISN'T REALLY A PROBLEM, BECAUSE THE POINT OF THE STUDY TENDS TO BE LOCATING SOMETHING WITHIN A BRAIN.

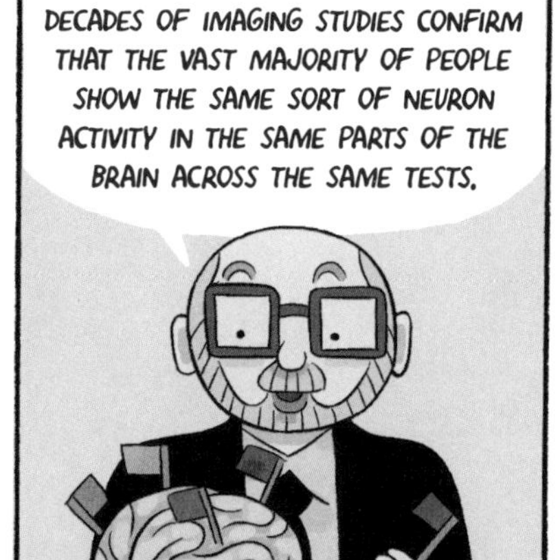
DECADES OF IMAGING STUDIES CONFIRM THAT THE VAST MAJORITY OF PEOPLE SHOW THE SAME SORT OF NEURON ACTIVITY IN THE SAME PARTS OF THE BRAIN ACROSS THE SAME TESTS.

SO, ALTHOUGH THE RED SPLODGE IMAGES AREN'T 100% TRUTHFUL, THEY'RE ALSO NOT LIES, EITHER.
PHEW! WE LOVE USING THEM AS SLIDES IN LECTURES.

DRAWING A CONCLUSION FROM MULTIPLE TRIALS IS A LITTLE TRICKIER.
LIBET'S TEST ONCE AGAIN SERVES AS A HELPFUL MODEL.

HIS ORIGINAL PAPER DETAILED THE RESULTS HE FOUND ON EACH OF HIS FIVE TEST SUBJECTS.*

THIS IS RARE – USUALLY, PSYCHOLOGISTS DISCUSS THE RESULTS BASED ON SOME FORM OF AVERAGE ACROSS ALL SUBJECTS.

IT'S LIKELY LIBET WANTED READERS TO BE AWARE OF THE COMPLEXITY OF THE DATA HE FOUND.

REMEMBER, EEG MEASURES ELECTRICAL IMPULSES GENERATED BY THE BRAIN – BRAINWAVES.
You won't be surprised to learn that your brain generates lots of waves all the time.

*YES, FIVE IS A RATHER SMALL SAMPLE SIZE – BUT NOT OUTRAGEOUSLY SO. AND YES, WE DO STILL USE THE LABEL "TEST SUBJECT," ALTHOUGH "PARTICIPANT" OR "VOLUNTEER" IS PREFERRED THESE DAYS.

IT'S PRETTY MUCH IMPOSSIBLE TO SWITCH OFF YOUR BRAIN TO FOCUS ENTIRELY ON ONE THING.
JUST ASK A BUDDHIST MONK.

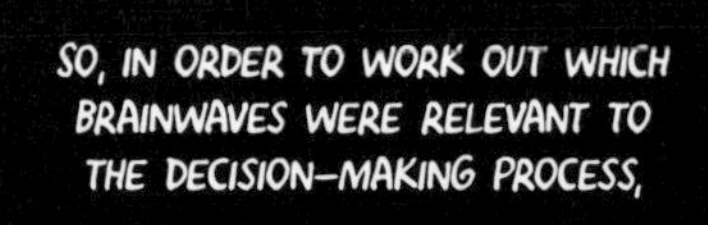
SO, IN ORDER TO WORK OUT WHICH BRAINWAVES WERE RELEVANT TO THE DECISION-MAKING PROCESS,

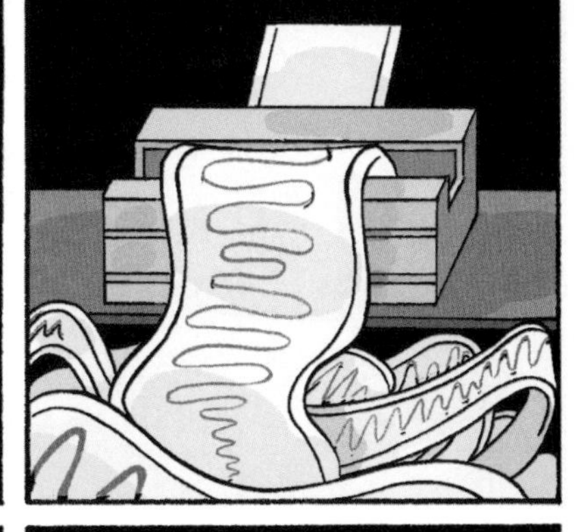
LIBET HAD TO RUN THE SAME TEST ON EACH SUBJECT REPEATEDLY, AND THEN COMPARE THE GRAPHS SPAT OUT BY THE EEG MACHINE.

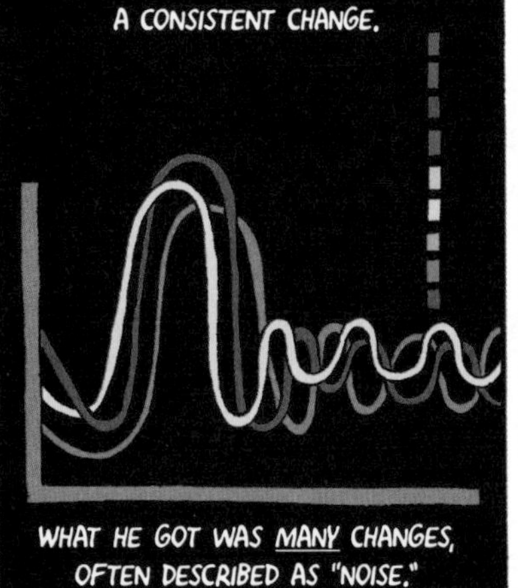
WHAT HE WANTED WAS TO SEE A CONSISTENT CHANGE.
WHAT HE GOT WAS MANY CHANGES, OFTEN DESCRIBED AS "NOISE."

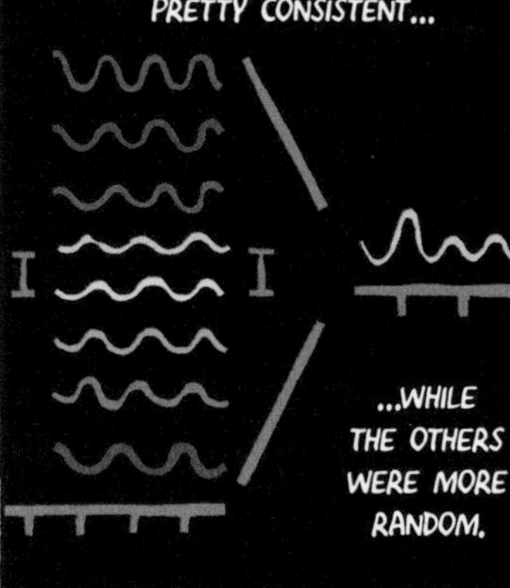
AFTER ABOUT 12–14 REPEATS, LIBET COULD START TO SEE THAT ONE CHANGE IN PARTICULAR WAS PRETTY CONSISTENT...
...WHILE THE OTHERS WERE MORE RANDOM.

THE FLUCTUATING LINES ON THE PAGE REPRESENT WANDERING THOUGHTS AND BRAIN PROCESSES THE SUBJECT FELT DURING THE EXPERIMENT.

THE LINE THAT SHOWED A CONSISTENT, REPEATED CHANGE ACROSS ALL REPEATS MUST, SURELY, SHOW THE BRAINWAVES THAT RELATE TO THE TASK AT HAND. IT'S A LITTLE LIKE TRYING TO IGNORE A CLUTTER OF IMAGES IN ONE'S MIND TO FOCUS ON JUST ONE THING.

THIS ONE THING, OR RATHER, ONE LINE ON THE EEG GRAPH, IS NOW CALLED THE SIGNAL – AND THIS IS WHERE WE START TO SEE TRUTH.

THAT CLAIM I MADE, THAT IF I COULD SEE INTO YOUR BRAIN I COULD PREDICT WHEN YOU'RE ABOUT TO TURN THE PAGE?

NOT ACTUALLY TRUE.

ALTHOUGH, IF I COULD STICK AN ELECTRODE DIRECTLY INTO THE CORRECT PART OF YOUR BRAIN...
...THUS ELIMINATING ALL THE NOISE FROM OTHER PARTS OF YOUR BRAIN...

...THEN I COULD DO IT.
PROBABLY.
IF LIBET WAS RIGHT.

UM, CAN I HAVE THE TOP OF MY HEAD BACK NOW?

IN THE WORLD OF BRAIN IMAGING, ONE THING EVERY EXPERIMENTER NEEDS TO DO IS CALLED "CORRECTING FOR MULTIPLE COMPARISONS."

IF YOU DON'T SEE THIS PHRASE SOMEWHERE IN A REPORT, ACTIVATE YOUR SKEPTICAL MODE.
Science.

IN ANY EXPERIMENT, ON ANY TEST SUBJECT, BRAIN SCANNERS WILL SHOW ACTIVITY HAPPENING ALL OVER THE BRAIN.

BUT IN MOST CASES, THE ORIGINAL HYPOTHESIS IS INTERESTED IN JUST ONE SPECIFIC PART OF THE BRAIN.

(THE SMALLEST POSSIBLE PART OF A BRAIN, IN THIS SENSE, IS DEFINED AS A 3MM X 3MM X 3MM CUBE OF BRAIN MATTER CALLED A VOXEL.)

HELLO.

SO, WHEN RUNNING AN EXPERIMENT, YOU NEED TO RUN MANY REPEATS TO BE SURE THAT THE ACTIVITY IN THE REGION – OR SET OF VOXELS – YOU WERE HOPING TO SEE ISN'T JUST RANDOM NOISE.
CLICK
CLICK
CLICK

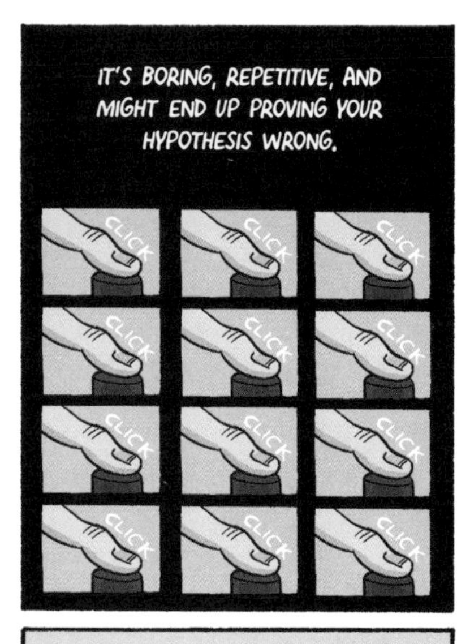
IT'S BORING, REPETITIVE, AND MIGHT END UP PROVING YOUR HYPOTHESIS WRONG.
CLICK
CLICK
CLICK
CLICK
CLICK
CLICK
CLICK
CLICK
CLICK
CLICK
CLICK
CLICK

BLURGH.

BUT ONCE YOU'VE DONE IT, YOU MIGHT DISCOVER A NEW TRUTH!

LET'S GET INTO A TECHNICAL ISSUE ABOUT PROBABILITY. DID YOU KNOW THAT, WHEN YOU'RE RUNNING AN EXPERIMENT...
FLIP
...THERE'S A 1/20 POSSIBILITY THAT SOMETHING THAT LOOKS INTERESTING WILL IN FACT BE A RANDOM COINCIDENCE.

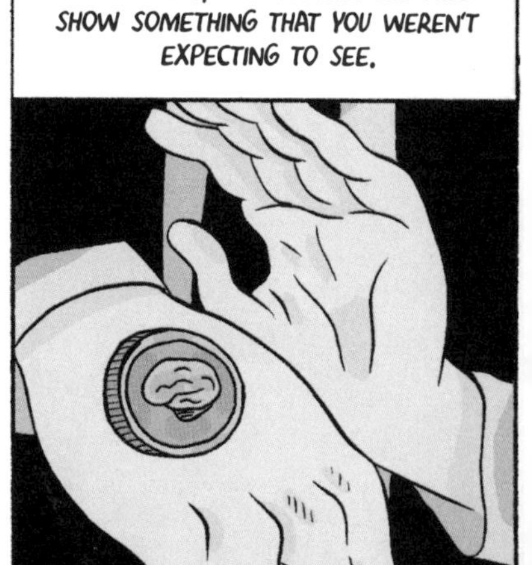
SPECIFICALLY, THIS COINCIDENCE WILL SHOW SOMETHING THAT YOU WEREN'T EXPECTING TO SEE.

THIS DOESN'T MEAN THAT YOU HAVE TO RUN 20 EXPERIMENTS BEFORE YOU CAN EXPECT TO FIND ANYTHING WORTH WRITING ABOUT.

IT MEANS THAT, IF YOU'RE RUNNING AN IMAGING TEST ON A SET OF INDIVIDUALS MORE THAN 20 TIMES, YOUR DATA IS PROBABLY GOING TO SHOW SOMETHING ACROSS ALL THEIR BRAINS THAT...
FLIP
FLIP

...A) YOU WEREN'T EXPECTING TO FIND
AND B) APPEARS TO BE HAPPENING SO OFTEN THAT THERE MUST BE SOMETHING INTERESTING GOING ON
—RIGHT?

IN FACT, NO, THIS 1 IN 20 EFFECT IS GENUINELY AN EXPRESSION OF COINCIDENCE.

HERE'S WHY.
IMAGINE TOSSING A COIN 100 TIMES IN A ROW AND GETTING HEADS EACH TIME.

IN ITSELF, THIS IS REMARKABLE, BUT BY NO MEANS IMPOSSIBLE.
AND, MATHEMATICALLY, GETTING HEADS 100 TIMES IN A ROW IS NO MORE OR LESS LIKELY THAN GETTING ANY PARTICULAR SEQUENCE OF HEADS/TAILS.

IT'S JUST THAT OUR BRAINS ARE PRIMED TO PICK OUT SOME SEQUENCES (SUCH AS ALL HEADS OR ALL TAILS) AS LOOKING UN-RANDOM.

SO, IN THE IMAGING STUDY, IF ONE VOXEL ACTIVATES ACROSS ALL TEST SUBJECTS AFTER HUNDREDS OF REPEATS, IT COULD ALSO BE THIS KIND OF RANDOM COINCIDENCE.

YOU NEED TO UNDERSTAND THAT SO MANY VOXELS ACTIVATE EVERY SINGLE TIME, IT'S ACTUALLY NOT SURPRISING THAT MAYBE SEVERAL OF THEM WILL ACTIVATE EVERY TIME.

ALL TOO OFTEN, EXPERIMENTERS CAN BLIND THEMSELVES AND OTHERS WITH DATA, ESPECIALLY IF THEY'RE LOOKING FOR A VERY PARTICULAR RESULT.

FOR EXAMPLE, THEY COULD JUST A) KEEP RUNNING THE EXPERIMENT ON MORE SUBJECTS UNTIL ENOUGH OF THEM CONFORM TO THE HYPOTHESIS.

OR B) USE A VARIETY OF DIFFERENT STATISTICAL ANALYSIS METHODS TO EXAMINE THEIR RAW DATA – AND THERE ARE LOTS –
UNTIL ONE OF THEM PRODUCES AN INTERESTING RESULT. A TECHNIQUE CALLED "DATA DREDGING."

ANY PAPER SUBMITTED TO AN ACADEMIC JOURNAL IS MEANT TO BE REVIEWED BY PEERS – WORKING PROFESSIONALS IN THE SAME FIELD,
WHO HAVE NO STAKE IN THE CLAIMS OF ANY EXPERIMENT THEY'RE REVIEWING.

PEER REVIEWERS CAN (AND SHOULD) DETECT ALL KINDS OF DATA TRICKERY.
BAD SCIENCE

MORE DIFFICULT IS HELPING PEOPLE TO SEE THAT WHEN DATA DISPROVES A HYPOTHESIS, IT CAN BE JUST AS VALUABLE, IF LESS FUN.
HYPOTHESIS WRONG!
"We're thrilled," says research team. "Now we can move forward with new ideas."

IT'S UP TO EDITORS TO BE BOLD ENOUGH TO PRINT THOSE PAPERS, EVEN KNOWING THAT FINDING NOTHING TENDS TO BE A LESS DRAMATIC STORY TO READ THAN FINDING SOMETHING.
Science.
Nothing discovered after experimental study

ANYWAY, ABOUT SIX PAGES AGO I WAS EXPLAINING THAT THE LIBET EXPERIMENT IS ONE THAT CAN BE AND HAS BEEN REPLICATED BY OTHER NEUROSCIENTISTS.

IN FACT, THIS IS SOMEWHAT UNUSUAL IN OUR FIELD (AND IN SOME OTHERS, FOR EXAMPLE CANCER RESEARCH).

JOURNALS HAVE BEEN REPORTING A "REPLICATION CRISIS."
ONE EXPERIMENTAL TEAM FINDS AN INTERESTING – AND SEEMINGLY GENUINE – RESULT. BUT OTHER TEAMS ARE UNABLE TO REPLICATE THE EXPERIMENT.

HERE'S AN EXAMPLE: IN 1996, ONE GROUP FOUND THAT, IF THEY PRIMED A ROOM FULL OF STUDENTS TO THINK ABOUT OLD AGE, THOSE STUDENTS WOULD THEN WALK MEASURABLY SLOWLY ONCE THEY GOT UP TO LEAVE THE LECTURE HALL.
THE EFFECTS OF OLD AGE: A DOCUMENTARY

LIKE A LOT OF PSYCHOLOGY FINDINGS, IT SOUNDS SO PLAUSIBLE THAT YOU MAY BE TEMPTED TO SHRUG AND SAY "SURE, OF COURSE."
UCL
UCL

BUT, IN FACT, WHEN OTHER GROUPS HAVE TRIED IT, THEY HAVE IN SOME CASES FOUND THE OPPOSITE RESULT...
...STUDENTS STARTED WALKING FASTER...
UCL
UCL

...OR, MORE COMMONLY, FOUND NO MEASURABLE RESULT AT ALL.
UCL
UCL

IN 2012, A GROUP IN BELGIUM, LED BY AXEL CLEEREMANS, ADDED A NEW WRINKLE:
WE PRIMED OUR EXPERIMENTERS BEFORE ASKING THEM TO CARRY OUT THE TEST.

ONE SET OF EXPERIMENTERS WAS TOLD THAT THE ORIGINAL 1996 STUDY WAS DEFINITELY TRUE.
ANOTHER SET WAS TOLD THE OPPOSITE.

BOTH SETS FOUND EXACTLY WHAT THEY HAD BEEN PRIMED TO FIND.

IT APPEARS THAT WHAT WE BELIEVE HAS A SIGNIFICANT EFFECT ON THE WAY WE INVESTIGATE THE WORLD – EVEN AS SUPPOSEDLY DISPASSIONATE SCIENTISTS.

THERE'S ANOTHER CONCERN EMBEDDED IN ALL THESE EXPERIMENTS: THE TEST SUBJECTS.
THEY'RE ALL WEIRD. THAT MEANS:
WESTERN,
EDUCATED,
INDUSTRIALIZED,
RICH,
DEMOCRATIC.

IT REFERS TO THE COUNTRY OF ORIGIN OF THE VAST MAJORITY OF ALL PEOPLE WHO HAVE EVER BEEN INVOLVED IN PSYCHOLOGY EXPERIMENTS.

BECAUSE MOST ARE PERFORMED IN UNIVERSITIES, WHERE THE EASIEST WAY TO FIND TEST SUBJECTS IS TO PAY STUDENTS – WHO ARE, ALMOST BY DEFINITION, WEIRD.

(EVEN STUDENTS IN NON-WEIRD COUNTRIES ARE MORE LIKELY TO BE AT THE WEIRD END OF THEIR LOCAL SPECTRUM)

IT MEANS THAT THESE EXPERIMENTS PRODUCE RESULTS THAT MAY, IN FACT, ONLY HOLD TRUE FOR OTHER WEIRD PEOPLE.

WEIRD

TO BE FAIR TO PSYCHOLOGY, FOR MOST OF ITS 150 YEARS THOSE HAVE BEEN THE ONLY PEOPLE WHO CARED TO READ AND CONSIDER SUCH STUDIES.
WEIRD

HANG ON, THAT'S A BIT LIKE SAYING THAT BECAUSE ONLY MEN WERE IN A POSITION TO STUDY SCIENCE FOR CENTURIES, IT'S OK THAT WOMEN WERE IGNORED THAT WHOLE TIME.
IS IT? SORRY.

IN ANY EVENT, IT MAY EXPLAIN PART OF THE REPLICATION CRISIS – LET'S NOT FORGET, BY USING WIDER POOLS OF SUBJECTS, WE'RE DOING BETTER SCIENCE NOW THAN EVER BEFORE.

ANOTHER WRINKLE: NOT ONLY ARE MOST TEST SUBJECTS FROM THE SAME SORT OF BACKGROUND – THEY'RE ALSO MORE LIKELY TO HAVE READ ABOUT PSYCHOLOGY EXPERIMENTS.

BUZZNEWS
TOP 10 PSYCHOLOGY EXPERIMENTS.
No. 7 will blow your mind. literally.

IT'S A RULE OF THUMB THAT TEST SUBJECTS MUSTN'T KNOW WHAT IT IS THAT YOU'RE LOOKING TO FIND IN AN EXPERIMENT, BECAUSE THIS WILL AFFECT THE RESULT.

BUT IF YOU'RE REPEATING AN OLD EXPERIMENT, ESPECIALLY A FAMOUS ONE, THE SORT OF PEOPLE WHO VOLUNTEER AS TEST SUBJECTS MAY WELL KNOW ABOUT IT ALREADY. AND THIS IS WHEN YOU RUN INTO PROBLEMS LIKE THE STORY OF PRIMING PEOPLE TO WALK FAST OR SLOW.
The Last of the Summer Wine
DID THIS IN HIGH SCHOOL. YAWN.

HERE'S A PHRASE THAT YOU* HAVE HEARD BEFORE:
Correlation is not causation!
*ERUDITE, INTELLIGENT PERSON WHO IS HALFWAY THROUGH A POPULAR SCIENCE BOOK.

WE'RE BRINGING IT UP BECAUSE IT'S A CONSTANT STUMBLING BLOCK, ESPECIALLY WHEN STUDYING THE CAUSES OF DISORDERS SUCH AS SCHIZOPHRENIA AND AUTISM.

LET'S LOOK AT AN EXAMPLE OF CORRELATION IN WHICH IT'S ALMOST IMPOSSIBLE NOT TO THINK THERE'S CLEAR CAUSATION:

DRINKING COFFEE CAUSES AROUSAL AND ANXIETY.

IF YOU DRINK A LOT OF COFFEE, OR KNOW SOMEONE WHO DOES, IT'S HARD NOT TO MAKE THIS LINK.
BUT WHAT IS REALLY GOING ON?

DOES COFFEE MAKE PEOPLE ANXIOUS?
OR, ARE ANXIOUS PEOPLE MORE LIKELY TO DRINK COFFEE?

HERE'S THE SCIENTIFIC WAY TO FIND OUT:
SELECT FOUR RANDOM GROUPS OF PEOPLE (WHERE RANDOM MEANS "A BROAD CROSS-SECTION OF SOCIETY").

OVER A SERIES OF DAYS, GIVE EACH OF THE FOUR GROUPS A PARTICULAR DRINK:
GROUP 1: GET ACTUAL COFFEE, AND ARE TOLD THAT IT IS COFFEE
GROUP 2: GET DECAF COFFEE, BUT ARE TOLD THAT IT IS REAL COFFEE
GROUP 3: GET ACTUAL COFFEE, BUT ARE TOLD THAT IT IS DECAF COFFEE
GROUP 4: GET DECAF COFFEE, AND ARE TOLD IT IS DECAF COFFEE
THE RANDOM ALLOCATION ENSURES THAT IT IS NOT DUE TO DIFFERENCES IN ANXIETY BETWEEN THE PARTICIPANTS. IN THEORY, THIS EXPERIMENT WOULD DEMONSTRATE IF COFFEE MAKES PEOPLE ANXIOUS AND/OR IF BELIEVING THEY ARE DRINKING COFFEE MAKES THEM ANXIOUS.

WITH A LARGE ENOUGH GROUP OF PEOPLE, AND TRULY RANDOM ASSIGNMENT OF COFFEE OPTIONS, IT SHOULD BE POSSIBLE TO CHECK THE HYPOTHESIS REGARDLESS OF WHAT OTHER FOOD AND DRINK EVERY PERSON CHOOSES TO EAT ALONGSIDE THEIR COFFEE RATION. YES, THE RESULTS WILL BE RATHER "NOISY," BUT THE CENTRAL "SIGNAL" SHOULD STILL SHOW THROUGH.

CHOC

AND TRY TO CORRECT FOR ANY OTHER VARIABLES THAT MAY AFFECT AROUSAL AND ANXIETY.
WE'RE IN BAD MOODS TODAY!

YOU CAN SEE WHY THIS KIND OF STUDY IS ALMOST IMPOSSIBLE TO CONDUCT IN A CONCLUSIVE WAY.
CHOC

BUT YOU WANT TO KNOW AN ANSWER, DON'T YOU!

SAD TO SAY, ALTHOUGH TEAMS AROUND THE WORLD HAVE STUDIED COFFEE-ANXIETY LINKS, THERE'S STILL NO CONCLUSIVE ANSWER.

MOST HAVE FOUND THAT SOME PEOPLE ARE MORE SENSITIVE TO THE EFFECTS OF CAFFEINE THAN OTHERS. MEN MORE THAN WOMEN, FOR EXAMPLE.

AND THERE'S A CONSISTENT CORRELATION THAT PEOPLE WHO SCORE HIGH ON ANXIETY TESTS SEEM TO DRINK MORE COFFEE.

BUT WE BOTH SLEEP FINE AT NIGHT AFTER DRINKING POST-DINNER ESPRESSOS, THANK YOU VERY MUCH.*

*ALTHOUGH WE MAY BE KEPT AWAKE FRETTING ABOUT SCIENCE THAT RELIES ON ANECDOTAL EVIDENCE.

WHAT IT BOILS DOWN TO IS THAT IT'S ENTIRELY POSSIBLE THAT THE COFFEE ITSELF IS NOT THE CAUSE OF ANXIETY.
BUT RATHER, THERE'S AN OVERLAP SHARED BY MANY COFFEE DRINKERS, E.G., BEING MIDDLE-CLASS WORRIERS.

THIS BRINGS US BACK TO RESEARCH INTO SCHIZOPHRENIA AND AUTISM.

IT'S HARD TO PROGRESS OUR UNDERSTANDING OF THESE CONDITIONS THROUGH EXPERIMENTS, BECAUSE WE CAN'T ASSIGN PEOPLE INTO THOSE FOUR NEAT GROUPS.

YOU CAN RANDOMLY ASSIGN A PERSON TO BE GIVEN COFFEE,

BUT YOU CAN'T RANDOMLY ASSIGN A PERSON TO BE AN INPATIENT IN A HOSPITAL DIAGNOSED WITH SCHIZOPHRENIA.

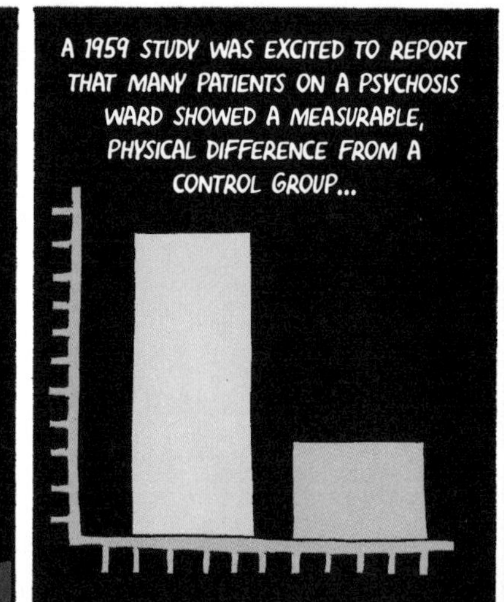
A 1959 STUDY WAS EXCITED TO REPORT THAT MANY PATIENTS ON A PSYCHOSIS WARD SHOWED A MEASURABLE, PHYSICAL DIFFERENCE FROM A CONTROL GROUP...

...PATIENTS DIAGNOSED AS SCHIZOPHRENIC ALL EXCRETED HIGH LEVELS OF SUBSTANCES CALLED PHENOLIC ACIDS.

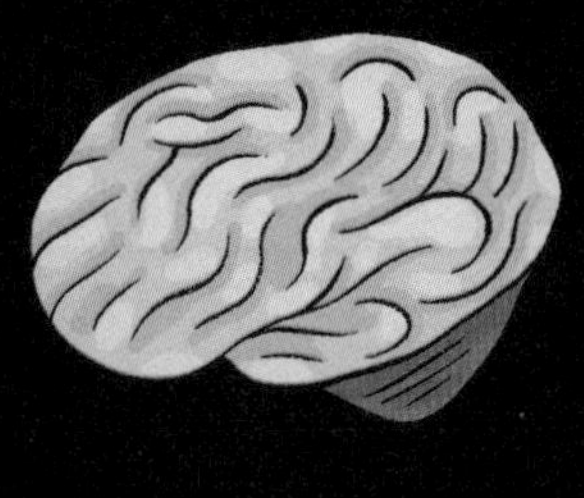

COULD THIS POINT THE WAY TO SOME BRAIN OR HORMONE-BASED DIFFERENCE?

IN FACT, IT DIDN'T.

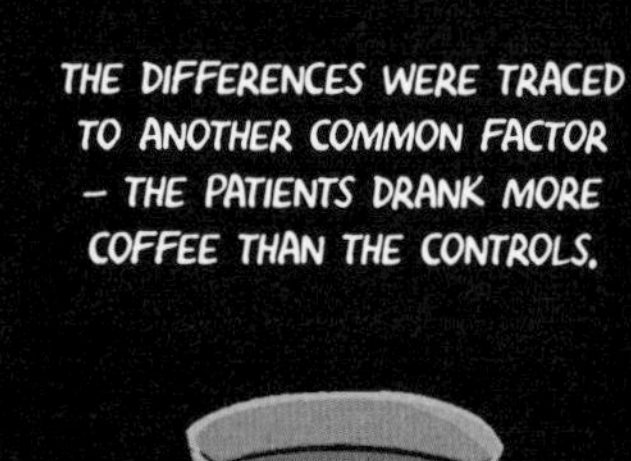
THE DIFFERENCES WERE TRACED TO ANOTHER COMMON FACTOR – THE PATIENTS DRANK MORE COFFEE THAN THE CONTROLS.

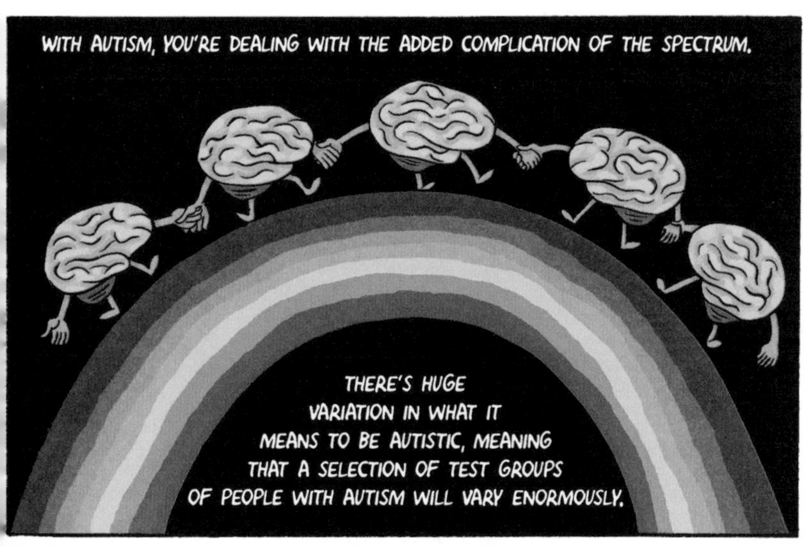
WITH AUTISM, YOU'RE DEALING WITH THE ADDED COMPLICATION OF THE SPECTRUM.
THERE'S HUGE VARIATION IN WHAT IT MEANS TO BE AUTISTIC, MEANING THAT A SELECTION OF TEST GROUPS OF PEOPLE WITH AUTISM WILL VARY ENORMOUSLY.

THE GOOD NEWS IS THAT OVER THE LAST COUPLE OF DECADES, THE EMERGENCE OF LARGE TEAMS OF RESEARCH GROUPS, ALL PERFORMING THE SAME EXPERIMENTS AROUND THE WORLD...

...HAS HELPED TO ELIMINATE MUCH OF THE NOISE, CORRECTING FOR THAT 1 IN 20 RANDOM COINCIDENCE, AND GRADUALLY HOMING IN ON A SIGNAL.

IF WE EVER IDENTIFY A CLEAR SIGNAL, IT WILL FINALLY HELP A) TO AGREE ON A BETTER DEFINITION OF WHAT AUTISM IS. B) MAKE IT EASIER TO DIAGNOSE PEOPLE CORRECTLY.

C) MAYBE EVEN BEGIN TO PROVIDE SOME SORT OF TREATMENT.
DAMMIT, DOING GOOD SCIENCE TAKES TIME!
NEWS

AND, BEING HUMAN, WE RESEARCHERS ARE LIABLE TO GET DISTRACTED BY OTHER PROJECTS.

WHICH TAKES US BACK TO THE ACTUAL THEME OF THIS BOOK.

INTERLUDE OVER. GO ON, TURN THE PAGE! WE'RE FINISHED WITH THIS BIT.

Chapter 7

Think, and think again.

BEFORE THE INTERLUDE, WE WERE TALKING ABOUT FEEDBACK LOOPS IN THE BRAIN. I PROPOSED THAT: WE DECIDE TO DO SOMETHING; THEN DO THE THING; AND THEN OUR BRAINS CLOSE THE LOOP BY MAKING A RECORD THAT WE DECIDED TO DO IT.

IF THIS IS WHAT IS GOING ON IN HUMAN BRAINS, IT'S LIKELY TO BE TRUE IN ANIMAL BRAINS, TOO.

BUT MIGHT HUMANS BE THE ONLY ANIMAL TO KNOW THAT THIS IS WHAT WE'RE DOING?

MANY PSYCHOLOGISTS SUSPECT HUMANS ARE THE ONLY ANIMALS THAT INDULGE IN SOMETHING CALLED METACOGNITION...

...THE SCIENTIFIC TERM FOR "THINKING ABOUT THINKING."

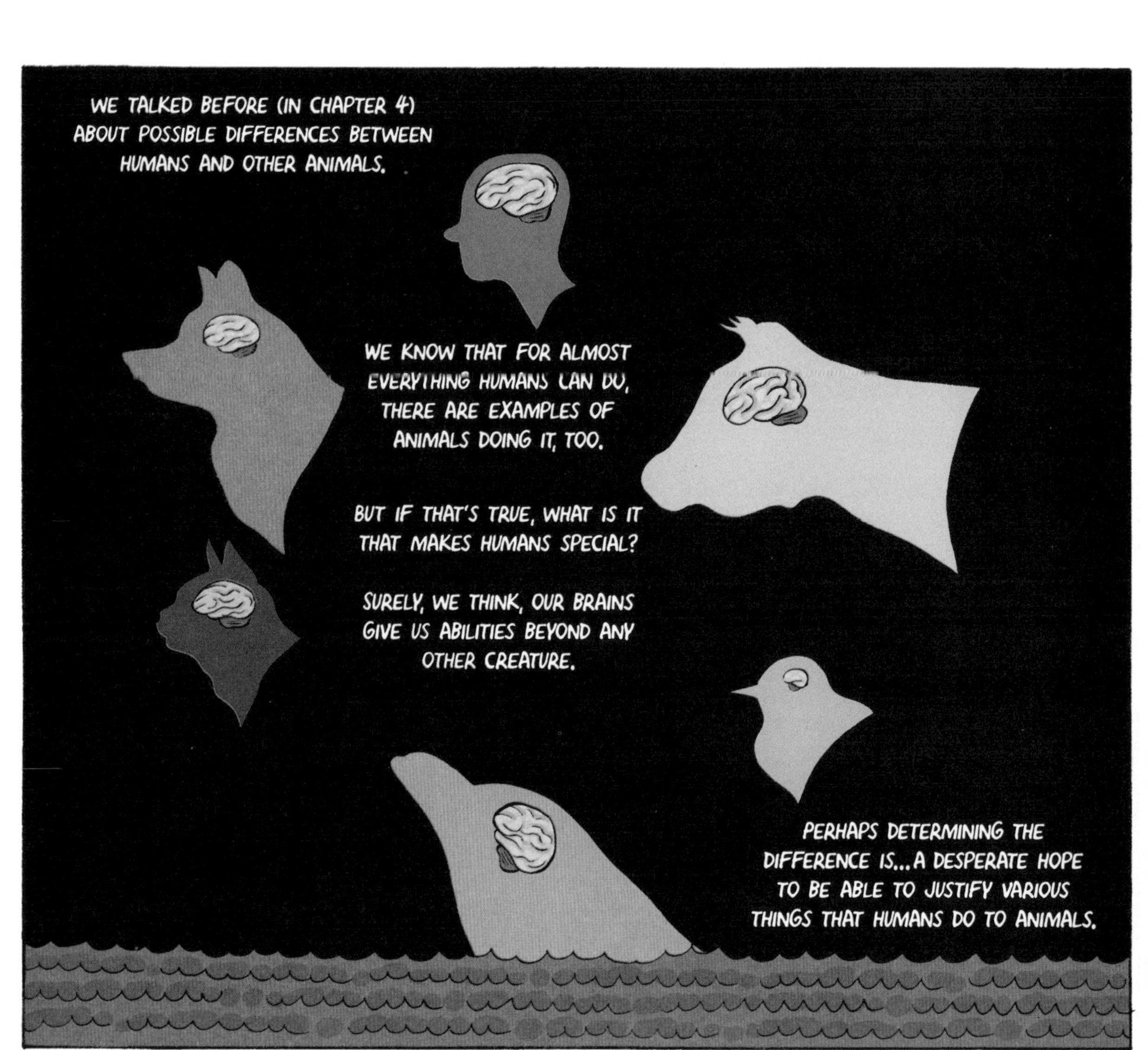
WE TALKED BEFORE (IN CHAPTER 4) ABOUT POSSIBLE DIFFERENCES BETWEEN HUMANS AND OTHER ANIMALS.
WE KNOW THAT FOR ALMOST EVERYTHING HUMANS CAN DO, THERE ARE EXAMPLES OF ANIMALS DOING IT, TOO.
BUT IF THAT'S TRUE, WHAT IS IT THAT MAKES HUMANS SPECIAL?
SURELY, WE THINK, OUR BRAINS GIVE US ABILITIES BEYOND ANY OTHER CREATURE.
PERHAPS DETERMINING THE DIFFERENCE IS...A DESPERATE HOPE TO BE ABLE TO JUSTIFY VARIOUS THINGS THAT HUMANS DO TO ANIMALS.

WE KEEP ANIMALS AS PETS.

WE TRAIN ANIMALS TO PERFORM TRICKS.

AND THE BIG ONE – WE PERFORM EXPERIMENTS ON ANIMALS THAT ARE INTENDED, ULTIMATELY, TO BRING ABOUT SOME BENEFIT FOR US HUMANS.

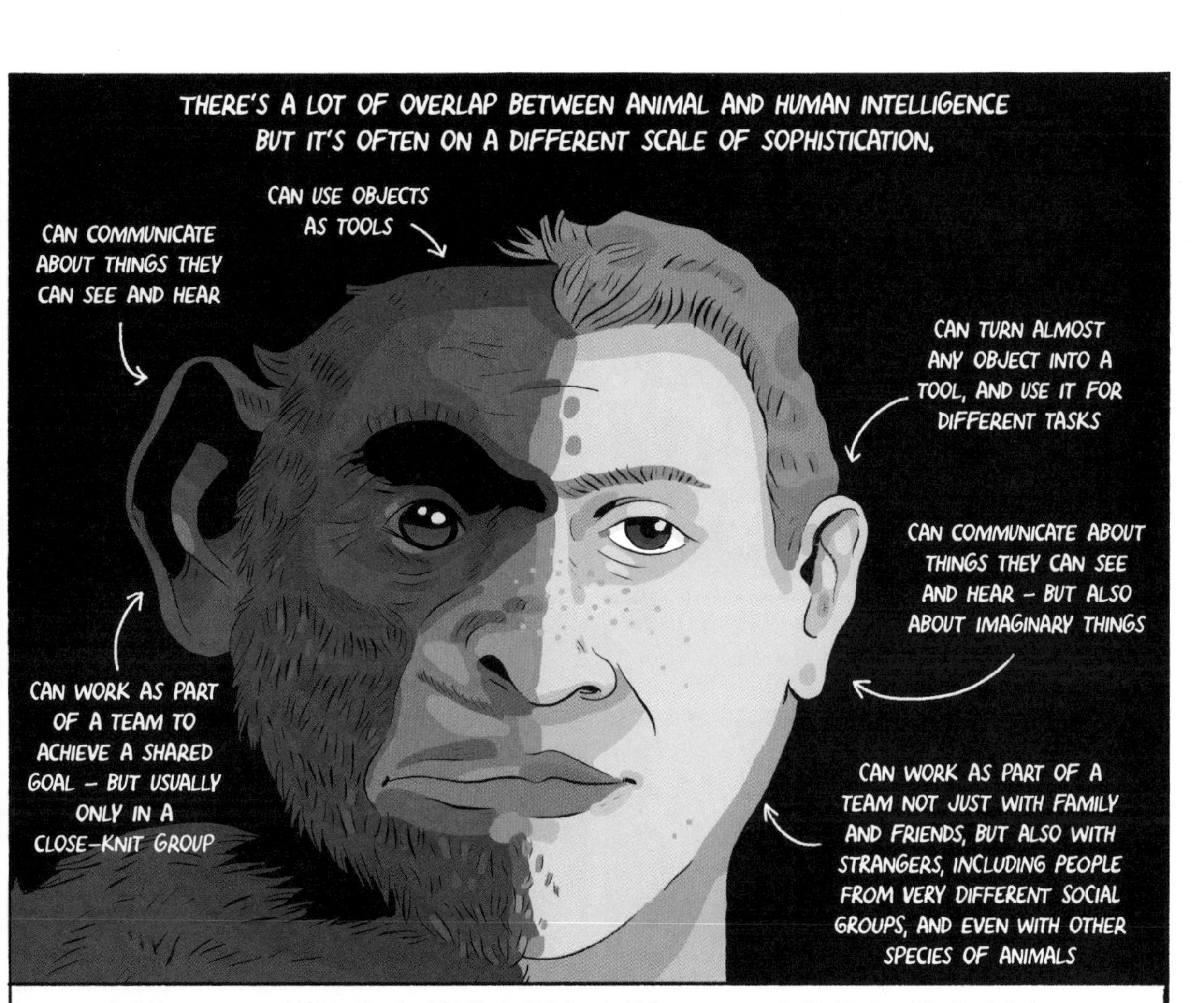
THERE'S A LOT OF OVERLAP BETWEEN ANIMAL AND HUMAN INTELLIGENCE BUT IT'S OFTEN ON A DIFFERENT SCALE OF SOPHISTICATION.
CAN COMMUNICATE ABOUT THINGS THEY CAN SEE AND HEAR
CAN USE OBJECTS AS TOOLS
CAN TURN ALMOST ANY OBJECT INTO A TOOL, AND USE IT FOR DIFFERENT TASKS
CAN COMMUNICATE ABOUT THINGS THEY CAN SEE AND HEAR – BUT ALSO ABOUT IMAGINARY THINGS
CAN WORK AS PART OF A TEAM TO ACHIEVE A SHARED GOAL – BUT USUALLY ONLY IN A CLOSE-KNIT GROUP
CAN WORK AS PART OF A TEAM NOT JUST WITH FAMILY AND FRIENDS, BUT ALSO WITH STRANGERS, INCLUDING PEOPLE FROM VERY DIFFERENT SOCIAL GROUPS, AND EVEN WITH OTHER SPECIES OF ANIMALS
FOR ALL THE INTELLIGENCE POSSESSED BY ANIMALS – IN THEIR PROPER HABITATS AND CONTEXTS – ONE THING WE HAVEN'T YET SEEN IS CLEAR EVIDENCE OF METACOGNTION.

ONE EXAMPLE OF METACOGNITION WE'VE LOOKED AT ALREADY IS "EXPLICIT THEORY OF MIND" – THE ABILITY TO THINK ABOUT WHAT OTHER PEOPLE ARE THINKING ABOUT.*
SOMETHING PLAYERS OFTEN MAKE USE OF WHEN PLAYING GAMES SUCH AS CHESS.
*NOT TO BE CONFUSED WITH THE "IMPLICIT" FORM OF THEORY OF MIND, THE KIND THAT MAY WELL CAUSE THE SOCIAL COMMUNICATION PROBLEMS IN AUTISM. SEE CHAPTER 4.

WE'RE BRINGING UP CHESS BECAUSE IT ILLUSTRATES SOME OTHER TYPES OF METACOGNITION, TOO.
THINKING ABOUT A SEQUENCE OF MOVES YOU INTEND TO MAKE...
...WHILE ALSO THINKING ABOUT THE MOVES YOUR OPPONENT MIGHT MAKE.

TO DO BOTH THESE TASKS, YOU HAVE TO KEEP UPDATING A MENTAL PICTURE OF WHERE THE PIECES ON THE BOARD WILL BE.

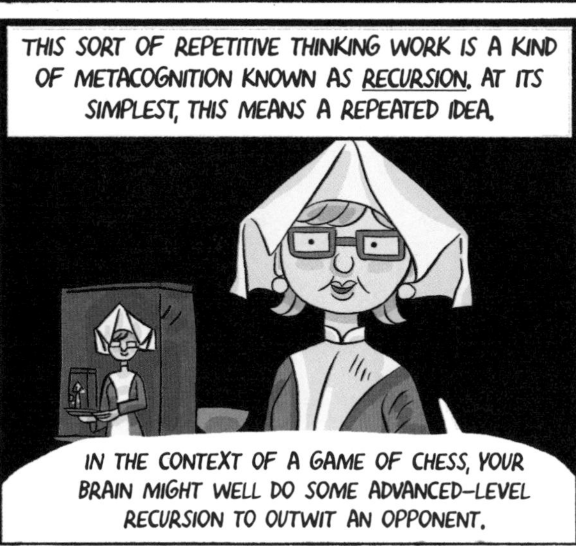
THIS SORT OF REPETITIVE THINKING WORK IS A KIND OF METACOGNITION KNOWN AS RECURSION. AT ITS SIMPLEST, THIS MEANS A REPEATED IDEA.
IN THE CONTEXT OF A GAME OF CHESS, YOUR BRAIN MIGHT WELL DO SOME ADVANCED-LEVEL RECURSION TO OUTWIT AN OPPONENT.

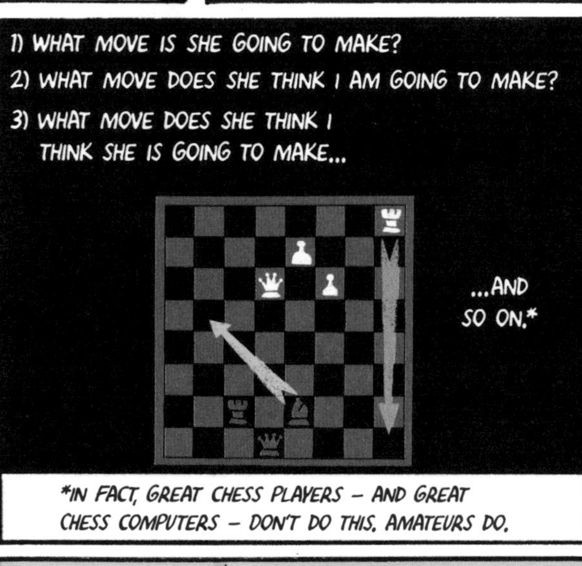
1) WHAT MOVE IS SHE GOING TO MAKE?
2) WHAT MOVE DOES SHE THINK I AM GOING TO MAKE?
3) WHAT MOVE DOES SHE THINK I THINK SHE IS GOING TO MAKE...
...AND SO ON.*
*IN FACT, GREAT CHESS PLAYERS – AND GREAT CHESS COMPUTERS – DON'T DO THIS. AMATEURS DO.

RECURSION COMES UP IN ALL SORTS OF SITUATIONS...
...AND IT'S NOT ALWAYS SOMETHING WE'RE CONSCIOUS OF.

FOR EXAMPLE, RECURSION IS A KEY PART OF LANGUAGE.
IT'S PART OF WHAT ALLOWS US TO START TO TALK WITHOUT KNOWING EXACTLY WHAT WE'RE GOING TO SAY, AND STILL GET OUR GRAMMAR CORRECT.

HERE'S AN EXAMPLE. IMAGINE YOU'RE DESCRIBING A HALF-REMEMBERED DREAM, AND VARIOUS DETAILS COME BACK TO YOU AS YOU DESCRIBE IT...
"LAST NIGHT I DREAMED ABOUT A DOG
A GERMAN SHEPHERD
ONE I HAD MET IN REAL LIFE THE DAY BEFORE
WHOSE OWNER WORE A RED JACKET
THAT REMINDED ME OF A POINSETTIA
A POPULAR CHRISTMAS PLANT
AND KNEE-LENGTH BOOTS
WITH A ZIP UP THE SIDE
AND IT CHASED ME DOWN THE ROAD."
THIS IS A CONVOLUTED YET GRAMMATICALLY CORRECT SENTENCE. IT RELIES ON YOU KNOWING IT'S OK TO REPEATEDLY EMBED NEW CLAUSES, ONE AFTER THE OTHER. GRAMMATICAL RECURSION.

SOME PEOPLE SAY THAT ONLY HUMANS CAN USE RECURSION DELIBERATELY. OTHERS SAY, WHAT ABOUT BIRDSONG?
TO AN UNTRAINED EAR, MUCH BIRDSONG SOUNDS LIKE REPETITION OF SNIPPETS OF SONG. BUT RECURSION IS MORE SOPHISTICATED THAN SIMPLY REPEATING SOMETHING.

IN PARTICULAR, STARLINGS USE AND RECOGNIZE EACH OTHER'S SONGS IN THE SAME WAY THAT I PUT TOGETHER THAT STRING OF CLAUSES WHEN DESCRIBING MY DREAM.
I DON'T BELIEVE THEY ARE THINKING ABOUT THEIR OWN THOUGHTS, BUT THEY ARE TAPPING INTO A MENTAL ABILITY THAT NOT ALL ANIMALS HAVE.

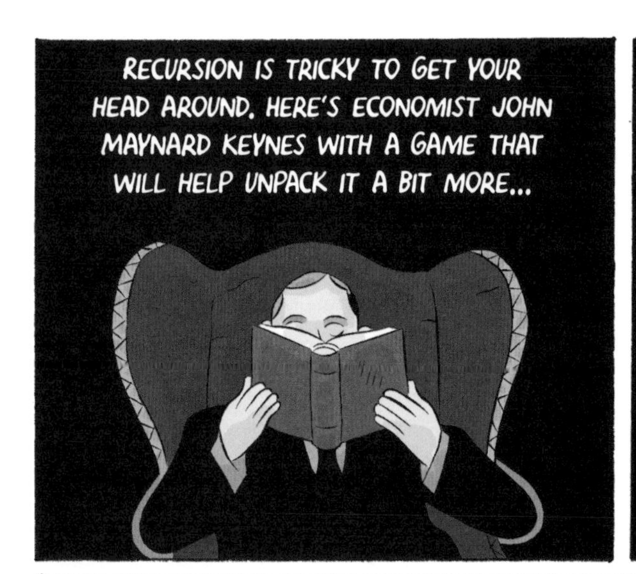
RECURSION IS TRICKY TO GET YOUR HEAD AROUND. HERE'S ECONOMIST JOHN MAYNARD KEYNES WITH A GAME THAT WILL HELP UNPACK IT A BIT MORE...

I CALL MY GAME "THE BEAUTY CONTEST." TO PLAY, I NEED A ROOM FULL OF PEOPLE. THEY HAVE TO LOOK AT EACH OTHER AND JUDGE EACH OTHER (IN THEIR HEADS), ON WHO THEY THINK IS BEAUTIFUL.

THE AIM OF THE GAME IS NOT TO DETERMINE WHO IS THE MOST BEAUTIFUL PERSON.
WHAT IS BEAUTY ANYWAY? SURELY IT IS IN THE EYE OF THE BEHOLDER, NOT AN OBJECTIVE TRUTH.

THE PLAYERS DO NOT CHOOSE THE FACES THEY PERSONALLY FIND MOST BEAUTIFUL...
OOH, HE'S PRETTY.

...INSTEAD, THEY MUST GUESS WHO THE MAJORITY OF PEOPLE WILL SAY IS MOST BEAUTIFUL.
I RECKON MORE PEOPLE WILL CHOOSE THAT BLANDER-LOOKING CHAP.

ALREADY, YOU NEED TO GIVE AS MUCH THOUGHT TO OTHER PEOPLE'S OPINIONS AS YOU DO TO YOUR OWN.
WHO WILL THE OTHERS THINK I WILL CHOOSE?

ECONOMISTS WHO STUDIED UNDER KEYNES DEVELOPED A VARIATION OF THIS GAME KNOWN AS THE P-BEAUTY CONTEST. IT'S NOT ABOUT BEAUTY, WHICH IS VERY HARD TO BE OBJECTIVE ABOUT. IT'S ABOUT NUMBERS.
INSTEAD OF LOOKING AT FACES, PEOPLE PLAYING THE GAME HAVE TO CHOOSE A NUMBER BETWEEN 1 AND 100. (100 MAKES THE MATH EASIER.)

THE WINNER IS THE PERSON WHO CHOOSES A PARTICULAR NUMBER...

...THE NUMBER THAT IS CLOSEST TO HALF OF THE AVERAGE OF ALL THE NUMBERS CHOSEN.

IF THAT SOUNDS CONFUSING, IT'S MEANT TO. IT'S NOT AN EASY GAME TO PLAY, LET ALONE WIN.

HERE'S HOW IT GOES. ALMOST EVERYBODY NOTICES THAT IF 100 IS THE BIGGEST POSSIBLE NUMBER, HALF OF THAT NUMBER IS 50. SO, NO ONE CHOOSES A NUMBER ABOVE 50.

MOST PEOPLE THEN APPLY A STEP OF RECURSION, AND DECIDE THAT THEY SHOULD CHOOSE HALF OF THAT NUMBER, NAMELY 25.

A HANDFUL OF PEOPLE APPLY ANOTHER STEP OF RECURSION, AND CHOOSE THE NUMBER 12 OR 13.
IF MOST PEOPLE WILL CHOOSE 25, I SHOULD CHOOSE HALF THAT NUMBER.

IF YOU EVER FIND YOURSELF PLAYING THIS GAME, CHOOSE 12 FOR THE BEST CHANCE OF WINNING.*

IT'S UNDERSTOOD THAT USING RECURSIVE THINKING IS A SIGN OF INTELLIGENCE, BUT IT'S INSTRUCTIVE TO LEARN THAT – ACCORDING TO THIS KIND OF TEST – MOST PEOPLE ONLY APPLY ONE AND A HALF LEVELS OF RECURSIVE THINKING.

*OR, IF YOU'RE PLAYING WITH A LECTURE HALL FULL OF ECONOMISTS, CHOOSE 6.

RECURSION IS NOT THE SAME THING AS ANOTHER SUBSET OF METACOGNITION, KNOWN AS REVISION.

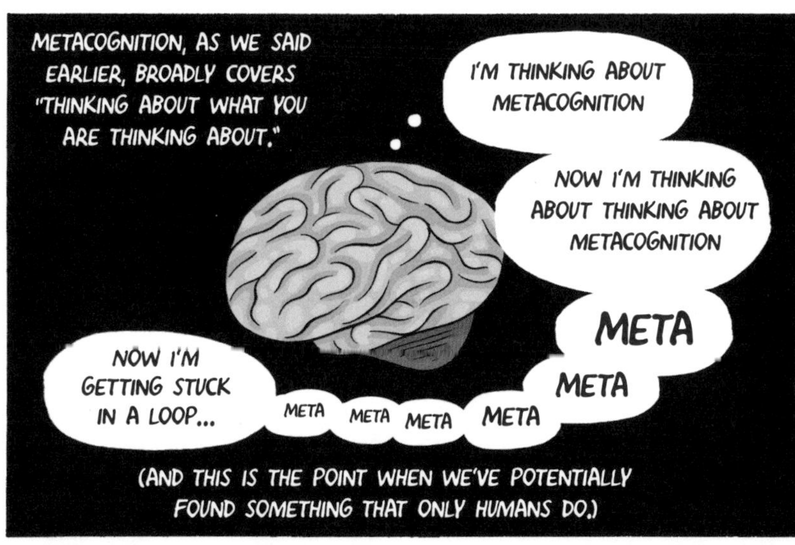
METACOGNITION, AS WE SAID EARLIER, BROADLY COVERS "THINKING ABOUT WHAT YOU ARE THINKING ABOUT."
I'M THINKING ABOUT METACOGNITION
NOW I'M THINKING ABOUT THINKING ABOUT METACOGNITION
META
META
META
META
META
META
NOW I'M GETTING STUCK IN A LOOP...
(AND THIS IS THE POINT WHEN WE'VE POTENTIALLY FOUND SOMETHING THAT ONLY HUMANS DO.)

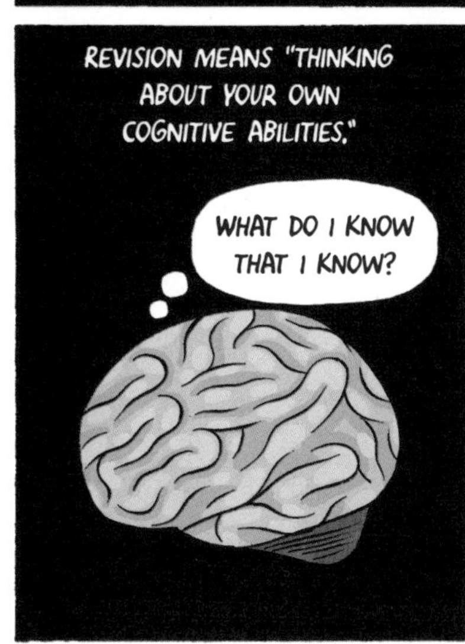
REVISION MEANS "THINKING ABOUT YOUR OWN COGNITIVE ABILITIES."
WHAT DO I KNOW THAT I KNOW?

DOESN'T "REVISION" MEAN "LOOKING OVER MY NOTES BEFORE TAKING AN EXAM"?*
THE MIND
*AND HAS A DIFFERENT EVERDAY MEANING IN US ENGLISH – EDITING YOUR WORK.

PRECISELY – THE POINT IS THAT YOU ONLY NEED TO REVISE THE SUBJECTS YOU KNOW YOU DON'T KNOW VERY WELL.
THE MIND
CUP O' JOE

IT'S POSSIBLE TO MEASURE ACTS OF REVISION BY ASKING HOW CONFIDENT A PERSON IS ABOUT WHAT THEY KNOW.

LOOK AT THIS JUMBLE OF OBJECTS.

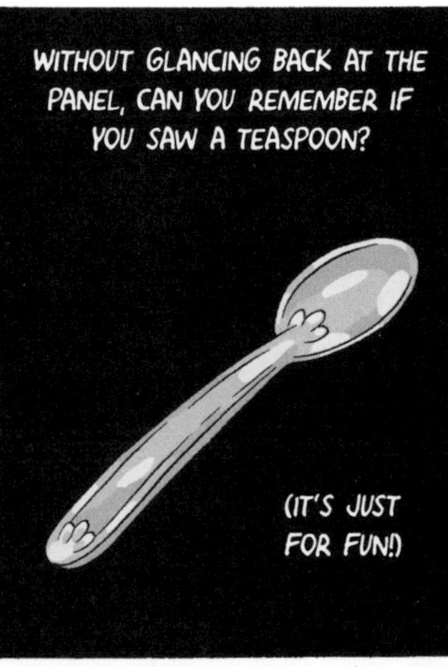
WITHOUT GLANCING BACK AT THE PANEL, CAN YOU REMEMBER IF YOU SAW A TEASPOON?
(IT'S JUST FOR FUN!)

NOW, HAVE A THINK ABOUT HOW CONFIDENT YOU ARE IN THE OPTION YOU CHOSE.
YES
NO
DON'T KNOW

PARISIAN COLLEAGUES SID KOUIDER AND LOUISE GOUPIL HAVE DEVELOPED A WAY TO DO THIS KIND OF TEST ON INFANTS (AGE 18 MONTHS), TO SHOW THAT THEY, TOO, EXAMINE CONFIDENCE IN THEIR MEMORIES – EVEN BEFORE THEY CAN TALK.

HERE'S ONE OF THEIR COLLEAGUES RUNNING THE EXPERIMENT.
STAGE 1: THE CHILD SITS WHERE SHE CAN SEE TWO BOXES.

A RESEARCHER PUTS AN OBJECT INTO ONE BOX.

THE RESEARCHER INVITES THE CHILD TO TAKE THE OBJECT OUT OF THE CORRECT BOX.

STAGE 2: THE RESEARCHER STARTS AGAIN, PUTTING THE OBJECT INTO ONE BOX, BUT THEN HIDES THE BOXES BEHIND A CURTAIN...

AND ASKS THE CHILD TO WAIT FOR SEVERAL SECONDS...
TICK
...BEFORE ASKING WHICH BOX HOLDS THE OBJECT.

15 SECONDS IS MORE THAN ENOUGH TIME FOR A HUMAN OF ANY AGE TO LOSE CONFIDENCE IN THEIR OWN MEMORY.

MEANWHILE, THE MEAN OL' RESEARCHER TAKES THE OBJECT OUT OF THE BOX (HIDDEN BEHIND THE CURTAIN).

THE CURTAIN IS REMOVED, AND THE CHILD IS INVITED TO FIND THE OBJECT.

SHE PUTS HER HAND INTO ONE BOX AND FEELS AROUND.

THE CHILD SPENDS MEASURABLY LONGER RUMMAGING AROUND IN THE BOX THAT SHE THOUGHT THE OBJECT SHOULD BE IN...

...THAN SHE DOES RUMMAGING IN THE OTHER BOX.

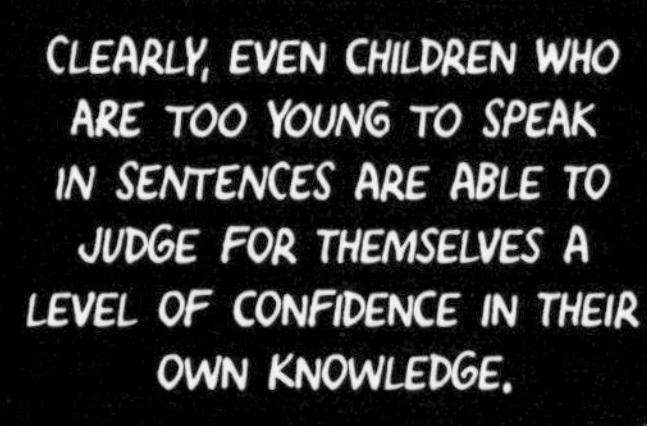
CLEARLY, EVEN CHILDREN WHO ARE TOO YOUNG TO SPEAK IN SENTENCES ARE ABLE TO JUDGE FOR THEMSELVES A LEVEL OF CONFIDENCE IN THEIR OWN KNOWLEDGE.

(DON'T WORRY, THE CHILD IS GIVEN THE OBJECT TO PLAY WITH AFTER THE EXPERIMENT IS FINISHED.)

AND WHILE WE'RE ON THE SUBJECT OF DOING EXPERIMENTS ON INFANTS...
...YES, IT'S TRUE THAT IF TWO NEUROSCIENTISTS MARRY AND HAVE CHILDREN, THEY WILL USE THOSE CHILDREN AS TEST SUBJECTS.
ALEXANDER
MARTIN

IN 1984, UTA WAS ABOUT TO BEGIN A LONG STRETCH OF WORK WITH THE MRC COGNITIVE DEVELOPMENT UNIT IN GORDON STREET, BLOOMSBURY...

...PART OF UNIVERSITY COLLEGE LONDON.

PRINCESS DIANA VISITED TO MARK THE OFFICIAL OPENING.

UTA DISPLAYED HER SONS AS EXHIBITS TO SHOW HOW AGE AFFECTS PROBLEM-SOLVING ABILITIES.

MARTIN, AGE 9, QUICKLY LEARNED TO MOVE BLOCKS ON A LEVER IN ORDER TO MAKE IT BALANCE.

ALEXANDER, AGE 6, COULDN'T DO IT.

NEITHER CHILD WAS HAPPY.
MARTIN COULDN'T SEE WHAT ALL THE FUSS WAS ABOUT GETTING TO MEET A FAMOUS PERSON.
?
ALEXANDER WAS TERRIBLY FRUSTRATED BY NOT BEING ABLE TO DO THE TASK.

ALEXANDER'S FRUSTRATION WAS AN EXAMPLE OF FAILURE OF SELF-ASSESSMENT.

INSPIRED BY MARTIN'S SUCCESS, ALEXANDER BELIEVED HE KNEW HOW TO BALANCE THE LEVER, EVEN THOUGH HE CLEARLY DID NOT.
HE SHOULD'VE PICKED A "DON'T KNOW" OPTION.

AS IN "I DON'T KNOW WHETHER I CAN BALANCE THE LEVER OR NOT."
BECAUSE THEN, EVEN THOUGH HE COULDN'T DO THE TASK...
...HIS BRAIN WOULD REWARD HIS CORRECT SELF-ASSESSMENT THAT HE DIDN'T KNOW.

IN MANY SITUATIONS, IT'S TYPICAL TO THINK YOU MIGHT KNOW SOMETHING, BUT NOT BE SURE ABOUT IT.
QUIZ
GENERAL
QUIZ
WHAT'S THE CAPITAL OF SLOVENIA?
I THINK IT'S... LJUBLJANA?

AS SOCRATES FAMOUSLY PUT IT:
TRUE "KNOWLEDGE" IS KNOWING THAT YOU KNOW NOTHING.

OFTEN, OF COURSE, PEOPLE ARE GENERALLY CONFIDENT IN WHAT THEY DO AND DON'T KNOW. BUT WE HAVE DISCOVERED THAT THE "DON'T KNOW" OPTION BECOMES AVAILABLE EARLY ON IN HUMAN BRAINS.
YES
NO
DON'T KNOW
IN FACT, IT'S EVEN AVAILABLE TO SOME ANIMALS.

LET'S FIND OUT HOW THAT WORKS IN A ROUND OF... THE BRAIN GAME!
IN THIS GAME, CONTESTANTS ARE GIVEN A BIG REWARD FOR A CORRECT "YES" OR "NO" ANSWER, BUT ALSO A SMALLER REWARD FOR A "DON'T KNOW."
FRITH
LET'S MEET OUR THREE COUPLES PLAYING TODAY: THE PROFESSORS FRITH...
...TWO RATS...
...AND TWO MACAQUES.
SQUEAK
SQUEAK
OOK
OOK

ASKING AN ANIMAL A QUESTION IS A BIT FIDDLY. THE QUESTIONS TEND TO BE VISUAL; FOR EXAMPLE, A COMPUTER SCREEN DISPLAYS A PATTERN OF MOVING DOTS.
The Brain Game
RATTUS
SIMIAN

THE ANIMAL HAS TO PRESS A BUTTON TO MATCH WHICH DIRECTION THE DOTS ARE MOVING.
BING
LEFT
DON'T KNOW
RIGHT

IF THE ANIMAL PRESSES THE CORRECT BUTTON, IT'LL GET A REWARD. RATS AND MONKEYS LEARN TO DO THIS TASK PRETTY FAST.

SOMETIMES THE DOTS ON THE SCREEN DON'T MOVE...
LEFT
DON'T KNOW
RIGHT

...WHICH IS WHEN THE ANIMALS ARE SEEN TO PRESS THE "DON'T KNOW" BUTTON.
RIGHT
DON'T KNOW
RIGHT
LEFT

SOME ARGUE THIS IS AN EXAMPLE OF METACOGNITION: THE ANIMAL KNOWS THAT IT DOESN'T KNOW. OTHERS SAY IT'S JUST THE ANIMAL LEARNING TO PRESS A PARTICULAR BUTTON WHEN GIVEN A CERTAIN CUE.

WHEN HUMANS DO THIS SORT OF TEST, IT'S ABSOLUTELY ABOUT META-COGNITION. THIS BROAD MENTAL TOOL IS WHAT HUMANS USE WHEN WE ARE EXAMINING OUR CONFIDENCE IN OUR OWN KNOWLEDGE.

RIGHT, TEAM FRITH, YOUR TURN! YOU CAN CONFER WITH EACH OTHER AS MUCH AS YOU LIKE BEFORE SELECTING YOUR ANSWER.
YOUR QUESTION IS: LJUBLJANA IS THE CAPITAL OF SLOVENIA. TRUE OR FALSE?

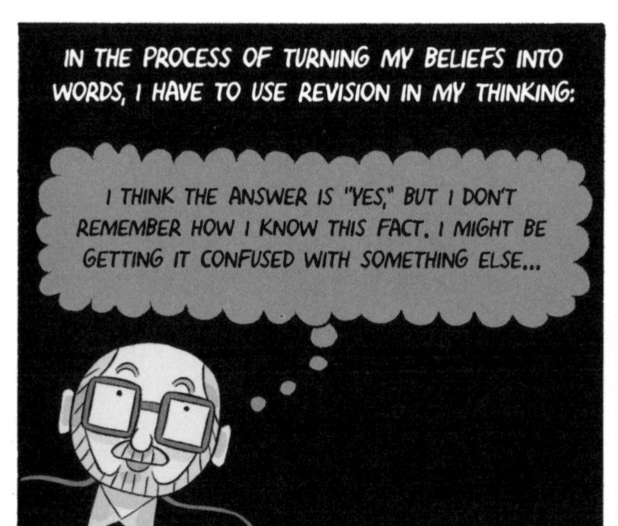
IN THE PROCESS OF TURNING MY BELIEFS INTO WORDS, I HAVE TO USE REVISION IN MY THINKING:
I THINK THE ANSWER IS "YES," BUT I DON'T REMEMBER HOW I KNOW THIS FACT. I MIGHT BE GETTING IT CONFUSED WITH SOMETHING ELSE...

AND, MORE IMPORTANTLY, THE PERSON LISTENING TO ME HAS TO THINK ABOUT CONFIDENCE – HER CONFIDENCE IN MY WORDS, AND HER ASSESSMENT OF MY CONFIDENCE IN MY OWN WORDS.
I WANT TO SAY YES, I THINK IT'S THE RIGHT ANSWER, BUT IT DOES SEEM PRETTY UNLIKELY THAT I WOULD KNOW.

FOR ALL OUR CLEVERNESS, HUMANS TEND TO OVERCOMPLICATE THINGS...
I'M STUCK FOR AN ANSWER, AND ALL THIS TALKING ISN'T HELPING!
SO I'M GOING TO SAY "DON'T KNOW."

THE ANSWER IS TRUE. LJUBLJANA IS THE CAPITAL OF SLOVENIA!

IT'S ALSO THE LOCATION OF A PARTICULARLY MEMORABLE FAMILY CAR JOURNEY, DRIVING ALONG THE ADRIATIC COAST AT MIDNIGHT DURING A THUNDERSTORM.
SO IN FACT YOU WERE WRONG TO SAY THAT YOU DON'T KNOW, BECAUSE YOU JOLLY WELL DID KNOW IT, SURELY?!

A QUICK RECAP: METACOGNITION MEANS, BROADLY, "THINKING ABOUT THINKING."

SOME ANIMALS MAY BE AWARE OF THEIR OWN THOUGHTS BUT WE DON'T KNOW FOR SURE.

WE DEFINITELY KNOW THAT WE HUMANS THINK ABOUT OUR OWN THOUGHTS!

SOMETIMES THIS IS ANNOYING, BUT IT'S VERY USEFUL TOO, ESPECIALLY WHEN DECIDING HOW CONFIDENT WE ARE IN OUR BELIEFS.

WE ALSO EVALUATE OTHER PEOPLE'S THOUGHTS – AND EVEN OTHER PEOPLE'S OWN LEVELS OF CONFIDENCE IN THEIR THOUGHTS.

SOMETIMES WE THINK ABOUT OUR THOUGHTS, AND OTHERS' THOUGHTS, IN A REPEATING PATTERN. THIS IS CALLED RECURSION – WE'LL GET BACK TO IT IN THE NEXT CHAPTER.

YOU MIGHT HAVE NOTICED THAT WE'VE FINALLY BEGUN TALKING ABOUT HOW BRAINS INTERACT WITH EACH OTHER.

IN OTHER WORDS, SOCIAL COGNITION.

BUT HOW ON EARTH CAN WE STUDY THIS? IT'S HARD ENOUGH TO EXAMINE A SINGLE BRAIN (IN REAL LIFE, GENERALLY WITHOUT CUTTING IT OPEN!), SO WHAT'S THE TRICK?

GLAD YOU ASKED! LET'S FIND OUT IN THE NEXT CHAPTER.

HANG ON A MINUTE, WHO WON THE BRAIN GAME??
DON'T KNOW.
DON'T KNOW.
FRITH
DON'T KNOW.
DON'T KNOW.
YOU'RE ALL CORRECT. NO ONE KNOWS IF ANY ANIMALS ARE, OBJECTIVELY, MORE INTELLIGENT THAN ANY OTHER.
SMALL REWARDS ALL ROUND!

Chapter 8

BRAINS ARE DESIGNED TO WORK WITH OTHER BRAINS.

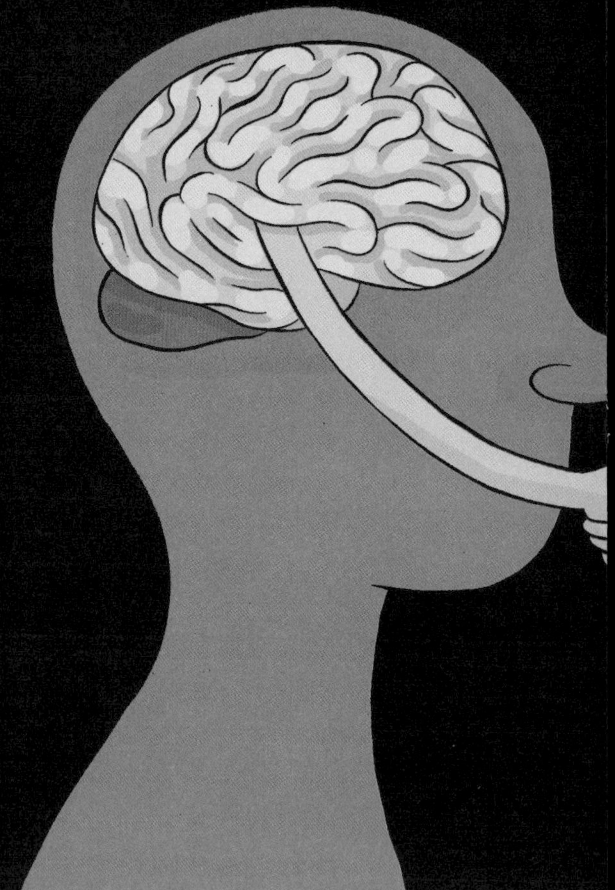

Watching brains at work.

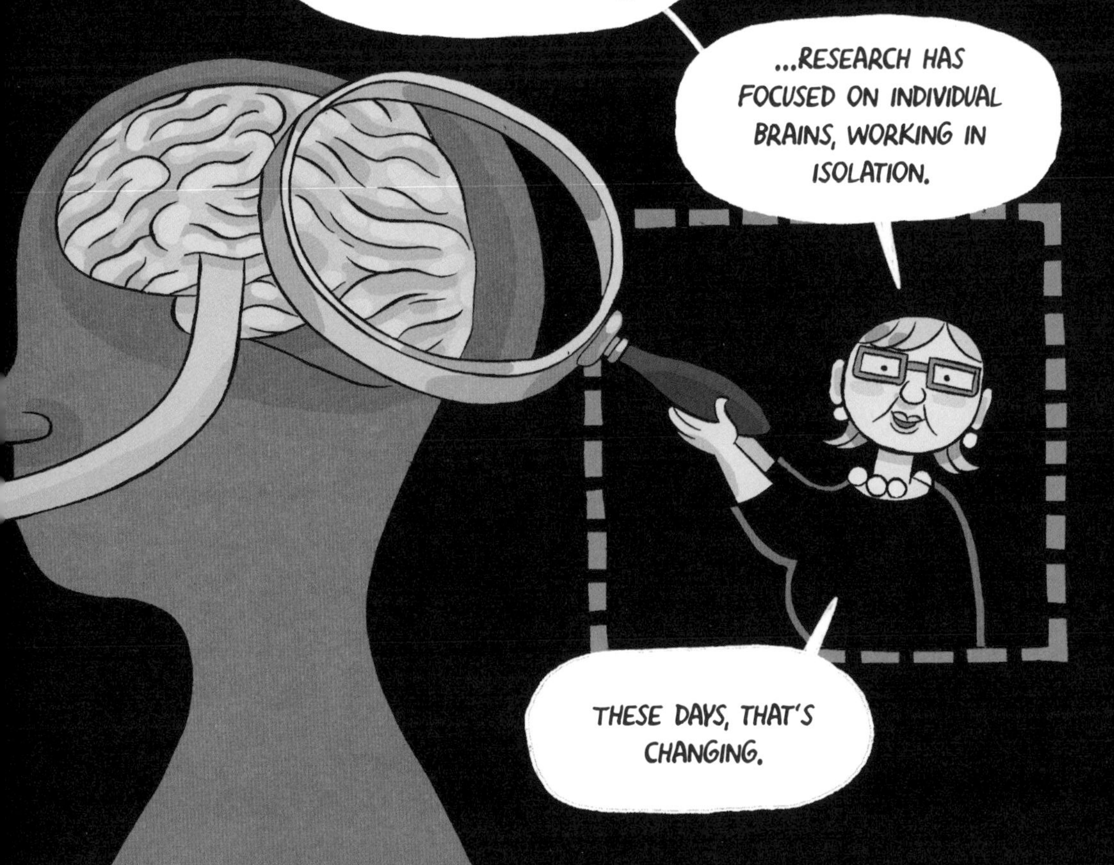
BUT, SO FAR, MUCH OF THE HISTORY OF NEUROSCIENCE HAS HAD A FUNDAMENTAL BARRIER...
...RESEARCH HAS FOCUSED ON INDIVIDUAL BRAINS, WORKING IN ISOLATION.
THESE DAYS, THAT'S CHANGING.

SCIENCE IS REALLY, REALLY HARD TO DO PROPERLY.

WE'VE ALREADY DESCRIBED SOME OF THE HIDDEN WORKINGS OF SCIENCE...

...INCLUDING ITS TRIALS AND, WE BELIEVE, ITS TRIUMPHS.

BUT PERHAPS YOU'RE STILL WONDERING WHAT EXACTLY RESEARCH PSYCHOLOGISTS DO ALL DAY...

THERE'S A LOT OF SITTING AT COMPUTERS STARING AT DATA.

AND A FAIR BIT OF SITTING AROUND TABLES DRINKING TEA AND DISCUSSING IDEAS.

NOT TO MENTION TRAVELING FROM COUNTRY TO COUNTRY TO TALK TO ROOMS FULL OF STRANGERS.

IT'S A GOOD LIFE.*
TO BE CLEAR, WE DO ACTUALLY PERFORM EXPERIMENTS. SOME OF THEM ARE EVEN DESIGNED TO HAVE A DIRECT, PRACTICAL EFFECT ON PEOPLE'S LIVES.
LET ME GIVE YOU AN EXAMPLE, FROM EARLY IN MY CAREER.
*ASIDE FROM THE CRIPPLING NERVES THAT PRECEDE EVERY. SINGLE. LECTURE.
I'M TALKING ABOUT THE EARLY 1970S, WHICH NOW FEELS SO LONG AGO THAT YOUNGER READERS WOULD BE FORGIVEN FOR PICTURING A HOSPITAL MENTAL HEALTH WARD AS IF IT'S FROM A GOTHIC HORROR MOVIE...
I'LL HAND YOU OVER TO MY COLLEAGUE EVE JOHNSTONE, WHILE I PREPARE THESE ELECTRODES.
AS YOU MAY RECALL, THE 20TH CENTURY IS NOT FONDLY REMEMBERED FOR ITS TREATMENT OF PEOPLE WITH MENTAL HEALTH DIFFICULTIES.

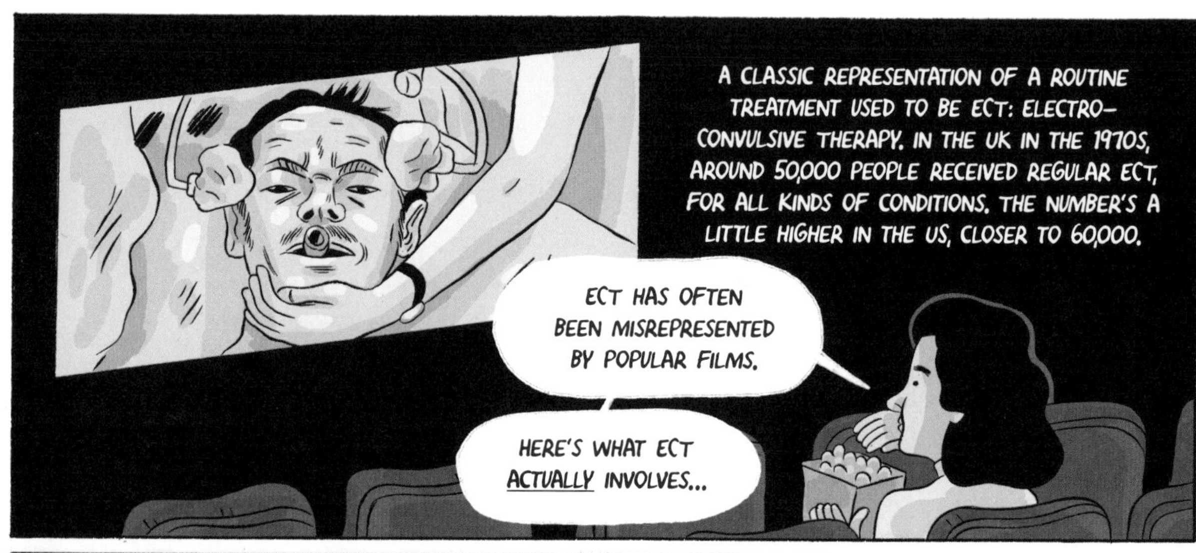
A CLASSIC REPRESENTATION OF A ROUTINE TREATMENT USED TO BE ECT: ELECTRO-CONVULSIVE THERAPY. IN THE UK IN THE 1970S, AROUND 50,000 PEOPLE RECEIVED REGULAR ECT, FOR ALL KINDS OF CONDITIONS. THE NUMBER'S A LITTLE HIGHER IN THE US, CLOSER TO 60,000.
ECT HAS OFTEN BEEN MISREPRESENTED BY POPULAR FILMS.
HERE'S WHAT ECT ACTUALLY INVOLVES...

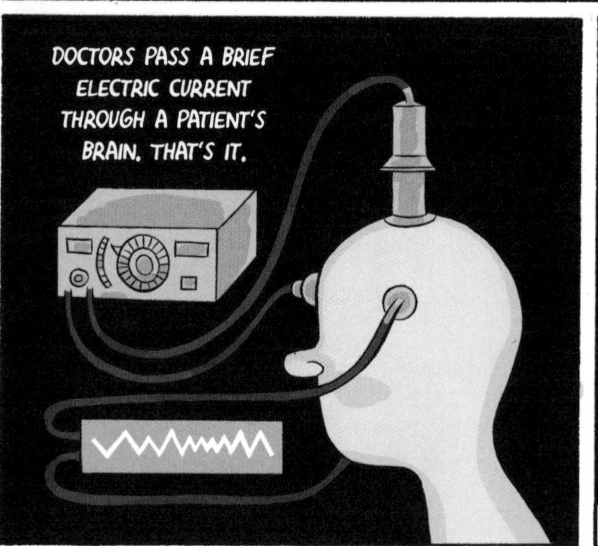
DOCTORS PASS A BRIEF ELECTRIC CURRENT THROUGH A PATIENT'S BRAIN. THAT'S IT.

THIS SOUNDS ALARMING, AS DOES THE INTENDED RESULT: INDUCING A CONTROLLED SEIZURE.
(PLEASE NOTE: YOU NEVER SEE ANY SPARKS)

REPORTEDLY, THIS HAS THE EFFECT OF TEMPORARILY RELIEVING ALL MANNER OF DIFFICULTIES.

– INSERT "HAVE YOU TRIED SWITCHING HIS BRAIN OFF AND ON AGAIN" JOKE HERE –

WHAT SENSATIONALISTIC FILMS DON'T EXPLAIN IS THAT PATIENTS ARE GIVEN A MUSCLE RELAXANT AND ANESTHETIC FIRST.
A REAL ECT SESSION IS NOT AT ALL PAINFUL.

NEVERTHELESS, ECT SHOULDN'T BE USED IF THE TREATMENT IS NOT ACTUALLY EFFECTIVE.

CHRIS AND I ATTEMPTED TO FIND AN ANSWER WHEN WE WORKED TOGETHER IN THE '70S.
OUR HYPOTHESIS: ECT MIGHT NOT BE EFFECTIVE.

SCIENTIFIC RESEARCH AT ITS PUREST, WITH A DIRECT CLINICAL APPLICATION, TOO.

TO TEST OUR HYPOTHESIS, WE NEEDED PATIENTS AND DOCTORS TO VOLUNTEER FOR A DOUBLE-BLIND TRIAL.

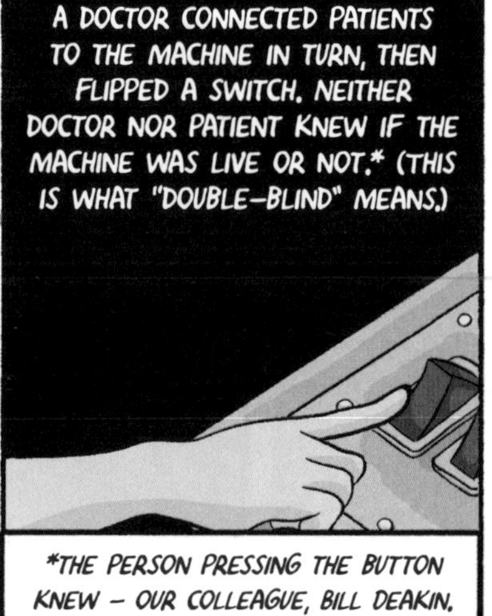
A DOCTOR CONNECTED PATIENTS TO THE MACHINE IN TURN, THEN FLIPPED A SWITCH. NEITHER DOCTOR NOR PATIENT KNEW IF THE MACHINE WAS LIVE OR NOT.* (THIS IS WHAT "DOUBLE-BLIND" MEANS.)
*THE PERSON PRESSING THE BUTTON KNEW – OUR COLLEAGUE, BILL DEAKIN.

(REMEMBER, THE PATIENTS ARE SEDATED, AND IN REAL LIFE, THERE'S NO GIANT CRACKLE OF ENERGY AROUND THEIR HEADS, EITHER.)

PURSUING THIS STUDY PROVED TOUGH ON ETHICAL GROUNDS.

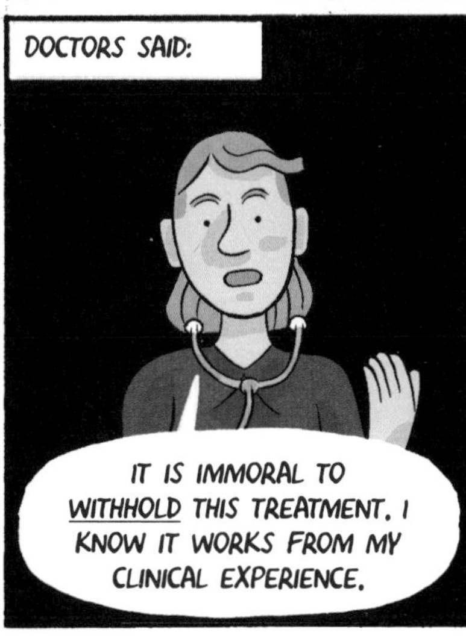
DOCTORS SAID:
IT IS IMMORAL TO WITHHOLD THIS TREATMENT. I KNOW IT WORKS FROM MY CLINICAL EXPERIENCE.

BUT SOCIAL WORKERS SAID:
IT IS IMMORAL TO GIVE THIS TREATMENT. IT DAMAGES THE BRAIN AND DOES NOT IMPROVE ANY ILLNESS.

AS IT WAS, WE DID MANAGE TO GET THE STUDY GOING...

...AND FOUND THAT, ON BALANCE, ECT DID HAVE A POSITIVE EFFECT ON PATIENTS WITH DEPRESSION – IN THE SHORT TERM.
I FEEL BETTER.

BUT WE ALSO FOUND THAT HAVING REAL OR SHAM ECT MADE NO DIFFERENCE TO HOW WELL PATIENTS FELT IN THE LONG TERM.
HERE'S HOW OUR RESULTS WERE RECEIVED...

DOCTORS SAID:
THE SIGNIFICANT REDUCTION IN DEPRESSION AT THE END OF THE COURSE SHOWS THAT THE TREATMENT HELPS PATIENTS GET THROUGH THE MOST DANGEROUS PHASE OF THE ILLNESS.
THIS SHORT-TERM BENEFIT CAN BE LIFE-SAVING.
THE TREATMENT MUST CONTINUE!

BUT SOCIAL WORKERS SAID:
ONE MONTH AFTER THE COURSE, REAL ECT WAS NO BETTER THAN SHAM. THE TREATMENT IS WORTHLESS AND SHOULD BE STOPPED.

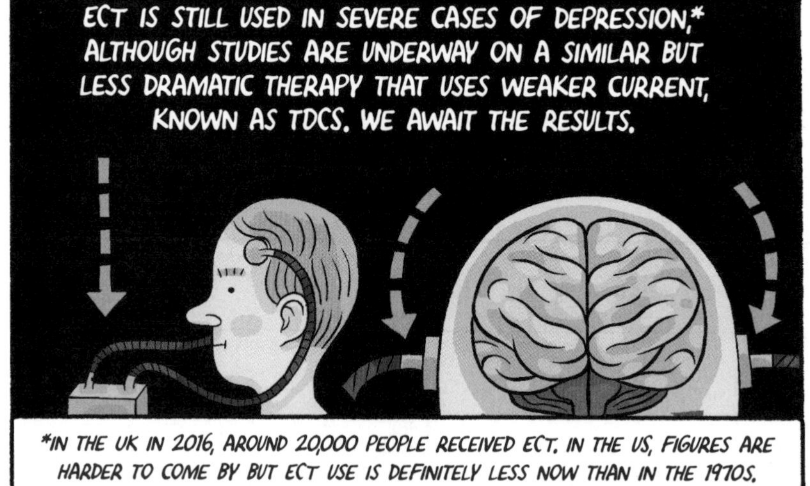
ECT IS STILL USED IN SEVERE CASES OF DEPRESSION,* ALTHOUGH STUDIES ARE UNDERWAY ON A SIMILAR BUT LESS DRAMATIC THERAPY THAT USES WEAKER CURRENT, KNOWN AS TDCS. WE AWAIT THE RESULTS.
*IN THE UK IN 2016, AROUND 20,000 PEOPLE RECEIVED ECT. IN THE US, FIGURES ARE HARDER TO COME BY BUT ECT USE IS DEFINITELY LESS NOW THAN IN THE 1970S.

AND I'LL REMIND MYSELF THAT EXPERIMENTAL DATA OFTEN ISN'T ENOUGH TO CHANGE PEOPLE'S MINDS!

I FOUND THAT WORKING ALONGSIDE PRACTICING PSYCHIATRISTS WAS AMAZINGLY PRODUCTIVE. BUT I COULDN'T ESCAPE A CERTAIN AMOUNT OF OFFICE POLITICS PULLING ME IN TWO DIRECTIONS.
YOU SHOULD PERSUADE PATIENTS TO COME TO OUR PRIVATE CLINIC, WHERE THEY'LL GET THE BEST TREATMENT!
NO, YOU SHOULD PERSUADE THEM TO JOIN OUR FREE EXPERIMENTAL TRIALS!

PUSHED AS WELL AS PULLED, I MOVED AWAY FROM THE PSYCHIATRY DEPARTMENT AT NORTHWICK PARK TO A RESEARCH DEPT AT THE HAMMERSMITH HOSPITAL...

...WORKING WITH A BRAND NEW MACHINE CALLED A PET* SCANNER.

*PET STANDS FOR "POSITRON-EMISSION TOMOGRAPHY," IN CASE YOU'RE WONDERING.

AT THE TIME (THE EARLY 1990S), THIS WAS THE CUTTING EDGE OF SEEING A BRAIN AT WORK. VARIOUS BRAIN REGIONS COULD BE HIGHLIGHTED, IN REAL TIME, ON THE SCANNING MACHINE – REVEALING WHERE BLOOD FLOW WAS MOST RAPID. IN TURN, THIS INDICATED WHICH BITS OF THE BRAIN WERE USING THE MOST ENERGY TO DO A TASK.

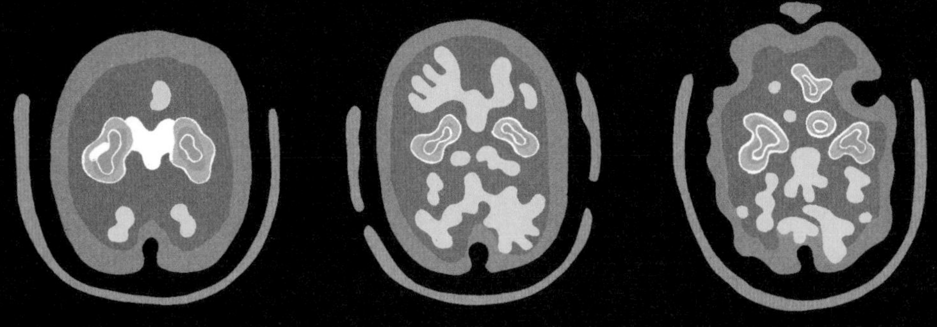

TO GET THIS EFFECT, PEOPLE IN THE SCANNER FIRST HAD TO BE INJECTED WITH A MILD DOSE OF RADIATION – THIS IS WHAT EMITS THE POSITRONS THAT THE SCANNER PICKS UP.

YOU'RE NOT ALLOWED TO DO PET SCANS ON CHILDREN OR WOMEN (OF AN AGE WHERE THEY MIGHT BE PREGNANT).

SOME MIGHT SAY THESE RESTRICTIONS LIMIT THE VALUE OF THE DATA YIELDED BY PET EXPERIMENTS.

IN ANY EVENT, PET TECHNOLOGY WAS SOON SUPERSEDED BY MRI (TURN BACK TO THE INTERLUDE FOR A REMINDER ON THAT).

BUT MRI SCANNERS PROMPTED ANOTHER ROUND OF OFFICE POLITICS.
LEADING RESEARCHERS SIMPLY REFUSED TO BELIEVE THAT A VERSION OF MRI – FUNCTIONAL MRI (OR FMRI) – WAS USEFUL.
THEY RECKONED THAT TINY HEAD MOVEMENTS BY PEOPLE IN FMRI SCANNERS WERE RUINING ANY POSSIBLE DATA. I DISAGREED.*

*FMRI SCANS ARE ALL ABOUT OBSERVING BRAINS OVER A PERIOD OF TIME, AND REQUIRE PEOPLE TO STAY STILL. PEOPLE DON'T ALWAYS STAY STILL – BUT IT HAS NOW BEEN PROVEN THAT FMRI SCANNERS CAN TAKE ACCOUNT OF THIS.

AND SO I WAS PROPELLED INTO A STATE-OF-THE-ART FUNCTIONAL IMAGING LABORATORY IN QUEEN SQUARE, CENTRAL LONDON.
NOW, JUST TO BOAST A LITTLE ABOUT THE IMAGING TEAM...

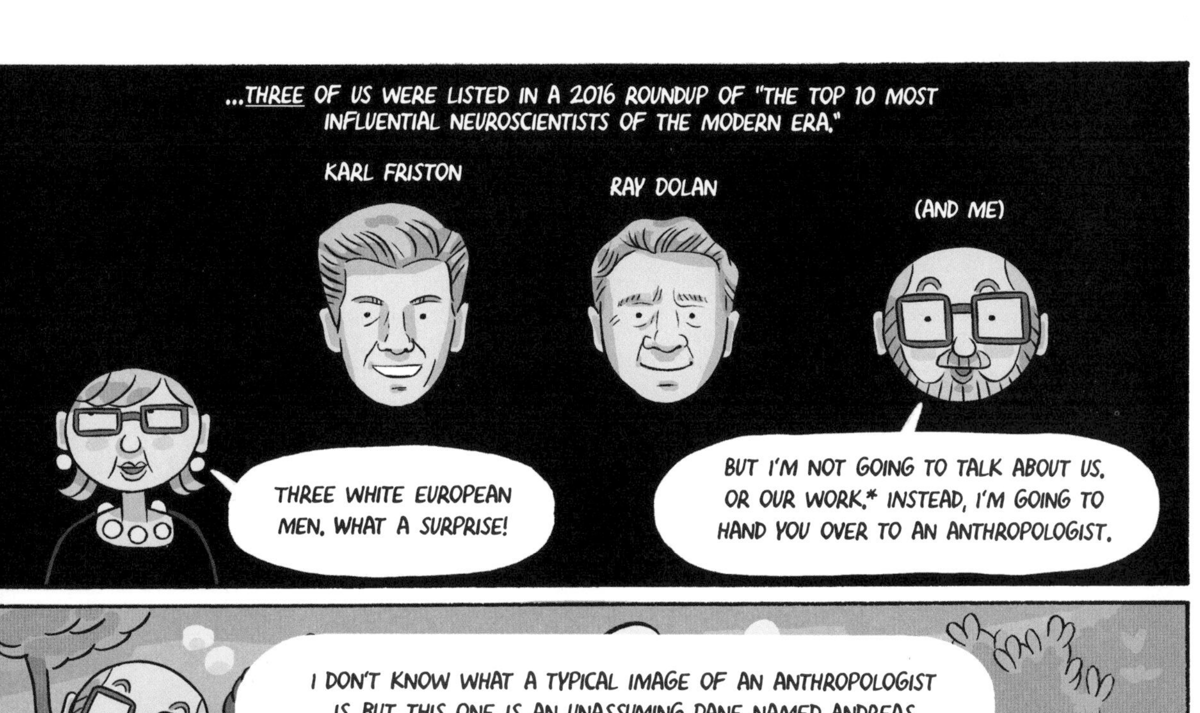
...THREE OF US WERE LISTED IN A 2016 ROUNDUP OF "THE TOP 10 MOST INFLUENTIAL NEUROSCIENTISTS OF THE MODERN ERA."
KARL FRISTON
RAY DOLAN
(AND ME)
THREE WHITE EUROPEAN MEN. WHAT A SURPRISE!
BUT I'M NOT GOING TO TALK ABOUT US. OR OUR WORK.* INSTEAD, I'M GOING TO HAND YOU OVER TO AN ANTHROPOLOGIST.

I DON'T KNOW WHAT A TYPICAL IMAGE OF AN ANTHROPOLOGIST IS, BUT THIS ONE IS AN UNASSUMING DANE NAMED ANDREAS ROEPSTORFF. (WHO HAS EVEN MORE CHILDREN THAN SHOWN HERE.)
HI, EVERYBODY, LOVELY TO MEET YOU ALL. NOW, DO YOU KNOW WHAT ANTHROPOLOGY IS?

*THE BEST BIT WAS ALREADY COVERED IN CHAPTER 1: KARL FRISTON'S WORK ON WHY BRAINS ARE BAYESIAN.

CLASSIC ANTHROPOLOGY INVOLVES PEOPLE FROM ONE CULTURE (USUALLY A RICH, WESTERN ONE) GOING TO LIVE WITH A SMALL GROUP FROM ANOTHER CULTURE (OFTEN A NON-RICH, NON-WESTERN ONE) TO OBSERVE AND REPORT.
BUT I CHOSE TO EMBED MYSELF WITH A TEAM OF EMINENT EUROPEAN NEUROSCIENTISTS AT A UNIVERSITY IN LONDON, WITH THE STATED AIM OF ANALYZING THEIR GROUP DYNAMICS.
DO YOU LIKE MY DISGUISE?

HERE'S ME, SAT IN A CORNER, TAKING NOTES.
THESE PEOPLE ARE CALLED "PRINCIPALS" – THEY SEEM TO BE THE MOST IMPORTANT IN THE BUILDING.
PEOPLE HERE SIT AROUND AND TALK A LOT.
HERE'S AN ACTUAL SCIENCE LAB, WITH BRAIN SCANNING EQUIPMENT IN.
HERE ARE THE FELLOWS. THESE PEOPLE SEEM TO WORK VERY HARD ALL THE TIME.
THE PERSON LYING DOWN IS REFERRED TO AS A "TEST SUBJECT."
THIS PERSON IS CALLED THE "ADMINISTRATIVE DIRECTOR." HE (AND IT IS A HE) THINKS HE'S IMPORTANT.
THE PEOPLE IN LAB COATS ARE SOMETIMES CALLED "TECHNICIANS."
HE LEADS MEETINGS IN THE BOARDROOM AT WHICH THE PRINCIPALS ARE FORCED TO TALK ABOUT FUNDING, AND OFTEN END UP GETTING CROSS.
(IN REALITY, TEST SUBJECTS ENTER ON THE GROUND FLOOR, AND ARE TAKEN TO SCANNERS IN THE BASEMENT.)

I COULD TELL YOU WHAT I THOUGHT IT WAS LIKE WORKING AT THE LAB, BUT ANDREAS'S VERSION IS INFINITELY MORE ENTERTAINING.
THE LAB RESEMBLES A GEORGIAN HOUSE, WHERE THERE IS A CLEAR HIERARCHY OF PEOPLE, AND SET ROOMS FOR SET TASKS. THE OVERALL TASK OF THE HOUSE IS TO TURN NATURE INTO SCIENCE...
TO SEE WHAT'S GOING ON, START AT THE BOTTOM. "NATURE," WE SEE, TAKES THE FORM OF A HUMAN BEING...
HUMAN TEST SUBJECTS ENTER AT THE GROUND FLOOR. THEY GO DOWN TO THE BASEMENT TO BE SCANNED.
THEIR DATA IS PROCESSED AS IT MOVES UP THROUGH THE HOUSE...
...UNTIL IT EMERGES IN THE SEMINAR ROOM ON THE TOP FLOOR AS "OBJECTIVE FACTS."
AT THIS POINT, IT'S SENT OFF FOR PUBLICATION – IDEALLY IN THE MAGAZINE CALLED NATURE.
SEMINAR ROOM
PRINCIPALS
FELLOWS
BOARDROOM
RECEPTION
(THE DIFFERENCE BETWEEN A BOARD, THE FELLOWS, AND THE PRINCIPALS MATTERS TO THEM – BUT NOT TO ANYONE ELSE.)
Nature

AS SCIENTISTS, WE'RE PRIMED TO LOOK DOWN ON ANTHROPOLOGISTS.

WHAT YOU'RE DOING IS INTERESTING, BUT IT'S NOT REAL SCIENCE.
(IS SOMETHING WE MAY HAVE THOUGHT BUT NEVER SAID OUT LOUD)

AS PSYCHOLOGISTS, WE'RE USED TO BEING LOOKED DOWN ON BY OTHER SCIENTISTS.

WHAT YOU'RE DOING IS INTERESTING, BUT IT'S NOT REAL SCIENCE.

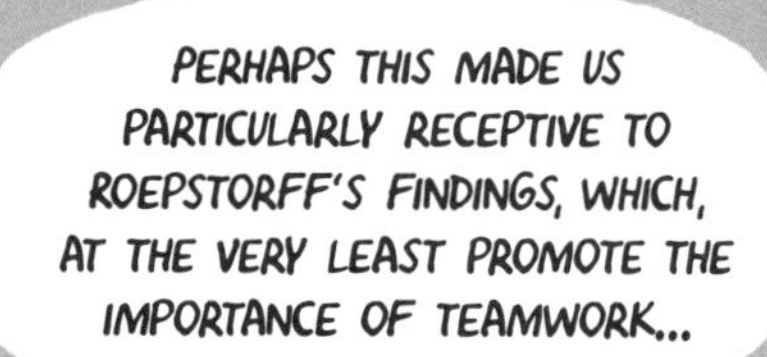
PERHAPS THIS MADE US PARTICULARLY RECEPTIVE TO ROEPSTORFF'S FINDINGS, WHICH, AT THE VERY LEAST PROMOTE THE IMPORTANCE OF TEAMWORK...

...AND MAKE THE INVALUABLE POINT THAT TEAMWORK IS NOT ONLY POSSIBLE BUT DESIRABLE.

"TEAMWORK IS GOOD" IS JUST THE SORT OF COMMONSENSE IDEA THAT MANY PEOPLE ALREADY THINK IS TRUE...
...AND THEREFORE SEE NO NEED TO USE THE TOOLS OF SCIENCE TO PROVE IT.

I REFER YOU BACK 6 PAGES, WHERE PSYCHIATRISTS "KNEW" THAT ECT WAS A USEFUL TREATMENT, WHILE SOCIAL WORKERS "KNEW" IT WAS USELESS.

SO, WHERE DOES THAT LEAVE US TODAY? WELL, WE'RE BACK AT THE PROBLEM RAISED ON THE FIRST PAGE OF THIS CHAPTER!
YOU COULD SAY THAT THE LAST 50 YEARS OR SO OF NEUROSCIENCE HAS BEEN ONE LONG PROJECT OF WATCHING BRAINS AT WORK...
...AS THEY PERFORMED IDENTICAL TASKS, ONE AFTER ANOTHER.
AND THERE'S GOOD, HARD SCIENCE HAPPENING HERE.
IN SUM, NEUROSCIENTISTS AROUND THE WORLD HAVE BEEN ABLE TO COMPARE ALL THE RESULTS TO DETERMINE WHAT AN "AVERAGE" BRAIN IS LIKE.
WE LIKE TO THINK WE CAME UP WITH SOME PRETTY CLEVER TESTS TO MAKE THIS PROCESS WORK WELL.

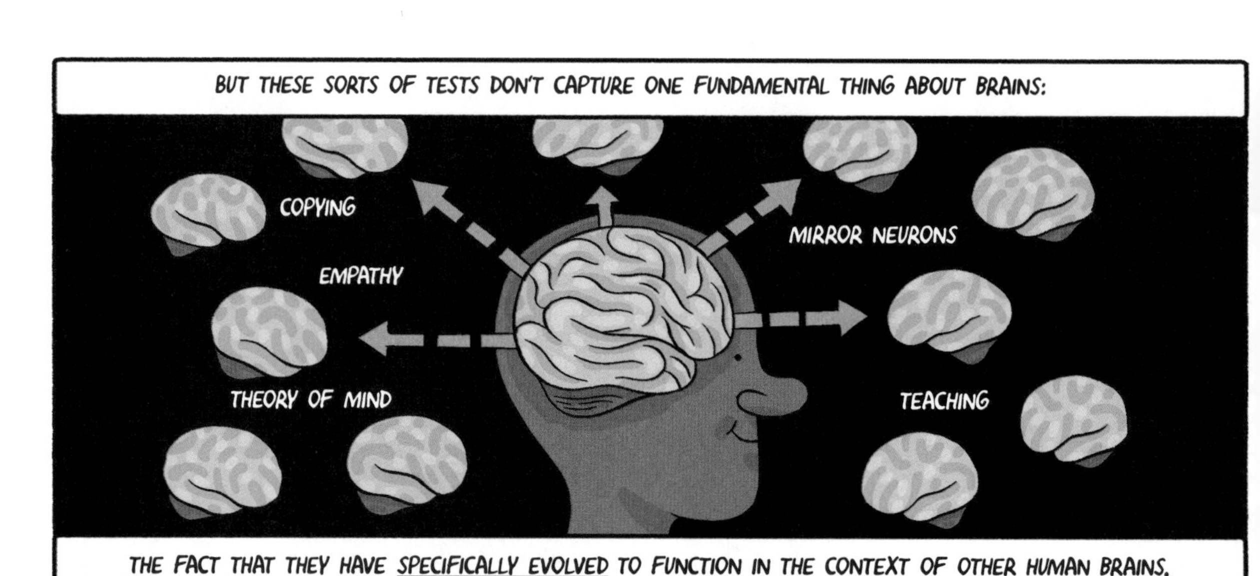
BUT THESE SORTS OF TESTS DON'T CAPTURE ONE FUNDAMENTAL THING ABOUT BRAINS:
COPYING
EMPATHY
THEORY OF MIND
MIRROR NEURONS
TEACHING
THE FACT THAT THEY HAVE SPECIFICALLY EVOLVED TO FUNCTION IN THE CONTEXT OF OTHER HUMAN BRAINS.

MRI SCANNERS BUILT FOR TWO DO EXIST,
BUT THEY HAVEN'T BEEN USED OFTEN ENOUGH – OR, WE SUSPECT, WELL ENOUGH – AS YET.
WE EAGERLY AWAIT THE DATA!*
*AND DO PLEASE NOTE THAT AS CREATING 300-PAGE COMICS TAKES YEARS, WE APPRECIATE THAT SEVERAL GOOD STUDIES MAY HAVE BEEN DONE SINCE THIS WAS WRITTEN...

ALL THIS SAID, EVEN THE TYPICAL "ONE BRAIN" TESTS ARE COLORED TO AN EXTENT BY INTERACTION...
...BECAUSE ALL TESTS INEVITABLY INVOLVE AT LEAST ONE BRAIN INTERACTING WITH THE BRAIN BEING TESTED.

EVEN IF I, THE EXPERIMENTER, PUT A TEST SUBJECT ON HER OWN IN AN ISOLATED ROOM...

...I STILL HAVE TO GREET HER, AND EXPLAIN THE TEST.
CLICK THE BUTTON WHEN YOU SEE RED DOTS.

AND OF COURSE HER BRAIN IS PRIMED BY THE NOTION THAT SHE IS "DOING A PSYCHOLOGY EXPERIMENT."

ONE EXPECTS SHE WILL NOT DELIBERATELY DO THE OPPOSITE OF WHAT I'VE ASKED.
I'M GOING TO CLICK WHEN I SEE GREEN DOTS.

A LIFETIME'S WORTH OF MY OWN DATA IS ALL BASED ON AN ASSUMPTION THAT PEOPLE HAVEN'T BEEN DOING THIS SORT OF THING!
THAT SAME LIFETIME OF WORK ALSO GIVES ME HOPE THAT IT HASN'T ALL BEEN A WASTE OF TIME, EVEN AS THE LAST DECADE OR SO HAS BEEN GIVEN OVER TO THE NEW CHALLENGE OF STUDYING BRAINS WORKING TOGETHER.
IN ESSENCE, WE'RE STUDYING THAT MOST VALUED CONCEPT, TEAMWORK.

DOING GOOD SCIENCE IS ALWAYS WORTHWHILE!
IT'S NOT JUST ABOUT PROVING A SIMPLE HYPOTHESIS. IN THIS INSTANCE, "TEAMWORK" ALONE ISN'T ENOUGH TO DESCRIBE HOW TO GET THE BEST FROM A GROUP.

WE THINK WE'VE BEEN ABLE TO DRILL INTO SOME OF THE DETAIL ABOUT WHAT KINDS OF TEAMS WORK BEST, AND WHAT IT MEANS TO WORK TOGETHER WITH OTHER PEOPLE.
CLATTER

THE WAY TO DO IT, OF COURSE, IS TO STUDY PEOPLE'S BRAINS. SOMETIMES INDIVIDUALLY, BUT INCREASINGLY, IN PAIRS OR EVEN GROUPS.

!
DEVISING TESTS TO FIT THESE SITUATIONS HAS BEEN AS TRICKY AS ANYTHING WE'VE DONE.

BUT, LUCKILY, AND INDEED APPROPRIATELY, WE HAVEN'T HAD TO DO IT ALONE, OR EVEN BY RESTRICTING OUR TEAMS TO NEUROSCIENTISTS.

I'M STICKING AROUND, THIS SOUNDS REALLY INTERESTING!

OUR HYPOTHESIS IS THAT GROUPS WORK BEST TOGETHER WHEN THE PEOPLE IN THEM HAVE A ROUGHLY SIMILAR LEVEL OF COMPETENCE...
...BUT YIELD THE BEST RESULTS WHEN THAT TEAM IS MADE UP OF A DIVERSE GROUP, WHO USE DIFFERENT WAYS TO APPROACH PROBLEMS.
WE'RE GOING TO TEST THIS HYPOTHESIS OVER THE NEXT FEW CHAPTERS.
DO FOLLOW US!

Chapter 9
ANDREAS ROEPSTORFF MADE ONE CLASSIC "ANTHROPOLOGIST'S ERROR" DURING HIS PERIOD OF FIELD OBSERVATION...

Two heads really are better than one.
...HE WENT NATIVE.

I FOUND BRAIN IMAGING SO FASCINATING THAT I DECIDED TO FOCUS ON THAT, INSTEAD OF FURTHER ANTHROPOLOGICAL STUDIES.

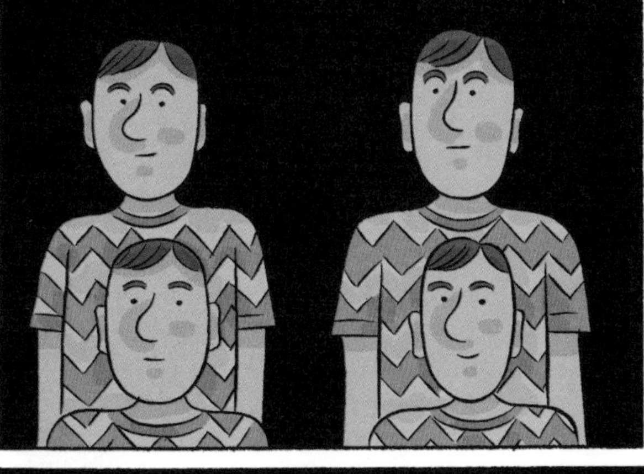
TRADITIONAL ANTHROPOLOGY INVOLVES INVESTIGATING CLOSED GROUPS OF PEOPLE.

I WANTED TO USE BRAIN IMAGING SPECIFICALLY TO FOCUS ON HUMAN DIVERSITY.

THIS LED TO THE CREATION OF THE INTERACTING MINDS CENTER IN AARHUS, DENMARK.
MOST DEPARTMENTS ARE THE PROVINCE OF ONE DISCIPLINE. BRAIN IMAGING CENTERS, FOR EXAMPLE, TEND TO INCLUDE ONLY MEDICINE, PHYSIOLOGY, AND PSYCHOLOGY.
MY CENTER DRAWS PEOPLE FROM LINGUISTICS, SEMIOTICS, POLITICAL SCIENCE, ARCHEOLOGY, AND EVEN THEOLOGY, AS WELL AS BRAIN IMAGING.
(NO SPACE TO TALK ABOUT IT HERE, BUT WE DID A STUDY ON THE EFFECTS OF PRAYER ON PAIN, AND EXAMINED GROUP WORK USING LEGOS. THIS IS DENMARK, AFTER ALL.)
MIXING DEPARTMENTS IN THIS WAY IS AN EXAMPLE OF DIVERSITY OF STUDY – BUT, IN TIME, WE WILL BE TALKING ABOUT SOCIAL AND ETHNIC DIVERSITY.

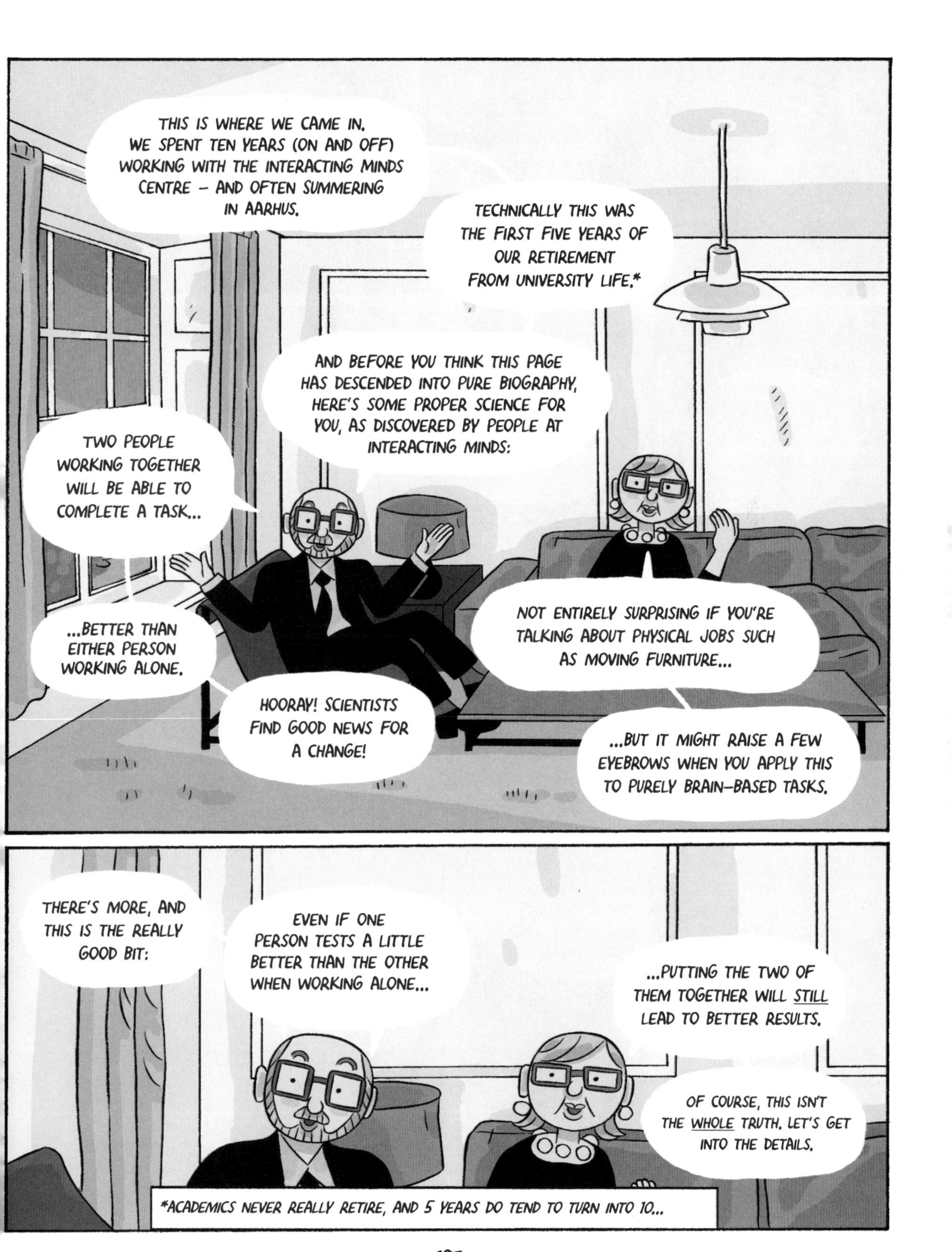
THIS IS WHERE WE CAME IN. WE SPENT TEN YEARS (ON AND OFF) WORKING WITH THE INTERACTING MINDS CENTRE – AND OFTEN SUMMERING IN AARHUS.
TECHNICALLY THIS WAS THE FIRST FIVE YEARS OF OUR RETIREMENT FROM UNIVERSITY LIFE.*
AND BEFORE YOU THINK THIS PAGE HAS DESCENDED INTO PURE BIOGRAPHY, HERE'S SOME PROPER SCIENCE FOR YOU, AS DISCOVERED BY PEOPLE AT INTERACTING MINDS:
TWO PEOPLE WORKING TOGETHER WILL BE ABLE TO COMPLETE A TASK...
...BETTER THAN EITHER PERSON WORKING ALONE.
HOORAY! SCIENTISTS FIND GOOD NEWS FOR A CHANGE!
NOT ENTIRELY SURPRISING IF YOU'RE TALKING ABOUT PHYSICAL JOBS SUCH AS MOVING FURNITURE...
...BUT IT MIGHT RAISE A FEW EYEBROWS WHEN YOU APPLY THIS TO PURELY BRAIN-BASED TASKS.
THERE'S MORE, AND THIS IS THE REALLY GOOD BIT:
EVEN IF ONE PERSON TESTS A LITTLE BETTER THAN THE OTHER WHEN WORKING ALONE...
...PUTTING THE TWO OF THEM TOGETHER WILL STILL LEAD TO BETTER RESULTS.
OF COURSE, THIS ISN'T THE WHOLE TRUTH. LET'S GET INTO THE DETAILS.
*ACADEMICS NEVER REALLY RETIRE, AND 5 YEARS DO TEND TO TURN INTO 10...

MEET ONE OF THE KEY FIGURES BEHIND THIS DISCOVERY: BAHADOR BAHRAMI. HE DEVISED A FIENDISHLY DIFFICULT – SOME WOULD EVEN SAY PAINFUL – TASK TO GIVE TO PEOPLE WORKING ALONE OR IN PAIRS.

THESE PEOPLE ARE STARING AT A SCREEN.
A SERIES OF DOTS FLASHES UP, ONE AFTER THE OTHER. EACH PERSON HAS TO DECIDE IF THEY SEE MORE BLUE OR YELLOW DOTS.
SOME PAIRS MUST DECIDE INDIVIDUALLY. BUT OTHER PAIRS ARE ASKED TO COME UP WITH A JOINT DECISION.

MY DATA ROUTINELY SHOWED THAT TWO PEOPLE WATCHING THE SCREEN TOGETHER, AND DISCUSSING THEIR OPINIONS, CONSISTENTLY GOT THE ANSWER RIGHT MORE OFTEN THAN PEOPLE WORKING ALONE.

THE ABILITY TO COMPARE DOTS MAY NOT SEEM ESPECIALLY USEFUL. LET'S TURN IT INTO A REAL-WORLD EXAMPLE.

THERE ARE TWO SOLDIERS ON A WATCHTOWER...

AS THEY STARE OUT, OVER LONG, BOREDOM-FILLED HOURS...SUDDENLY IN THE DISTANT HILLS, THEY CAN MAKE OUT A FLURRY OF MOVEMENT.
DO THEY SEE MOSTLY BLUE FLAGS – THEIR OWN ARMY, RETURNING HOME TRIUMPHANT?

OR DO THEY SEE MOSTLY YELLOW FLAGS, AS THEIR ARMY RACES HOME, PURSUED BY THE ENEMY?

IF YOU'RE THE WATCHTOWER LEADER, YOU CAN CITE BAHRAMI'S PAPER TO BACK UP YOUR BUDGET REQUEST TO HAVE TWO WATCHERS ON DUTY AT ALL TIMES.
BECAUSE SENIOR MANAGEMENT ARE ALWAYS PERSUADED BY SCIENTIFIC EVIDENCE. (I WISH.)

MOVING ON, HERE'S KRISTIAN TYLÉN AND RICCARDO FUSAROLI, TWO MORE MEMBERS OF THE TEAM. THEY INVESTIGATED WHAT WAS GOING ON WHEN PEOPLE ENGAGE IN THIS KIND OF COOPERATIVE TASK.

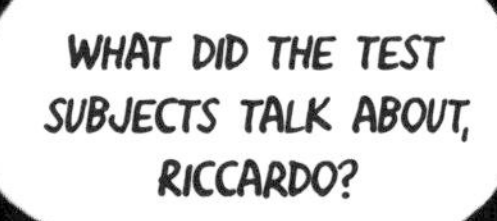

WHAT DID THE TEST SUBJECTS TALK ABOUT, RICCARDO?
IT'S ALL ABOUT CONFIDENCE, KRISTIAN.

HOW CONFIDENT IS EACH PERSON IN WHAT THEY HAVE SEEN?

HOW DO THEY COMPARE NOTES ON THEIR OWN CONFIDENCE?

MOST PEOPLE DOING THE TASK VERY QUICKLY DEVELOP A CONFIDENCE SCALE BETWEEN THEMSELVES.

I'M PRETTY SURE THERE WERE MORE BLUE DOTS.
I THINK THERE MAY HAVE BEEN MORE YELLOW DOTS.

WHO IS MORE CONFIDENT? WOULD YOU PICK BLUE OR YELLOW?

IN PRACTICE, IT'S RARE FOR BOTH OBSERVERS TO BE STRONGLY CONFIDENT IN OPPOSITE DIRECTIONS,

BUT IT'S EQUALLY RARE FOR ANY ONE OBSERVER TO BE 100% CONFIDENT IN AN ANSWER, TOO.

THIS MAY REMIND YOU OF SOMETHING YOU'VE READ ABOUT EARLIER IN THIS VERY BOOK...

DO YOU REMEMBER THE SWARM OF BEES SEARCHING FOR A NEW HOME?*
*SEE CHAPTER 1

SCOUTS EXPLORE, THEN COME BACK AND REPORT TO EACH OTHER ABOUT WHAT THEY'VE FOUND. THE ONES WHO ARE "MORE CONFIDENT" DANCE FOR LONGER, UNTIL THEY PERSUADE OTHER BEES TO FOLLOW THEM.

OR CONSIDER A SHOAL OF FISH, WHO CAN EACH DETECT A WEAK SIGNAL THAT FOOD IS NEARBY IN A PARTICULAR DIRECTION.
BY STICKING TOGETHER, THEY ARE ABLE TO FOLLOW A GROUP "CONFIDENCE REPORT" THAT SUGGESTS THE FOOD IS MORE LIKELY IN ONE DIRECTION THAN ANOTHER. IF THEY SEARCHED ALONE, THEY'D SOON GO SEPARATE WAYS AND HAVE VERY LITTLE SUCCESS.

COOPERATION HAS HELPED US TREMENDOUSLY. CHRIS'S ABILITY TO SOLVE CRYPTIC CROSSWORDS ON HIS OWN IS PRETTY HIGH...
...BUT HE NEEDS ME TO CONFIRM THE SPELLING OF WORDS.
AND, DESPITE MY GENERAL LACK OF INTEREST, IT INVARIABLY TAKES BOTH (OR, IF AVAILABLE, MORE PEOPLE) TO COMPLETE THE WHOLE PUZZLE.
(ALEX JOINS IN ON SUNDAY AFTERNOON VISITS.)

WORKING TOGETHER DOESN'T HELP IN PAIRINGS OF PEOPLE WHO ARE MISMATCHED IN TWO WAYS.
a
b
a
b
a
b
FIRST, IF ONE PERSON IS MUCH BETTER AT DOING THE TEST.
a
b
1
2
AND SECOND, IF BOTH PEOPLE ARE POOR AT ASSESSING THEIR OWN ABILITIES.
I'M SO BAD AT THIS.
I'M AMAZING.
a
b
1
(A PROBLEM THAT MOST OF US FACE IN REALITY...)*

*THIS IS SOMETIMES CALLED THE DUNNING-KRUGER EFFECT, ALTHOUGH TECHNICALLY THAT REFERS ONLY TO POORLY PERFORMING PEOPLE OVERASSESSING THEIR OWN ABILITIES, NOT THE OTHER WAY AROUND.

WILLIAM SHAKESPEARE PUTS IT BEST IN AS YOU LIKE IT...
THE FOOLE DOTH THINKE HE IS WISE, BUT THE WISE MAN KNOWES HIMSELFE TO BE A FOOLE.

AH, BUT DOES THE WISE MAN KNOW WHEN THE FOOLE IS ACTUALLY A FOOLE?

THAT'S ANOTHER DANISH COLLEAGUE, DAN BANG, WHO IS INTERESTED IN HOW PEOPLE WORK TOGETHER, TOO.
I'M GOING TO TALK ABOUT STARING AT DOTS AGAIN.
Denmark.

MY TESTS ARE AMONGST THE MOST PATIENCE-TESTING OF ALL TIME!

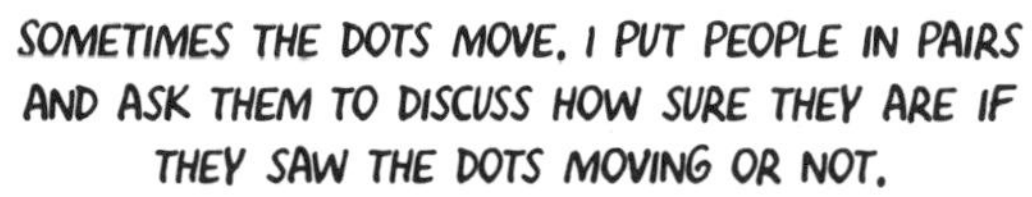
SOMETIMES THE DOTS MOVE. I PUT PEOPLE IN PAIRS AND ASK THEM TO DISCUSS HOW SURE THEY ARE IF THEY SAW THE DOTS MOVING OR NOT.

I'M STILL WAITING FOR CALLS FROM REALITY TV PRODUCERS.

I CONFIRMED BAHADOR'S RESULT: PEOPLE DO THE TEST BETTER IN PAIRS THAN WORKING ALONE.

MY INTEREST WAS IN HOW PEOPLE DISCUSS THEIR CONFIDENCE LEVELS.
I'M PRETTY SURE THEY DIDN'T MOVE.
I'M FAIRLY SURE THEY MOVED.

SAY WE HAVE A 6-POINT SCALE OF CONFIDENCE:
6. VERY SURE
5. PRETTY SURE
4. FAIRLY SURE
3. THINK SO
2. MAYBE
1. JUST MAYBE

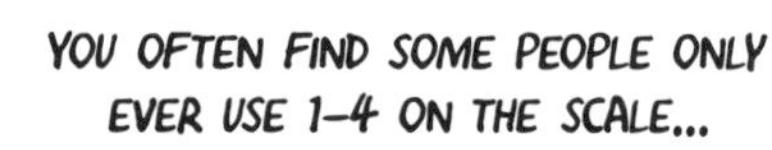
YOU OFTEN FIND SOME PEOPLE ONLY EVER USE 1–4 ON THE SCALE...

4. FAIRLY SURE
3. THINK SO
2. MAYBE
1. JUST MAYBE

...AND THERE ARE SOME WHO ONLY EVER USE 5 AND 6.
6. VERY SURE
5. PRETTY SURE

WE NEUROSCIENTISTS CALL THIS "REPORT BIAS" – THE PROBLEM OF PEOPLE USING DIFFERENT SCALES TO DESCRIBE THE SAME THING. I EXPECT YOU CAN CONJURE UP YOUR OWN REAL-WORLD EXAMPLES OF THIS. MY FAVORITE COMES FROM THE CIA...
NEWS

IN FEBRUARY 1951, THE CIA WERE PUZZLING OVER TENSIONS BETWEEN THE USSR AND YUGOSLAVIA.
WHAT'S THE LATEST INTELLIGENCE?
ASKED THE CHAIRMAN OF THE CIA PLANNING COMMITTEE.

ALTHOUGH IT IS IMPOSSIBLE TO DETERMINE WHICH COURSE THE KREMLIN IS LIKELY TO ADOPT, WE BELIEVE THAT AN ATTACK ON YUGOSLAVIA IN 1951 SHOULD BE CONSIDERED A SERIOUS POSSIBILITY.
SAID AN EXCERPT FROM CIA NOTE 29-51.

60%
25%
NEVER MIND "SERIOUS POSSIBILITY," JUST TELL ME THE ODDS!
20%
40%
SHRUGGED THE UNCERTAIN COMMITTEE MEMBERS.

YOU GUYS ARE USELESS! I CAN'T MAKE RECOMMENDATIONS BASED ON THAT MUCH VARIATION.
(THE SOVIETS DIDN'T INVADE IN THE END.)

BACK TO DAN BANG AND HIS MOVING DOTS.

THE PEOPLE I STUDIED SHOWED GREATER COLLABORATION THAN THE CIA...

...MEANING THEY SOON LEARN TO CALIBRATE THEIR SCALES.
AHA! YOUR "FAIRLY SURE" IS WORTH MORE THAN MY "PRETTY SURE."

THE IMPORTANT THING TO NOTE AT THIS POINT IS THAT WE'RE DESCRIBING AN UNCONSCIOUS PROCESS.

PEOPLE DON'T NOTICE THEMSELVES MAKING THESE ADJUSTMENTS.

EVEN MORE INTERESTING IS THAT WHEN PEOPLE ARE VERY MISMATCHED...

...MEANING ONE PERSON IS CLEARLY BETTER AT DOING THE TASK, WHILE THE OTHER IS INCLINED TO EXPRESS GREATER LEVELS OF CONFIDENCE...

...THEN THE MORE COMPETENT PERSON WILL START TO ALIGN THEIR OWN CONFIDENCE WITH THEIR LESS COMPETENT PARTNER.
YOU SEEM PRETTY SURE OF YOURSELF, I WON'T PRESS TOO HARD IF I DISAGREE.

AND YES, THIS MEANS THE PAIR DO LESS WELL ON THE TASK THAN THE COMPETENT PERSON WOULD HAVE DONE ON THEIR OWN.

WE WARNED YOU "TWO HEADS ARE BETTER THAN ONE" ISN'T THE WHOLE TRUTH.
COULD BANG'S FINDINGS BE THE RESULT OF BIAS IN HIS POOL OF TEST SUBJECTS? OR, IN OTHER WORDS...

...IS THIS A DANISH THING?
ACCORDING TO THE WORLD RANKINGS OF "HOW MUCH PEOPLE TRUST EACH OTHER," DENMARK IS NUMBER 3.
NORWAY AND SWEDEN ARE 1 AND 2, BUT WE DON'T LIKE TO TALK ABOUT THAT.

AT THE OPPOSITE END OF THE SCALE FOR "MUTUAL TRUST" RANKINGS, YOU'LL FIND CHINA AND IRAN.
MY COLLEAGUE ALI MAHMOODI AND I MADE A POINT OF RUNNING THE TEST ON PEOPLE SPECIFICALLY FROM THOSE TWO COUNTRIES.
SAME RESULT: AFTER A FEW TRIALS, THE MORE COMPETENT PERSON ALIGNED THEIR CONFIDENCE WITH THE LESS COMPETENT PERSON.

SO IF IT'S NOT ABOUT HOW MUCH PEOPLE TRUST EACH OTHER, WHAT MIGHT EXPLAIN DAN'S ORIGINAL FINDING?

1. IF SOMEONE ELSE EXPRESSES HIGH CONFIDENCE IN AN OPINION THAT IS DIFFERENT FROM YOUR OWN, YOU CAN'T HELP BUT THINK YOU MIGHT BE WRONG.
I DON'T THINK THEY MOVED?
THEY DEFINITELY MOVED.
OH, PERHAPS I'M WRONG.

2. IN SOME SETTINGS, PEOPLE MIGHT FEEL THAT AGREEING WITH EACH OTHER IS MORE IMPORTANT THAN SOLVING A PROBLEM CORRECTLY.

WHO CARES WHO'S RIGHT? LET'S JUST TRY TO GET ALONG.
YOU HAVE A TURN CHOOSING.

3. IT MIGHT NOT BE ABOUT GETTING ALONG WITH YOUR COMPANION – PERHAPS PEOPLE ARE MORE CONCERNED ABOUT AVOIDING BLAME IF THEY GET THE ANSWER WRONG.

THEY DEFINITELY MOVED.

I DON'T THINK THEY MOVED, BUT HE SEEMS PRETTY SURE. IF I AGREE WITH HIM, AND WE'RE WRONG, I CAN SAY IT'S HIS FAULT...

ONCE AGAIN, DURING THE TEST PEOPLE DON'T NOTICE THEY HAVE THESE THOUGHTS.
ACCORDING TO WHAT THEY TELL US, ANYWAY.

SO ALTHOUGH WE'VE FOUND SOME EXAMPLES WHEN TWO PEOPLE WORKING TOGETHER CAN AND DO GET THINGS WRONG...
...IN GENERAL, IT'S TRUE THAT TWO PEOPLE WORKING TOGETHER CAN SOLVE PROBLEMS BETTER THAN A PERSON WORKING ALONE.

THIS IS HARDLY A GROUNDBREAKING SCIENTIFIC DISCOVERY – BUT NEVER FORGET THAT UNTIL AN IDEA HAS BEEN TESTED, YOU CAN'T ASSUME THAT IT'S TRUE.

AND A WELL-CONSTRUCTED TEST CAN STILL YIELD A SURPRISING RESULT.

A BAT AND BALL TOGETHER COST $1.10. THE BAT COSTS $1 MORE THAN THE BALL.
QUESTION: HOW MUCH DOES THE BALL COST?

DID YOU SAY 10 CENTS? (OR $0.10, IF YOU PREFER?)

IF YOU DISCUSSED THE ANSWER WITH OTHER PEOPLE, ONE OF YOU MAY HAVE NOTICED THIS IS THE WRONG ANSWER.

IF THE BALL COSTS 10¢, AND THE BAT COSTS $1, THE TWO TOGETHER ADD UP TO $1.10.

BUT THE BAT IS SUPPOSED TO COST $1 MORE THAN THE BALL – $1.00 IS ONLY 90¢ BIGGER THAN $0.10.

THE CORRECT ANSWER IS: THE BALL COSTS 5¢ (OR $0.05).

YES, IT'S A QUESTION DELIBERATELY DESIGNED TO MAKE PEOPLE'S FIRST THOUGHT (USUALLY TRUSTWORTHY) BE THE WRONG ANSWER.*

THIS IS WHERE PEOPLE WORKING TOGETHER CAN DO BETTER – ONE PERSON SOWING A SEED OF DOUBT FORCES EVERYONE ELSE IN THE GROUP TO THINK AGAIN.

IN THIS CASE, JUST A LITTLE MORE CAREFUL THOUGHT IS ENOUGH TO SOLVE WHAT IS NOT AN ESPECIALLY TRICKY PROBLEM, EVEN IF IT'S NOT TOTALLY EASY, EITHER.

*IN CASE YOU'RE WONDERING, NEITHER PROFESSOR HAS EVER ACTUALLY PLAYED BASEBALL. OR CRICKET, COME TO THAT. CROQUET, MAYBE.

SO FAR, NOT SO SURPRISING. THE CLEVER BIT IS IN THE WHY.

IT SEEMS TO BE AN URGE THAT ONCE AN INDIVIDUAL HITS UPON AN ANSWER TO A PROBLEM—
THE BALL MUST COST 10 CENTS.

THEIR INSTINCT IS TO JUSTIFY THAT ANSWER.
THE BAT COSTS A DOLLAR AND THE BALL COSTS 10 CENTS. $1.10, SEE?

WHILE THE FIRST PERSON IS SATISFIED, THE SECOND PERSON HAS THE URGE TO ARGUE.
SOUNDS LIKE A TRICK QUESTION TO ME. THAT ANSWER'S TOO EASY.

THIS OPENS UP THE QUESTION OF CONFIDENCE...

...AND FORCES BOTH PEOPLE TO THINK ABOUT THE ANSWER AGAIN, IN MORE DETAIL.

WHICH WILL, IN MOST CASES, MEAN THEY FIND THE RIGHT ANSWER.
THE BALL COSTS 5 CENTS!

LET'S MOVE AWAY FROM SIMPLE ARITHMETIC PROBLEMS AND INTO THE WORLD OF THE SOCIAL.

GETTING FAR BEYOND THE SAFE CONFINES OF A LAB, AND INTO THE MURKY WATERS OF IDLE SPECULATION.

HERE'S A SIMPLE RULE OF LIFE: GROUPS BENEFIT FROM HAVING DIFFERENT KINDS OF PEOPLE.

WHEN IT COMES TO FOOD, FOR EXAMPLE, ALL ANIMALS MUST DEAL WITH THE EXPLOIT/EXPLORE DILEMMA.

IS IT BETTER TO EXPLOIT NEARBY FOOD SOURCES, OR TO EXPLORE IN THE HOPES OF FINDING NEW FOOD SOURCES?

UTA AND I TEND TO EXPLOIT THE FOOD SOURCES THAT REQUIRE LITTLE EFFORT, AND HAVE BEEN BOUNTIFUL IN THE PAST...

SupaFood

...BUT MAYBE WE'D BE BETTER OFF EXPLORING, IN CASE THE UNTHINKABLE HAPPENS.

SupaFoods
CLOSED
OH DEAR!
LET'S TRY BOROUGH MARKET FOR A CHANGE.

BUT EXPLORING IS POTENTIALLY DANGEROUS AND MAY NOT YIELD ANYTHING EDIBLE.

TOO TRENDY!
TOO CROWDED!

SOCIAL ANIMALS GET AROUND THIS DILEMMA BY VARIATION. MOST GROUPS INCLUDE INDIVIDUALS WHO INSTINCTIVELY PREFER EXPLOITING OR EXPLORING.

(YOU DON'T NEED EQUAL AMOUNTS OF EACH.)

FOR EXAMPLE, AROUND 4 IN EVERY 100 BEES ARE SCOUTS, GENETICALLY WIRED TO BE EXPLORERS.

IN OUR PERSONAL LIFE, WE THINK OF OURSELVES AS EXPLOITERS...

...PREFERRING TO STAY IN OUR COMFORTABLE HOUSE...
...FOLLOWING FAMILIAR ROUTINES.

BUT, PROFESSIONALLY, WE LIKE TO THINK OF OURSELVES AS EXPLORERS.

THIS MEANS LEARNING FROM WHAT HAS GONE BEFORE...

...AND THEN DEVELOPING IDEAS NO ONE HAS REALLY LOOKED AT.
RIP

IT'S A LITTLE LIKE TRYING TO CATCH A RARE BUTTERFLY.

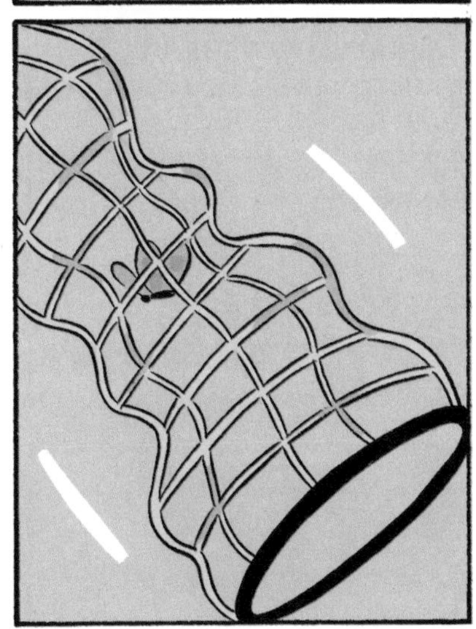

SOMETIMES THE BUTTERFLY SLIPS AWAY – AS WHEN A THEORY IS DISPROVED!

IN GENERAL, AND CERTAINLY IN THE WORLD OF NEUROSCIENCE, STUDY THRIVES ON DIVERSITY OF APPROACH.

SCIENTIFIC EXPLORERS IDENTIFY NEW AREAS OF RESEARCH TO TACKLE.

WHEREAS SCIENTIFIC EXPLOITERS BURY THEMSELVES IN STUDIES ALREADY CONDUCTED, TO SYNTHESIZE WHAT WE ACTUALLY KNOW SO FAR.

OR, MORE TO THE POINT, IDENTIFY WHAT WE DON'T KNOW.
PERHAPS ABOVE ALL, SCIENTIFIC EXPLOITERS PUSH AGAINST IDEAS THAT TURN OUT TO BE FALSE.
A VITAL AND UNDERVALUED TASK.

AS EXPLORERS OURSELVES, WE'VE WORKED ALONGSIDE OTHER EXPLORERS AND PLENTY OF EXPLOITERS, TOO. WE LIKE TO BELIEVE THAT THESE COLLABORATIONS HAVE BEEN FRUTIFUL – THAT WE'VE MADE SOME SPECIFIC IMPROVEMENTS TO PEOPLE'S UNDERSTANDING OF HOW THE BRAIN WORKS.

SUCH AS NOTING THE BRAIN'S INTERNAL FEEDBACK MECHANISM, WHICH ALLOWS YOU TO KNOW THAT IT IS YOU WHO HAS HAD A THOUGHT.

AND MY WORK WITH BRAIN SCANNERS, WHICH HAS HELPED SHOW THAT WHAT PEOPLE SAY THEY ARE THINKING ABOUT IS REFLECTED IN THEIR BRAIN ACTIVITY.
THIS, I THINK, PROVES TWO THINGS.
1. PEOPLE AREN'T LIARS.
2. PEOPLE DO HAVE SOME LEVEL OF SELF-AWARENESS.

HERE'S ONE OF MY PROUDEST ACHIEVEMENTS. IN MY TIME STUDYING AUTISM, I HELPED OVERCOME A NOTION FROM THE 1960S AND '70S, THAT THE CONDITION WAS CAUSED BY UNFEELING PARENTS.
AUTISM IS NOBODY'S FAULT!
SO-CALLED REFRIGERATOR MOTHERS. A HORRIBLE IDEA, AND 100% NOT TRUE!

THE BREAKTHROUGH IN THAT ACHIEVEMENT CAME FROM AN UNUSUAL EXAMPLE OF COLLABORATION – SCIENTISTS WORKING WITH NONSCIENTISTS. IN THIS CASE, MOTHERS OF CHILDREN WITH AUTISM.

AUTISM IS NOBODY'S FAULT!
NOBODY'S FAULT!
AUTISM IS NOBODY'S FAULT!

IN PARTICULAR, I LEARNED A LOT FROM MARGARET DEWEY, MOTHER TO AN AUTISTIC SON, WHO HELPED ME TO TRANSLATE FINDINGS FROM EXPERIMENTS INTO REAL-LIFE EXPERIENCES. SHE LITERALLY IMPROVED THE QUALITY OF MY WRITING.

I AND MY TEAM ALSO HELPED TO OVERTURN THE EARLY BELIEF THAT ALL PEOPLE WITH AUTISM ARE LEARNING DISABLED. ACTUALLY, MANY ARE NOT – BUT THEY ALL HAVE SPECIFIC COMMUNICATION PROBLEMS.

IS FAULT!

IN FACT, THEY TYPICALLY SHOW BETTER THAN AVERAGE SCORES ON CERTAIN TESTS, ESPECIALLY ONES INVOLVING ATTENTION TO DETAIL.

IT WAS AMITTA SHAH, THEN MY PHD STUDENT, WHO DISCOVERED THIS, BY SETTING THE CLASSIC TASK OF LOOKING FOR HIDDEN FIGURES.
YES, JUST LIKE THE POPULAR WHERE'S WALLY* BOOKS.
*OR WALDO, AS HE'S KNOWN IN THE STATES.

THE WAY THINGS WORK IN RESEARCH MEANT THAT WE HAD NUMEROUS PHD STUDENTS, POSTDOCS, AND VISITING RESEARCHERS, ALL MAKING IMPORTANT CONTRIBUTIONS. WE'VE NEVER BEEN A TEAM OF ONE, OR EVEN TWO.
30

THIS VERY BOOK IS A COLLABORATION BETWEEN FOUR PEOPLE, NOT TO MENTION LOTS OF OTHERS WHO'VE ALL WORKED WITHOUT GETTING DUE CREDIT.*
44
11
*WELL, THERE'S THE MINOR CONSOLATION OF AN ACKNOWLEDGMENTS SECTION ON PAGE 319.

ONE PART OF TEAMWORK WE'VE ALWAYS VALUED IS MIXING UP DISCIPLINES.
41

SO MANY OF OUR PROJECTS HAVE ONLY COME ABOUT BECAUSE OF INPUT FROM ALL OVER THE ACADEMIC WORLD:

PHYSICISTS, RADIOLOGISTS, NEUROLOGISTS, PSYCHOLOGISTS, NEURO-ANATOMISTS, STATISTICIANS, COMPUTER SCIENTISTS, ECONOMISTS, LINGUISTS, SEMIOTICIANS, ANTHROPOLOGISTS, ARCHAEOLOGISTS, SCIENCE OF RELIGION-ISTS, MUSICIANS, POLITICAL SCIENTISTS, COMPUTATIONAL MODELLERS, ENGINEERS, PHILOSOPHERS...

LOOK, HERE'S A VENN DIAGRAM OF JUST A FEW WAYS PEOPLE IN A RESEARCH TEAM CAN BE DIFFERENT.

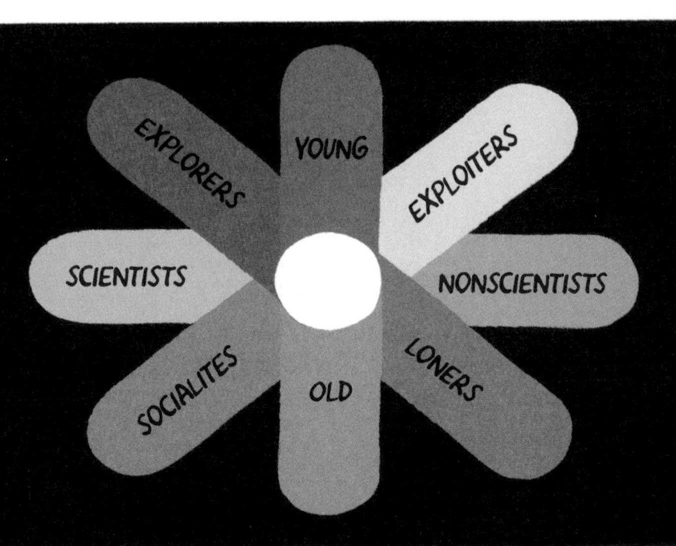
EXPLORERS
YOUNG
EXPLOITERS
SCIENTISTS
NONSCIENTISTS
SOCIALITES
OLD
LONERS

WE BELIEVE THAT THE BEST CHANCE OF FINDING TRUTH – IN SCIENCE, SOCIAL SCIENCE, OR ANY ENDEAVOR – HAPPENS WHEN YOU CAN GET AN OVERLAP OF MANY DIFFERENT QUALITIES.

BECAUSE, IF THERE'S ONE MESSAGE WE WANT YOU TO TAKE AWAY FROM THIS CHAPTER, IT'S THIS: TO SOLVE A DIFFICULT PROBLEM, ALWAYS WORK WITH OTHER PEOPLE...AS LONG AS YOU'RE ROUGHLY MATCHED IN ABILITY...

...AND ARE ABLE TO ASSESS YOUR OWN, AND EACH OTHER'S, LEVELS OF CONFIDENCE.

11
22
44
30
41
23

LISTEN TO DIFFERENT IDEAS AND ENCOURAGE VIGOROUS ARGUMENTS FOR AND AGAINST THEM. WHO BETTER TO ACCOMPLISH THIS THAN A DIVERSE GROUP* WITH A MIX OF EXPLOITERS AND EXPLORERS. OF COURSE, IF IT WAS THAT SIMPLE, EVERYONE WOULD DO IT ALREADY, WOULDN'T THEY? IN THE NEXT CHAPTERS, WE'LL TAKE A LOOK AT WHY DIVERSITY IS DIFFICULT, AND THE PSYCHOLOGY OF FAILED COOPERATION.

*WE MEAN DIVERSE IN EVERY POSSIBLE SENSE. WE HOPE IT'S SELF-EVIDENT THAT THE MORE DIVERSE YOUR TEAM IN TERMS OF GENDER, ETHNICITY, SEXUALITY, AND SHEER LIFE EXPERIENCES, THE MORE LIKELY THAT TEAM IS TO SHOW DIVERSITY OF APPROACH.

Chapter 10

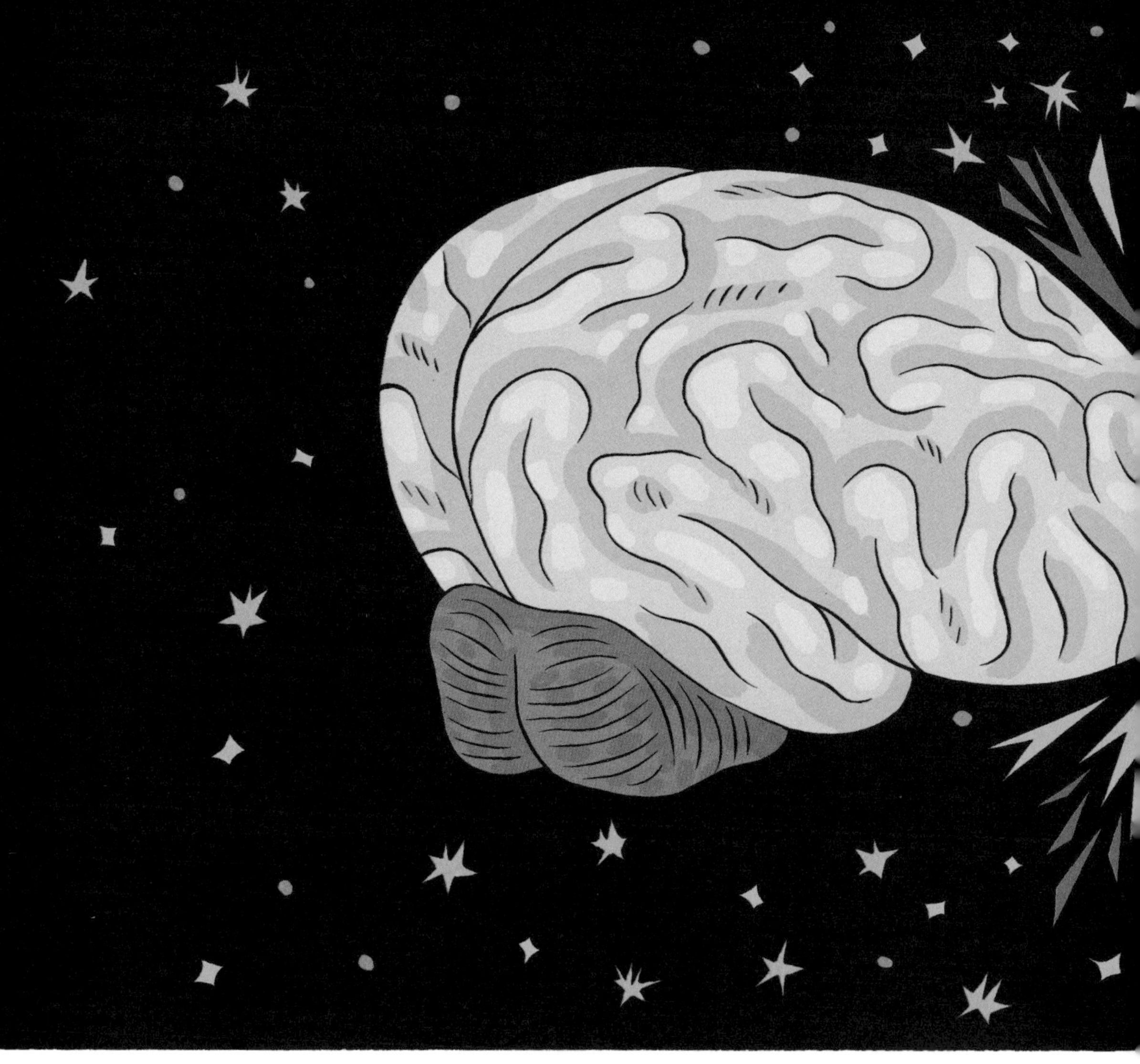

When cooperation breeds confusion.

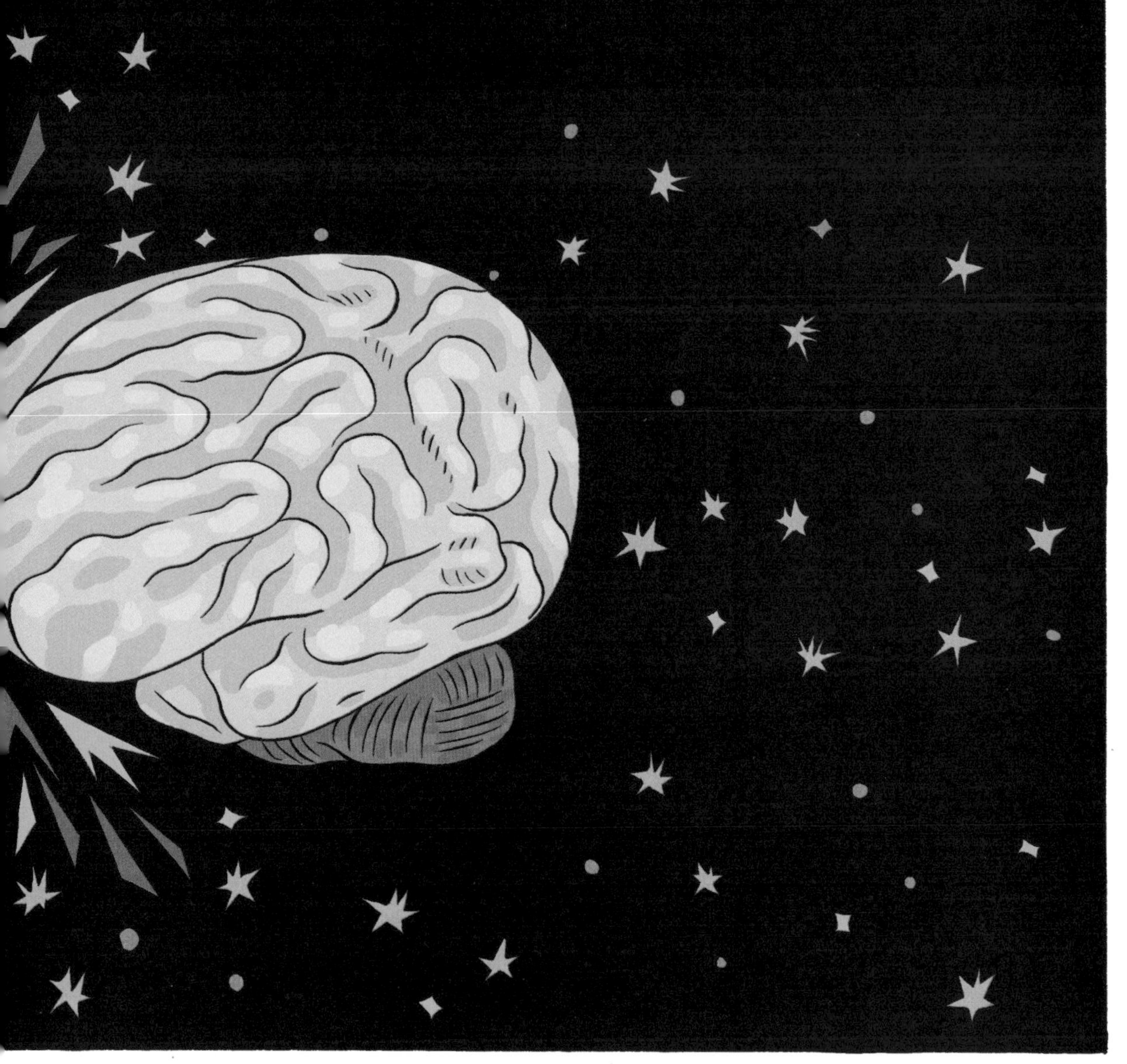

I IMAGINE YOU'VE EXPERIENCED THIS SITUATION YOURSELF.

IT WOULD BE NICE IF THERE WAS A RULE THAT EVERYONE UNDERSTOOD AND FOLLOWED WITHOUT THINKING ABOUT IT.

FOR EXAMPLE, ALWAYS STEP ASIDE TO THE LEFT.

UNSPOKEN RULES HELP BECAUSE WE DON'T HAVE TO THINK ABOUT WHAT WE'RE DOING.

AND OUR BRAINS OFTEN WORK EFFICIENTLY WHEN OUR CONSCIOUS SELVES ARE NOT INVOLVED.

AND WE DO HAVE A NUMBER OF WELL-KNOWN, IF SOMETIMES UNSPOKEN, RULES FOR CERTAIN THINGS IN LIFE.
SUCH AS WEARING RINGS TO SHOW WE ARE MARRIED.*
EXCEPT THAT WEDDING RINGS ARE LARGELY A WESTERN SYMBOL, NOT USED WORLDWIDE.

*THE PROFESSORS CELEBRATED THEIR PERSONAL UNION OF ENGLAND AND GERMANY IN, OF ALL TIMES, THE SUMMER OF 1966 – WHEN ENGLAND WON THE WORLD CUP. THE FRITHS WEREN'T WATCHING.

SOME SHARED RULES ARE VERY LOCAL. IN OSAKA, JAPAN, COMMUTERS STAND ON THE LEFT.

BUT IN KYOTO, COMMUTERS STAND ON THE RIGHT.

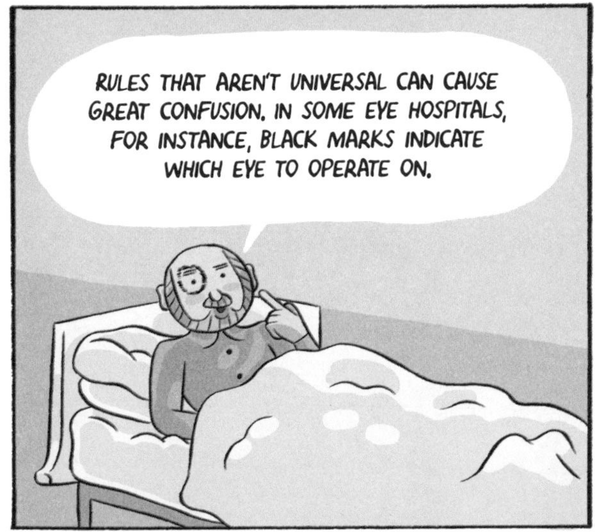
RULES THAT AREN'T UNIVERSAL CAN CAUSE GREAT CONFUSION. IN SOME EYE HOSPITALS, FOR INSTANCE, BLACK MARKS INDICATE WHICH EYE TO OPERATE ON.

IN OTHERS, MARKS INDICATE WHICH EYE NOT TO OPERATE ON.

IN FACT, OUR ABILITY TO REASON OUT WAYS TO INTERACT WITH OTHER PEOPLE IS SO IMPERFECT THAT PEOPLE HAVE A NAME FOR THE STUDY OF IT: GAME THEORY.

HERE'S ONE EXAMPLE OF THE TYPES OF GAMES STUDIED: IT'S CALLED "THE BATTLE OF THE SEXES."*

*IN CASE THE NAME DOESN'T GIVE IT AWAY, THIS WAS INVENTED IN THE 1950S.

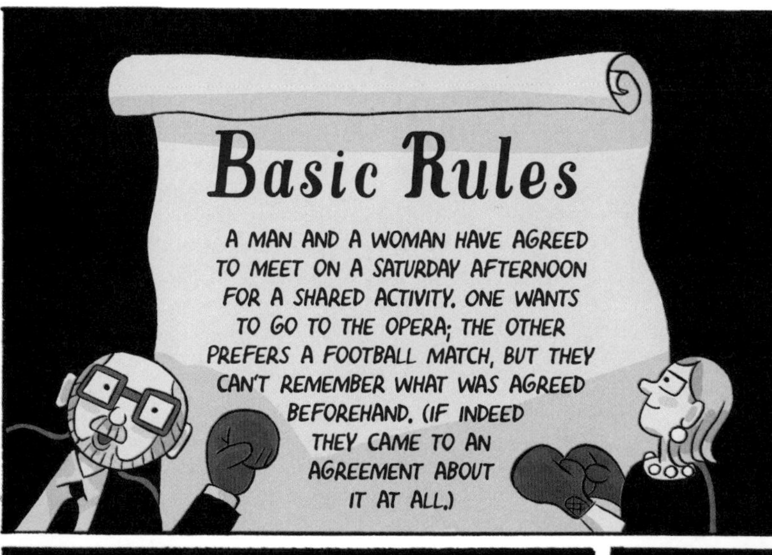
Basic Rules
A MAN AND A WOMAN HAVE AGREED TO MEET ON A SATURDAY AFTERNOON FOR A SHARED ACTIVITY. ONE WANTS TO GO TO THE OPERA; THE OTHER PREFERS A FOOTBALL MATCH, BUT THEY CAN'T REMEMBER WHAT WAS AGREED BEFOREHAND. (IF INDEED THEY CAME TO AN AGREEMENT ABOUT IT AT ALL.)

HOWEVER, THEY KNOW THEY HAVE TO BE AT ONE OR OTHER VENUE AT A SET TIME.

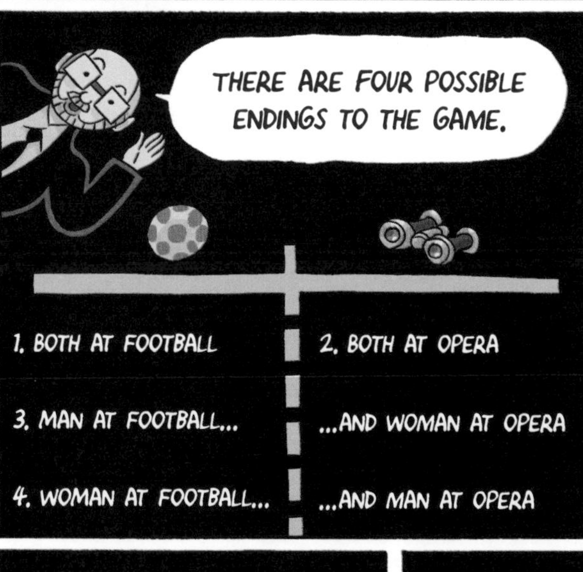
THERE ARE FOUR POSSIBLE ENDINGS TO THE GAME.
1. BOTH AT FOOTBALL
2. BOTH AT OPERA
3. MAN AT FOOTBALL...
...AND WOMAN AT OPERA
4. WOMAN AT FOOTBALL...
...AND MAN AT OPERA

TWO ARE HAPPY(ISH) OUTCOMES: BOTH MEET AT THE OPERA, OR AT THE FOOTBALL.

ONE PERSON IS AT THE EVENT HE OR SHE LIKES LESS, BUT THEY'LL BE TOGETHER, WHICH IS THE MOST IMPORTANT THING (IN THIS GAME).

ONE IS A MODERATE OUTCOME: BOTH ARE ALONE, BUT AT THE EVENT THEY THEMSELVES ENJOY.

ONE IS A TRAGIC OUTCOME: BOTH ARE ALONE, AND AT THE EVENT THEY HATE!

OF COURSE, WE'D WIN THIS GAME EASILY EVERY TIME – THE IDEA THAT EITHER OF US WOULD CHOOSE FOOTBALL OVER OPERA IS PATENTLY ABSURD.
BUT THERE'S A VARIANT OF THE GAME THAT IS EXACTLY THE SAME, ONLY IT'S CALLED "BACH OR STRAVINSKY."
THAT ONE'S A BIT MORE OF A CHALLENGE.

BACK TO THE POINT – IT TURNS OUT THAT THE BEST WAY TO PLAY THIS GAME, AND FOR BOTH PLAYERS TO CONSISTENTLY GET THE BEST POSSIBLE OUTCOME...

...IS TO TAKE TURNS.
OH NO – IT'S STRAVINSKY TODAY!
NEXT WEEK, BACH.

BUT WHEN PEOPLE ACTUALLY PLAY THE GAME (OR RATHER, AN IMAGINARY VERSION OF THE GAME)...

...THEY OFTEN DON'T TAKE TURNS.
I'M GOING TO CHOOSE BACH EACH TIME.

TO GET THE BEST RESULTS IN THE GAME, THE TACTIC IS TO INTRODUCE AN OUTSIDE PERSON TO POINT OUT THE "TAKING TURNS" STRATEGY.
SO HERE, TWO HEADS MAKE MORE TROUBLE THAN ONE, BUT THE SOLUTION IS TO ADD A THIRD HEAD.
IT HELPS, OF COURSE, IF THAT HEAD IS NEUTRAL.

REMEMBER CHAPTER 7? THAT'S WHERE WE DESCRIBED RECURSION – A TRICK PEOPLE USE TO THINK THROUGH WHAT THEY, AND OTHER PEOPLE, ARE THINKING ABOUT.
MY FAVORITE EXAMPLE OF THIS SORT OF RECURSION COMES FROM A SHORT STORY CALLED "THE GIFT OF THE MAGI."
(CHRIS OFTEN READS TO ME AT BATH–TIME.)

(IT FEELS LIKE THE SORT OF STORY THAT HAS ALWAYS EXISTED, BUT AS FAR AS ONE KNOWS AMERICAN AUTHOR O. HENRY CAME UP WITH IT IN 1905.)

JAMES AND DELLA ARE A HAPPY BUT POOR YOUNG COUPLE. CHRISTMAS IS COMING UP, AND EACH RESOLVES TO BUY THE OTHER A SECRET GIFT.

JAMES KNOWS THAT DELLA'S PROUDEST POSSESSION IS HER LONG, FLOWING, GOLDEN HAIR.

HE DECIDES TO BUY HER A SILVER COMB.

DELLA KNOWS HOW MUCH JAMES VALUES HIS POCKET WATCH. SHE DECIDES TO BUY A MATCHING GOLD CHAIN.

IN ORDER TO AFFORD THEIR GIFTS, DELLA CUTS OFF AND SELLS HER HAIR...

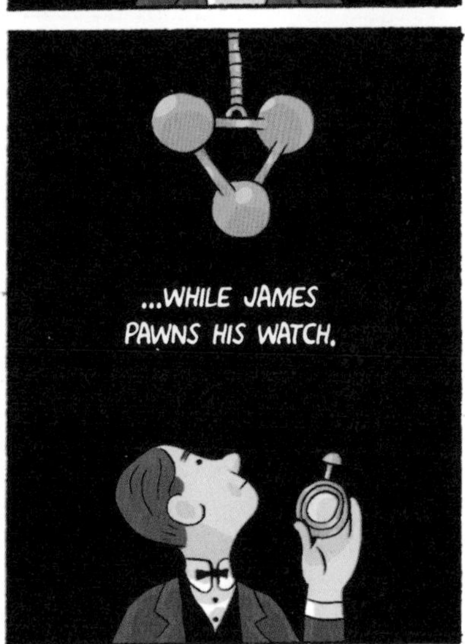
...WHILE JAMES PAWNS HIS WATCH.

ON CHRISTMAS MORNING, THEY EXCHANGE NOW USELESS GIFTS.

IN THIS STORY, RECURSIVE THINKING LED TO A SAD ENDING. BUT IS IT REALLY SAD?
O. HENRY NOTES THAT THE COUPLE ARE MOVED BY THEIR SACRIFICE FOR EACH OTHER. IT INDICATES THAT THEY ARE, IN FACT, BOTH HAPPY AND WISE PEOPLE.

THINK ABOUT THAT WHILE WE DISCUSS GAME THEORY, A FUN PART OF THE STUDY OF ECONOMICS.

PEOPLE IN THE WORLD OF GAME THEORY ARE INTERESTED IN IDENTIFYING THE STRATEGIES THAT BRING THE "BEST" OUTCOME, USUALLY DEFINED BY MONEY OR POWER.

THE BATTLE OF THE SEXES GAME, NOT UNLIKE THE O. HENRY STORY, ALSO INVOLVES PEOPLE THINKING ABOUT EACH OTHER.

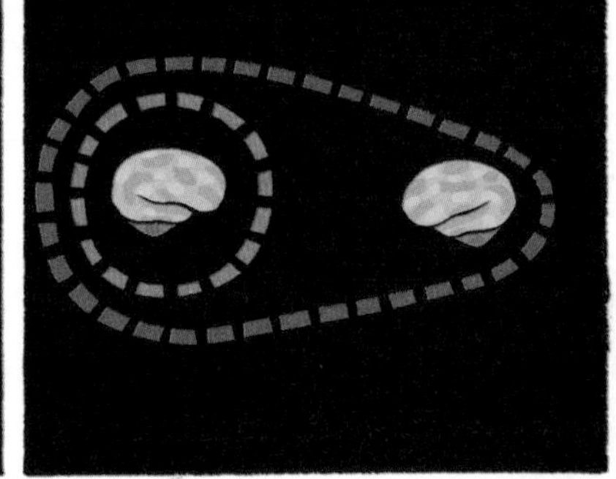
ARGUABLY, IT'S EASIER IF ONLY ONE PERSON THINKS ABOUT THE OTHER.

IF PERSON A DOES NOT THINK ABOUT WHAT PERSON B IS THINKING ABOUT...
I'D MUCH RATHER GO TO THE FOOTBALL – SO I GUESS I'LL CHOOSE TO GO THERE.
HE'D MUCH RATHER GO TO THE FOOTBALL – SO I GUESS I'LL CHOOSE TO GO THERE.
...WHILE PERSON B PUTS SOME EFFORT INTO THINKING ABOUT WHAT PERSON A IS THINKING ABOUT, THEN THEY END UP WITH AT LEAST A MODERATELY HAPPY OUTCOME.

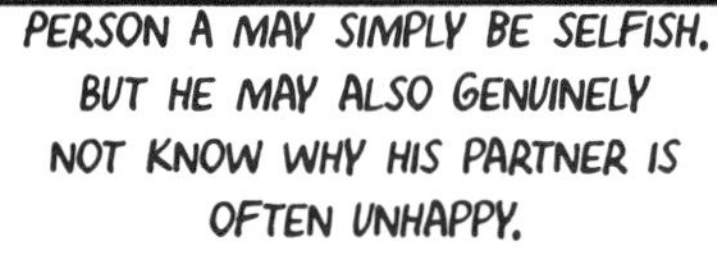
PERSON A MAY SIMPLY BE SELFISH. BUT HE MAY ALSO GENUINELY NOT KNOW WHY HIS PARTNER IS OFTEN UNHAPPY.

DOESN'T EVERYONE LIKE FOOTBALL BEST?

IT'S A FALSE STEREOTYPE THAT MEN ARE LESS LIKELY TO THINK ABOUT THEIR PARTNER'S DESIRES THAN THEIR OWN.
BUT IT'S TRUE THAT OUR BRAINS LEARN TO BELIEVE STEREOTYPES, AND ADJUST OUR BEHAVIOR TO MATCH THEM.
(WHICH IS WHY I ALWAYS ASK CHRIS TO CHANGE LIGHTBULBS, BECAUSE THAT'S A BOY'S JOB.)*
*ANOTHER FALSE STEREOTYPE

AS IT HAPPENS, REPEATED STUDIES SHOW THAT MOST PEOPLE, ESPECIALLY IN BIG GROUPS, ARE PRIMED TO BE UNSELFISH.
IT'S NOT A CONSCIOUS DECISION; WE'RE JUST WIRED TO WANT TO DO THE BEST FOR OUR GROUP MORE THAN FOR OURSELVES.
WE'LL TALK A LOT MORE ON THE CONCEPT OF "OUR GROUP" IN THE NEXT CHAPTER.

FOR NOW, TIME TO PLAY ANOTHER GAME!
FAR PAVILION

FAR PAVILION IS NOT REALLY A COMPUTER GAME (AS FAR AS WE KNOW).

GAME THEORISTS INVENTED THIS GAME AS A WAY TO EXAMINE HOW PEOPLE COMPETE AGAINST EACH OTHER, INSTEAD OF WORKING TOGETHER.

THERE ARE TWO PLAYERS:
ONE IS A RUNAWAY SLAVE, TRYING TO ESCAPE THE SULTAN.

...THE OTHER IS THE SULTAN'S GRAND VIZIER, WHO IS TRYING TO CATCH THE SLAVE.

THE RUNAWAY CAN MAKE TWO CHOICES:

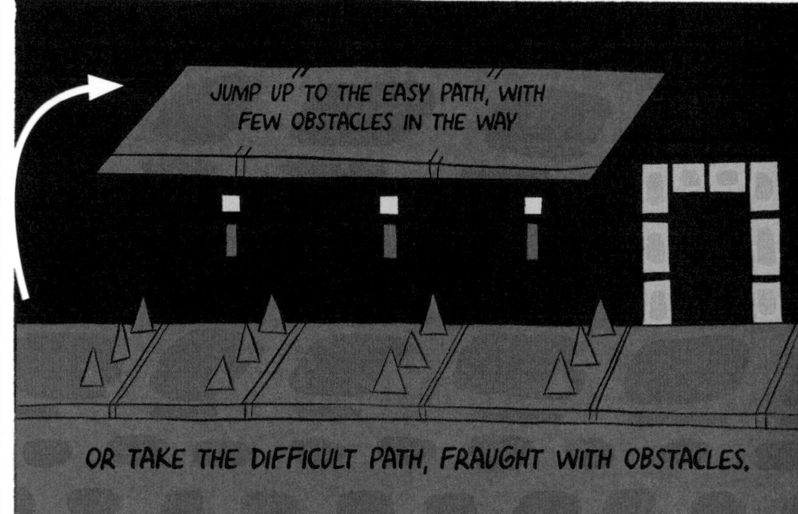
JUMP UP TO THE EASY PATH, WITH FEW OBSTACLES IN THE WAY
OR TAKE THE DIFFICULT PATH, FRAUGHT WITH OBSTACLES.

THE VIZIER MUST THEN MAKE THE SAME CHOICE:
?

DOES HE TAKE THE EASY PATH, AND THEN VERY LIKELY CATCH UP TO THE FUGITIVE SLAVE?

OR DOES HE ASSUME THE FUGITIVE CHOSE THE DIFFICULT PATH, AND MAKE PURSUIT THAT WAY?

THE TACTICS IN THIS GAME ARE ALL ABOUT RECURSIVE THOUGHT PROCESSES.
?
EACH PLAYER HAS TO THINK THROUGH WHAT THE OTHER PLAYER WILL DO.

HE'D CHOOSE THE DIFFICULT PATH, BECAUSE IT GIVES HIM A BETTER CHANCE TO ESCAPE.

BUT HE KNOWS I'D THINK THAT, SO MAYBE HE'S REALLY GOING TO TAKE THE EASY PATH.

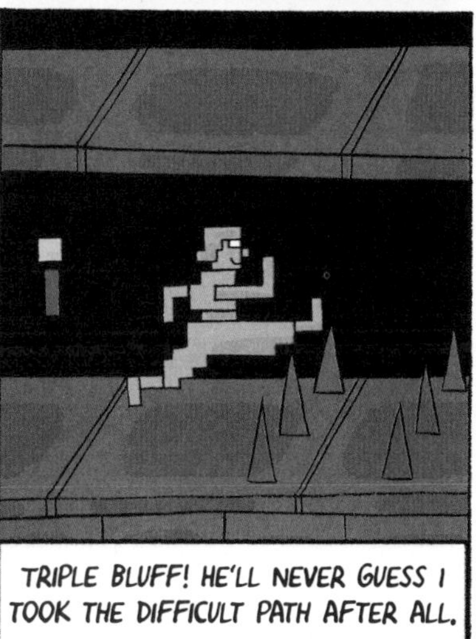
TRIPLE BLUFF! HE'LL NEVER GUESS I TOOK THE DIFFICULT PATH AFTER ALL.

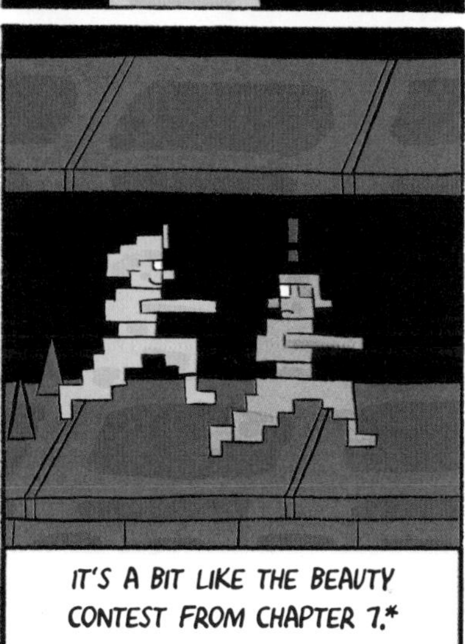
IT'S A BIT LIKE THE BEAUTY CONTEST FROM CHAPTER 7.*

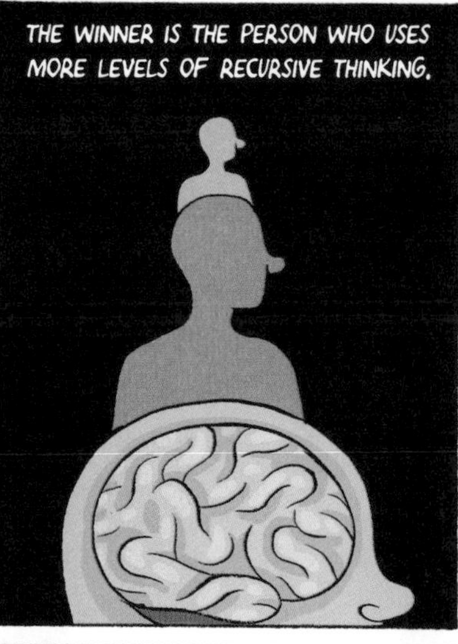
THE WINNER IS THE PERSON WHO USES MORE LEVELS OF RECURSIVE THINKING.

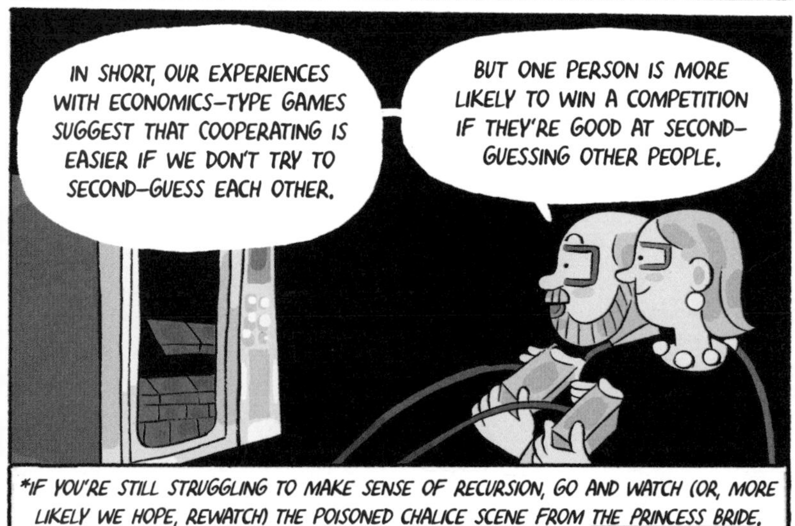
IN SHORT, OUR EXPERIENCES WITH ECONOMICS-TYPE GAMES SUGGEST THAT COOPERATING IS EASIER IF WE DON'T TRY TO SECOND-GUESS EACH OTHER.
BUT ONE PERSON IS MORE LIKELY TO WIN A COMPETITION IF THEY'RE GOOD AT SECOND-GUESSING OTHER PEOPLE.

*IF YOU'RE STILL STRUGGLING TO MAKE SENSE OF RECURSION, GO AND WATCH (OR, MORE LIKELY WE HOPE, REWATCH) THE POISONED CHALICE SCENE FROM THE PRINCESS BRIDE.

BUT IT'S NOT JUST "THEORY OF MIND" SKILLS AT WORK, EVEN IN COMPETITION GAMES.

PERHAPS THE MOST BASIC "COMPETITION GAME" OF ALL IS ONE PLAYED ALL OVER THE WORLD:

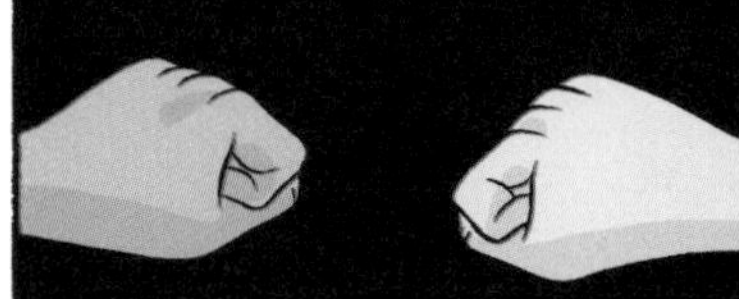

ROCK, PAPER, SCISSORS.

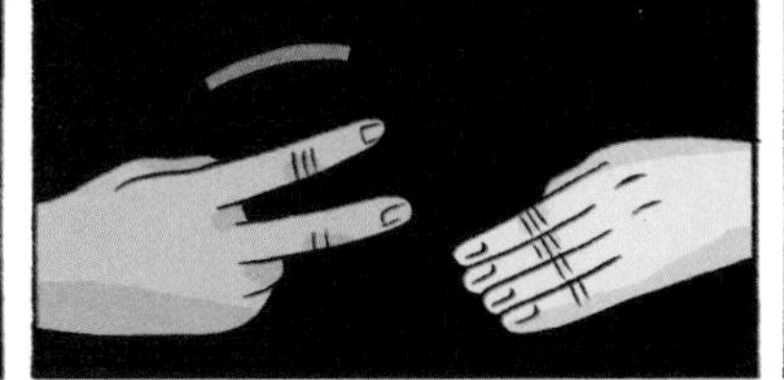

ECONOMISTS STRUGGLE WITH THE CONCEPT OF HAVING THREE CHOICES, SO THEY PREFER A VARIATION OF THE GAME CALLED "MATCHING SHAPES."

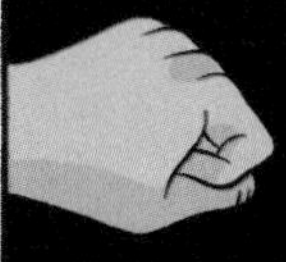

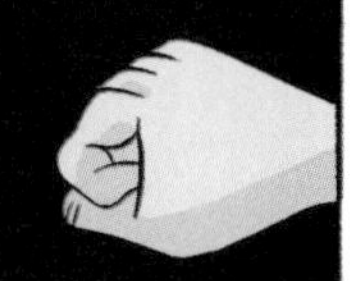

EACH PLAYER, A AND B, STARTS WITH ONE HAND IN THIS POSITION:

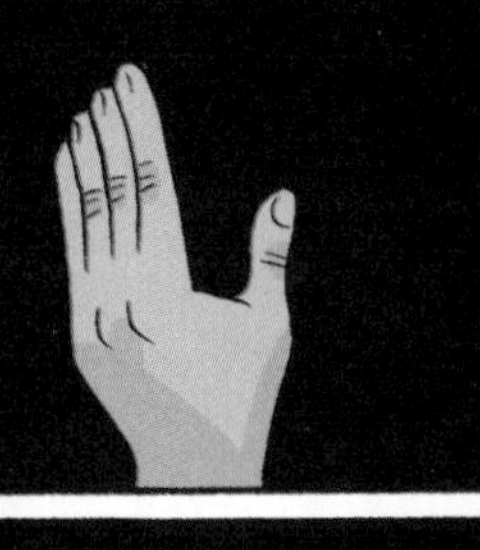

ON THE COUNT OF THREE, BOTH CHANGE THEIR HANDS TO REVEAL TWO POSSIBLE NEW SHAPES:

OPEN HAND

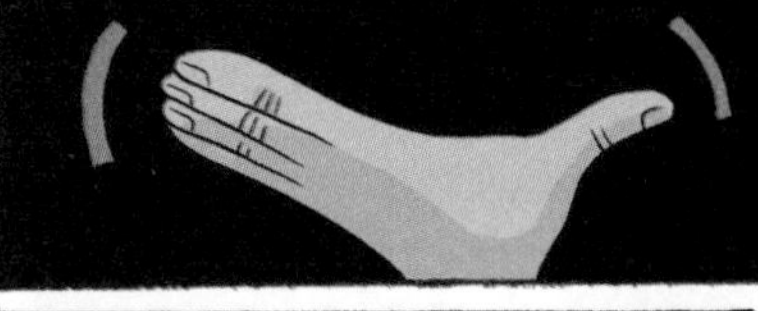

CLOSED FIST

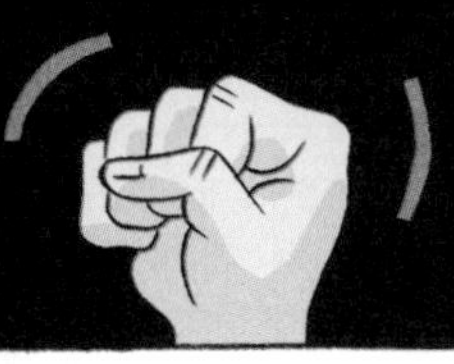

IF BOTH HANDS SHOW THE SAME SHAPE, PLAYER A WINS.

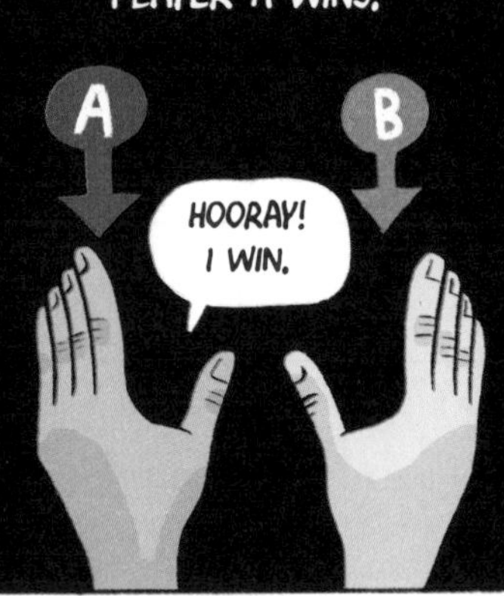
A
B
HOORAY! I WIN.

IF THE HANDS SHOW DIFFERENT SHAPES, PLAYER B WINS.

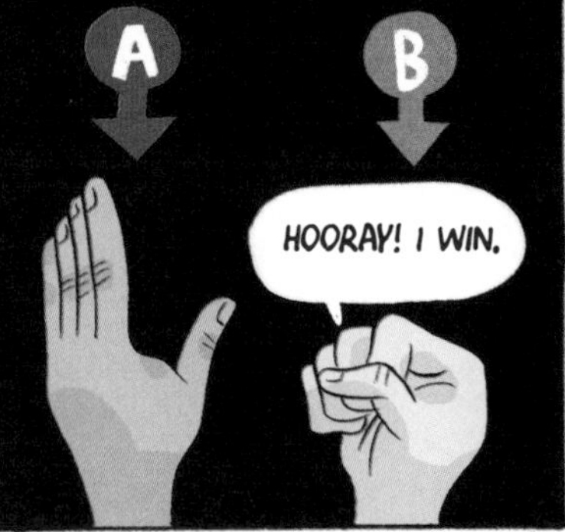
A
B
HOORAY! I WIN.

NEVER AGREE TO PLAY THIS GAME AS PLAYER B.

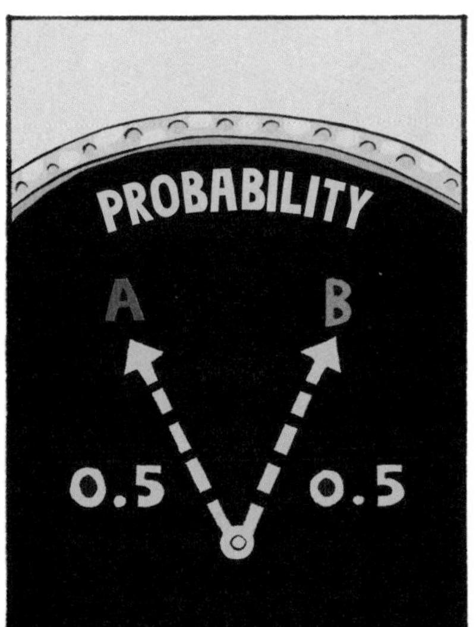
PROBABILITY
A
B
0.5
0.5

MATHEMATICS MAY TELL YOU THAT A AND B SHOULD WIN EQUALLY, BUT NEUROSCIENCE TELLS YOU THAT...
...PLAYER B WILL AUTOMATICALLY WANT TO COPY PLAYER A, EVEN THOUGH THE POINT OF THE GAME IS NOT TO COPY!

IN THEORY, THE PLAYERS MAKE SHAPES SIMULTANEOUSLY. BUT THE BRAIN HAS ENOUGH TIME FOR THE COPYING INSTINCT TO KICK IN. THIS DESIRE IS ANOTHER UNCONSCIOUS PROCESS...

...PLAYERS DON'T NOTICE IT. BUT THEY DO NOTICE THAT, OVER TIME, PLAYER A WILL ALWAYS WIN THIS GAME MORE OFTEN.

WHEN ECONOMISTS ASK PEOPLE TO PLAY THESE KINDS OF GAMES, THEY'RE USUALLY CURIOUS ABOUT...

...WHICH STRATEGIES ACHIEVE THE GREATEST PROFIT.

NEUROSCIENTISTS ARE MORE INTERESTED IN HOW PEOPLE MAKE DECISIONS WHEN PLAYING.

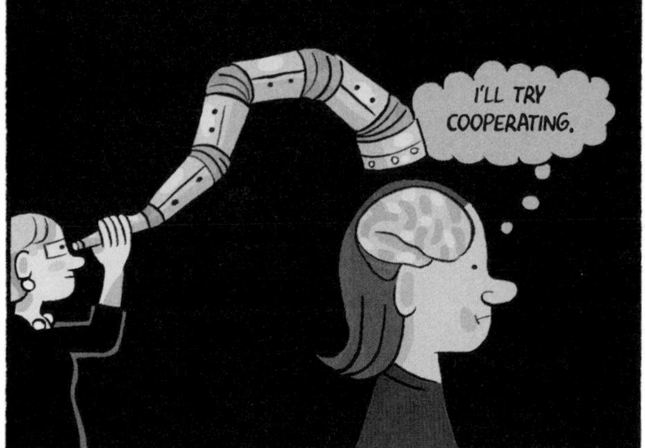
I'LL TRY COOPERATING.

AND ESPECIALLY IN HOW THEY EVALUATE THEIR OWN DECISIONS AFTER THE FACT.

COOPERATION GOOD!
COOPERATION BAD!

LOSERS, FOR EXAMPLE, MAY REGRET THE DECISIONS THEY'VE MADE DURING THE GAME, AND WALLOW IN THEIR REGRET.
(REGRET IS AMAZINGLY INTERESTING. WE'RE GOING TO TALK ABOUT IT A LOT MORE IN THE NEXT CHAPTER.)

WINNERS ARE LESS LIKELY TO FEEL REGRET, BUT THEY TOO MAY DWELL ON THE DECISIONS THEY MADE TO ACHIEVE VICTORY –

– FOR EXAMPLE, TO JUSTIFY BEHAVIOR THAT MIGHT BE TAKEN AS CRUEL OR TOO COMPETITIVE.

BOTH EXAMPLES OF POST-GAME ANALYSIS CAN BE OVERWHELMING AT TIMES,

TO THE EXTENT THAT PEOPLE OFTEN WORRY, BEFORE PLAYING A GAME, ABOUT THE MENTAL TURMOIL THEY MAY EXPERIENCE AFTERWARDS, WHETHER THEY WIN OR NOT!

IN FACT, ONE OF THE HIDDEN TRICKS BEHIND COOPERATION IS ABOUT TIME.
THE MORE TIME YOU GIVE PEOPLE TO DECIDE WHETHER OR NOT TO COOPERATE –

– THE MORE CHANCE THERE IS THAT THEY WILL DECIDE NOT TO.

THAT SAID, MOST OF OUR INTERACTIONS WITH OTHER PEOPLE MAKE IT WORTHWHILE TO COOPERATE.

IN GENERAL, WE ALL WANT OTHER PEOPLE TO COOPERATE WITH US. (AND WE WANT PEOPLE WHO CAN SEE US TO SEE THAT WE ARE BEING COOPERATIVE.)
WE BELIEVE THIS DESIRE IS INTUITIVE. TO DEMONSTRATE HOW INTUITIVE THIS CAN BE, LET'S TAKE A TRIP TO NIAGARA FALLS.

NOT TO THE FALLS THEMSELVES!
WE'RE HERE TO VISIT A NOTABLE HAUNTED HOUSE ATTRACTION, IN THE TOWN OF NIAGARA FALLS, NEW YORK.

NIGHTMARES.
NOT IN REAL LIFE, DON'T BE RIDICULOUS! WE HAVE NO STOMACH FOR THRILL–SEEKING.

COLLEAGUES FROM SWITZERLAND AND PARIS STUDIED IMAGES OF PEOPLE WHO HAD BRAVED THE HAUNTED HOUSE, AND OBSERVED A VERY COMMON BEHAVIOR: GRIPPING.

IN PAIRS, PEOPLE CONFRONTED WITH FEAR WILL TYPICALLY CLING ONTO EACH OTHER.

AS OPPOSED TO, SAY, RUNNING AWAY FROM THE DANGER, OR TRYING TO PUSH THE OTHER PERSON INTO DANGER.

WHILE IN BIGGER GROUPS, THE MOST SCARED WILL GRIP OTHERS, BUT THE LEAST SCARED WON'T GRIP ONTO PEOPLE, OR EVEN OFFER THEM THE COMFORT OF RETURNING THEIR GRIP.
THE POINT REMAINS: OUR INSTINCT, WHEN SCARED, IS TO FIND ANOTHER PERSON, NOT TO RUN AWAY FROM OTHER PEOPLE.

BACK IN THE WORLD OF ECONOMIC GAMES, A STUDY BY DAVID RAND ASKED PEOPLE TO MAKE ANONYMOUS DONATIONS TO A GROUP.

WHATEVER THEY GAVE WOULD BE DOUBLED, AND THEN SHARED OUT. IF EVERYONE GIVES EVERYTHING, ALL PLAYERS DO WELL.

BUT A SINGLE FREE RIDER, WHO GIVES NOTHING, STANDS TO GET EVEN MORE MONEY.

TYPICALLY, SOME PEOPLE GIVE LITTLE OR NOTHNG, MEANING ANY GENEROUS DONORS END UP WITH LESS THAN THEY EACH GAVE.

WHEN PLAYING THE GAME WITH NO TIME TO THINK, ALL PLAYERS TEND TO BE GENEROUS.

BUT WHEN ASKED TO THINK ABOUT IT CAREFULLY FIRST, THEY BECOME LESS GENEROUS.

MORE THAN THIS, THEY ALSO BECOME LESS COOPERATIVE IN OTHER GAMES PLAYED IMMEDIATELY AFTERWARDS.

PRIMING IN ACTION AGAIN! (AS WE SAW IN OUR INTERLUDE CHAPTER.)

OVERALL, THE PICTURE SEEMS TO BE THAT PEOPLE GENERALLY AND INSTINCTIVELY WANT TO COOPERATE WITH EACH OTHER.
COULD IT BE THAT COOPERATION IS A LEARNED HABIT RATHER THAN HARDWIRED?

IN CERTAIN SITUATIONS, THIS DESIRE TO COOPERATE CAN GET IN THE WAY OF CREATING A HAPPY OUTCOME.

IN FACT, THE MORE PEOPLE THINK ABOUT COOPERATING WITH EACH OTHER...

...THE MORE LIKELY THEY ARE TO GET IT WRONG, OR, IN THE RIGHT SITUATION, CHOOSE NOT TO COOPERATE AT ALL, OFTEN IN THE HOPE OF GAINING SOME SELFISH ADVANTAGE.

SO, WHAT'S GOING ON IN OUR HEADS WHEN WE THINK ABOUT OUR DECISIONS? YOU WON'T "REGRET" TURNING TO THE NEXT CHAPTER TO FIND OUT.
HA HA.

Chapter 11

Free will and regret.
WE TALKED A LITTLE IN CHAPTER 6 ABOUT WHETHER OR NOT PEOPLE HAVE FREE WILL.
THE SCIENCE IS NOT AND MAY NEVER BE CONCLUSIVE – BUT I BELIEVE WE DO HAVE FREE WILL.
SO DO I. (AND I'M NOT JUST COPYING!)

BUT WE'RE PREPARED TO BELIEVE THAT: WE MIGHT ONLY THINK WE HAVE FREE WILL...
POP

...BECAUSE FREE WILL MIGHT BE AN ACCIDENTAL BY-PRODUCT OF HAVING A CONSCIOUS BRAIN.
POP

IN OTHER WORDS, OUR MENTAL LIVES COULD BE PURELY DETERMINED BY THE MOLECULAR MAKEUP OF OUR BRAINS, AS GUIDED THROUGH LIFE FOLLOWING THE LAWS OF PHYSICS. THIS INCLUDES BOTH THE CHOICES WE MAKE EACH DAY, AND OUR PERCEPTION THAT THOSE CHOICES ARE MADE FREELY.

DOES THIS BOTHER YOU AT ALL?
IT'S ONLY A POSSIBILITY.

WE COULD PUT IT ANOTHER WAY: DOES THE BODY RULE THE MIND?
...OR DOES THE MIND RULE THE BODY?
WE DON'T KNOW.*

*YES, THESE ARE LYRICS FROM A SONG BY THE SMITHS, BUT YOU'D HAVE TO ASK ALEX ABOUT THAT.

AS YOU CAN IMAGINE, THE QUESTION OF FREE WILL IS NOT AT ALL NEW.
BEFORE I STUDIED PSYCHOLOGY, I WAS SOMETHING OF A CLASSICIST. I READ A CERTAIN AMOUNT OF GREEK PHILOSOPHY.

MEET EPICURUS, BORN 341 BC IN ATHENS.
χαιρε*
*GREETINGS AND SALUTATIONS

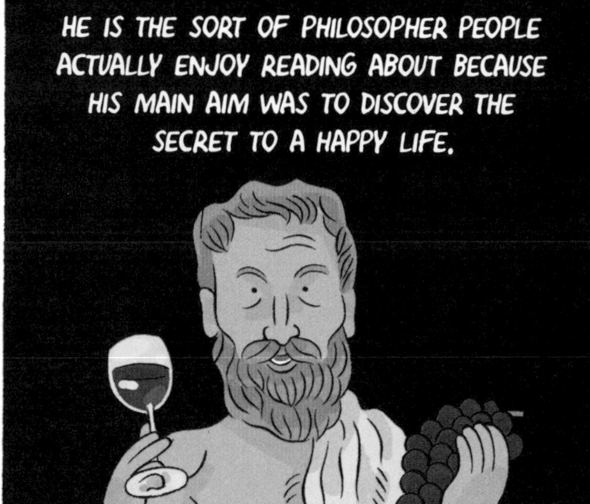

HE IS THE SORT OF PHILOSOPHER PEOPLE ACTUALLY ENJOY READING ABOUT BECAUSE HIS MAIN AIM WAS TO DISCOVER THE SECRET TO A HAPPY LIFE.

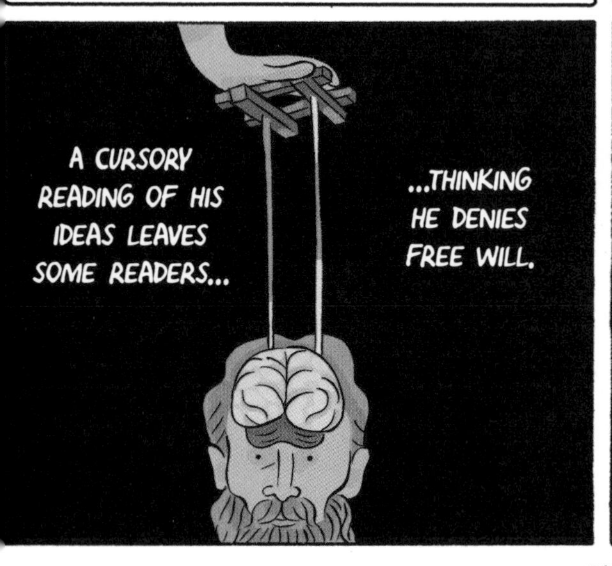

A CURSORY READING OF HIS IDEAS LEAVES SOME READERS...
...THINKING HE DENIES FREE WILL.

HAPPINESS, FOR EPICURUS, WAS ALL ABOUT RELAXING AND LETTING LIFE AND THE UNIVERSE TAKE OVER.
EVERYTHING'S JUST ATOMS, MAN.*
*THE EXISTENCE OF ATOMS WAS FIRST POSITED BY DEMOCRITUS, A CENTURY BEFORE EPICURUS LIVED.

BUT NO, EPICURUS RECOGNIZED FREE WILL.

FIRSTWISE, I AM CAUSING THIS EVENT. SECONDWISE, I KNOW THAT I COULD HAVE DONE OTHERWISE.

WE'RE GOING TO FOCUS ON THE SECOND POINT FOR A BIT...
THE FEELING THAT YOU COULD HAVE MADE A DIFFERENT CHOICE ONLY ARISES BECAUSE YOU THINK YOU HAVE FREE WILL.*

*NOTICE THAT IT DOESN'T MATTER WHETHER OR NOT YOU ACTUALLY HAVE FREE WILL.

THERE ARE TWO RESPONSES TO A BAD OUTCOME.

1) DISAPPOINTMENT. I EXPECTED A GOOD THING TO HAPPEN, BUT IT DIDN'T.
FOR EXAMPLE, IF SOMEONE ELSE WAS CHOSEN TO RECEIVE A PRIZE THAT I THOUGHT MAYBE I DESERVED.

2) REGRET. IF I HAD MADE A DIFFERENT CHOICE, A BETTER THING WOULD HAVE HAPPENED – BUT I DIDN'T.

AMUSINGLY, A CLASSIC EXAMPLE OF REGRET IS BACKING THE WRONG HORSE IN A RACE.

BETTING SLIP
OF COURSE, YOU HAVE NO CONTROL OVER THE HORSE'S PERFORMANCE – BUT YOU CAN REGRET NOT HAVING CHOSEN DIFFERENTLY WHEN YOU PLACED YOUR BET.

WHERE THIS REALLY GETS INTO FREE WILL IS THE CONCEPT OF ANTICIPATED REGRET.
I KNOW THAT I WILL FEEL REGRET IN THE FUTURE, IF I GET THE WRONG OUTCOME. SO, I WILL FACTOR THIS FEELING INTO MY DECISIONS. I WILL NOT BET ON ANY HORSE, TO AVOID LOSING.

IT SEEMS TO US THAT THE VERY CONCEPT OF "ANTICIPATING THAT YOU WILL FEEL REGRET IN THE FUTURE" IS A PURELY MENTAL STATE.
AND YET IT'S SUCH A POWERFUL SENSATION THAT IT IMPACTS HOW OUR PHYSICAL BODY FEELS AND ACTS.
IN OTHER WORDS, A RARE BUT CLEAR EXAMPLE OF THE MIND RULING THE BODY.

IT IS A FACT OF LIFE THAT REGRET FEELS WORSE AFTER MAKING A PROACTIVE CHOICE...

...COMPARED TO HOW IT FEELS WHEN YOU CHOOSE TO DO NOTHING.

POP

POP

I REGRET THAT I CHOSE TO DRIVE TO THE SHOPS (BECAUSE I GOT STUCK AT SOME ROADWORKS). I COULD HAVE WALKED!

DEEPLY ANNOYED WITH MYSELF.

I REGRET THAT I DON'T HAVE ANY MILK LEFT IN THE HOUSE (BECAUSE I CHOSE NOT TO GO SHOPPING).

ONLY SLIGHTLY ANNOYED WITH MYSELF.

THERE'S A SUBTLE DIFFERENCE BETWEEN THESE TWO EXAMPLES: CHOOSING TO DO SOMETHING, VERSUS CHOOSING NOT TO DO SOMETHING.
HERE'S OUR FRIEND DEAN MOBBS TO PRESENT A SCIENTIFIC STUDY INTO THIS.

HELLO! NOW, I RAN AN EXPERIMENT* BASED ON A CLASSIC ETHICS EXPERIMENT, THE TROLLEY PROBLEM.
THIS KIND OF TROLLEY?
PRECISELY.
*WELL, I COWROTE THE PAPER. YOU'LL MEET THE FIRST AUTHOR, ORIEL FELDMANHALL, SHORTLY.

IN THIS CLASSIC VERSION OF THE PROBLEM, A TROLLEY CARRYING A GROUP OF PEOPLE IS RUNNING ALONG A RAILWAY TRACK, WITH LIMITED CONTROL.
A LITTLE DISTANCE AHEAD, THE TRACK SPLITS INTO TWO FORKS.
COLLECTIVELY, THE GROUP OF PEOPLE HAVE TO MAKE A MAJOR DECISION. THE ASSUMPTION IS THAT THEY HAVE FREE WILL, AND THE DECISION THEY WILL MAKE IS A FREE ONE.
IT'S NOT AN EASY DECISION, MIND!

A DASTARDLY VILLAIN HAS TIED UP A SELECTION OF PEOPLE.
THE GROUP ON THE TROLLEY HAVE THE CHANCE TO PULL A LEVER OR NOT.
IF THEY DON'T PULL IT, THE TROLLEY WILL GO LEFT, AND RUN OVER TWO PEOPLE TIED TO THE TRACKS.
IF THEY DO PULL IT, THE TROLLEY WILL GO RIGHT, RUNNING OVER A SINGLE PERSON TIED TO THE TRACKS.
HNN HNN HNN!

THIS IS WHAT CHRIS'S LAUGH SOUNDS LIKE, IN CASE YOU WERE WONDERING.
HNN HNN HNN!

THE PEOPLE ON BOARD THE TROLLEY ARE FREE TO CHOOSE WHETHER TO PULL THE LEVER OR NOT.
?
?
?

HANG ON! THAT'S NOT HOW THE PROBLEM GOES IN MY ORIGINAL FORMULATION!*

*THIS IS PHILIPPA FOOT, WHO DEVISED "THE TROLLEY PROBLEM" IN 1967. AND YES, WE HAVE ADJUSTED SOME DETAILS, TO MAKE IT EASIER TO DRAW.

THE POINT IS, IN THIS SCENARIO, EVERYONE HAS TO SPEND TIME DEBATING AND, ULTIMATELY, TO MAKE A CHOICE. CRUCIALLY, THERE'S ALSO TIME FOR THEM TO JUSTIFY THAT CHOICE TO BOTH THEMSELVES AND EACH OTHER...
...AND THERE'LL BE TIME TO REGRET THEIR DECISION LATER.

I DIDN'T SET UP A FIENDISH TRAIN TRACK AND ASK PEOPLE TO COMMIT MURDER.
THIS IS ORIEL FELDMANHALL.

I USED THOSE OLD PSYCHOLOGY TEST STANDBYS, HARD CASH AND ELECTRIC SHOCKS.

I INVITED EACH TEST SUBJECT INTO A BOOTH...

...WHERE THEY WITNESS A TORTURER ADMINISTERING SHOCKS TO A MAN STRAPPED TO A CHAIR.

(THE TESTEE DOESN'T KNOW THAT BOTH ARE REALLY ACTORS PRETENDING.)

THE TESTEE IS GIVEN A BAG OF MONEY, AND TOLD THAT SHE CAN PAY THE TORTURER.

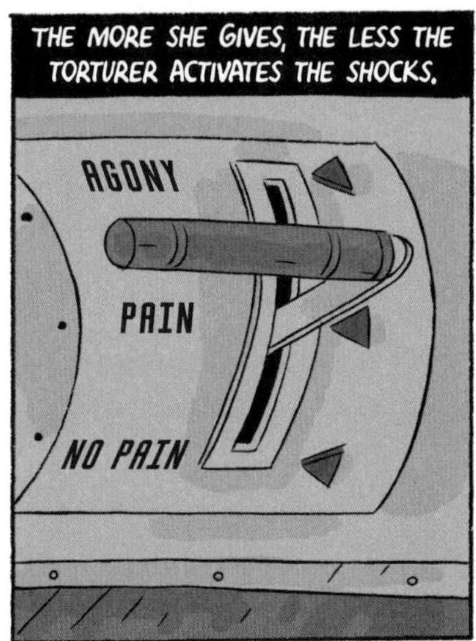
THE MORE SHE GIVES, THE LESS THE TORTURER ACTIVATES THE SHOCKS.
AGONY
PAIN
NO PAIN

WHEN THIS TEST IS SIMPLY DESCRIBED, 93% OF PEOPLE SAY THEY'D GIVE ALL OF THEIR MONEY TO ENSURE THE SHOCKS STOPPED.

WE WERE THE FIRST GROUP TO PERFORM THIS TEST "FOR REAL" (USING ACTORS)...

...AND DISCOVERED THAT, IN FACT, 53% OF PEOPLE CHOOSE TO KEEP SOME OF THE MONEY.
THAT PEOPLE EXAGGERATE THEIR OWN KINDNESS IS NOT UNEXPECTED.

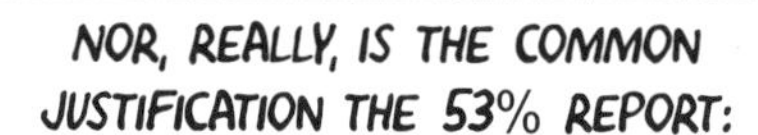
NOR, REALLY, IS THE COMMON JUSTIFICATION THE 53% REPORT:

I THOUGHT I COULD KEEP A LITTLE OF THE MONEY, AND HE COULD HAVE JUST A LITTLE BIT OF PAIN.

HERE'S A VARIATION ON THE TEST THAT HASN'T ACTUALLY BEEN CONDUCTED YET.

BEFORE YOU GO INTO THE BOOTH, HERE'S A PUZZLE TO SOLVE.
PLEASE WORK ON IT WHILE THE OTHER EXPERIMENT IS GOING ON.

AND HERE'S OUR HYPOTHESIS OF WHAT WOULD HAPPEN.

HERE, TAKE THE MONEY, I'VE GOT TO FIX THIS.
?

TURNS OUT, WHEN A BRAIN IS BUSY SOLVING A PUZZLE, IT HAS LESS CAPACITY TO PONDER MORAL DILEMMAS.*

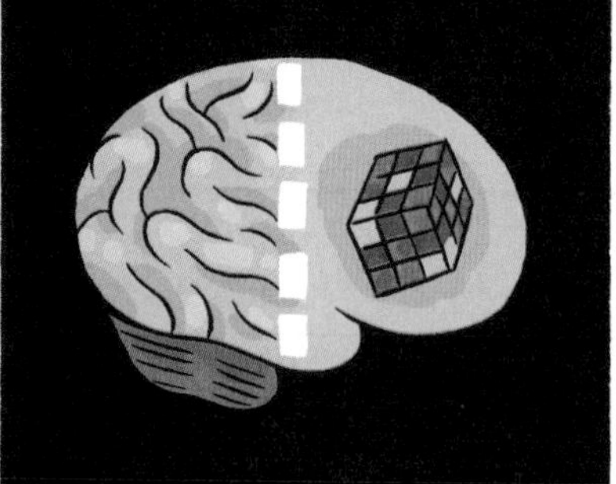

AND THE DEFAULT POSITION IS TO BE NICE. WE PREDICT THAT ALMOST EVERYONE IN THIS SECOND TEST WOULD VERY QUICKLY PAY ALL THEIR MONEY TO STOP THE SHOCKS.

*REMEMBER THAT STUDY FROM PAGE 237, WHICH SHOWED THAT PEOPLE ARE MORE SELFISH WHEN THEY HAVE TIME TO THINK.

SO PERHAPS HAVING FREE WILL ISN'T A GOOD THING...
...IF PEOPLE WHO HAVE MORE TIME TO THINK END UP MAKING SELFISH CHOICES.

BUT WE STILL LIKE HAVING FREE WILL! FOR EXAMPLE, LIKE MOST PEOPLE WE'D BE RELUCTANT TO GIVE UP FREE WILL WHEN IT COMES TO MONEY.
WE LIKE TO FEEL WE'RE IN CONTROL OF HOW WE SPEND IT!

IMAGINE YOU'RE BIDDING IN AN AUCTION.
NO, NOT THIS KIND.
THE KIND WHERE EVERYONE WRITES DOWN THEIR MAXIMUM BID, AND THE HIGHEST AMOUNT WINS.

IF YOU BID $500, BUT THE WINNING BID TURNS OUT TO BE $1,000, YOU'D LIKELY SHRUG YOUR SHOULDERS AND MOVE ON.

BUT IF THE WINNING BID WAS JUST $505, YOU'D EXPERIENCE GREAT PANGS OF REGRET FOR NOT HAVING BID A HIGHER AMOUNT.

IF PEOPLE IN THIS KIND OF "SEALED BID" AUCTION KNOW THAT THEY WILL FIND OUT WHAT THE WINNING BID IS...

...EVERYONE ENDS UP BIDDING A LITTLE BIT MORE – PRECISELY HOPING TO AVOID THE KIND OF REGRET MENTIONED ABOVE.
500
525
545
550

IT IS NOT INSIGNIFICANT THAT THIS EXAMPLE INVOLVES ONE PERSON DEALING WITH A GROUP OF OTHER PEOPLE.
11
12

IN THEORY, EACH INDIVIDUAL IS SIMPLY DECIDING HOW MUCH AN OBJECT IS WORTH TO HIM– OR HERSELF.
?

BUT IN REALITY, EVERYONE IS ALSO THINKING "HOW MUCH IS IT WORTH TO THESE OTHER PEOPLE?"
$
$
$
YES, IT'S RECURSIVE THINKING AT WORK AGAIN!

PARTLY BECAUSE THEY WANT TO WIN THE AUCTION, OF COURSE...
...BUT PARTLY ALSO BECAUSE ALMOST EVERYTHING TAKES ON GREATER VALUE WHEN YOU KNOW HOW MUCH OTHER PEOPLE COVET IT.

AND NOW WE'RE BACK TO THE MORE EXPLICIT FORM OF SOCIAL COGNITION.

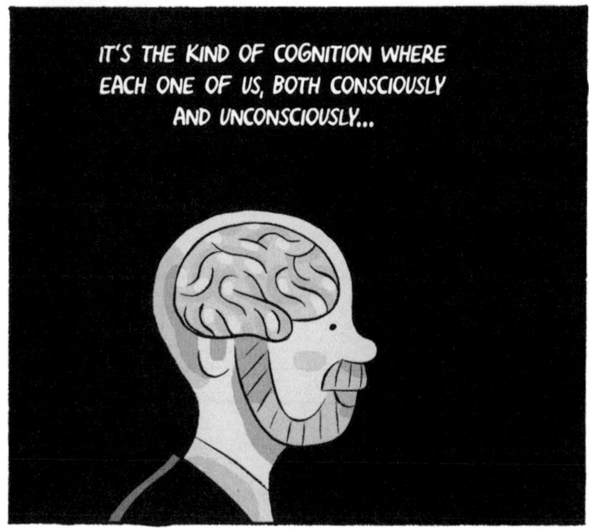
IT'S THE KIND OF COGNITION WHERE EACH ONE OF US, BOTH CONSCIOUSLY AND UNCONSCIOUSLY...

...SPENDS OUR TIME THINKING ABOUT WHAT OTHER PEOPLE ARE THINKING ABOUT.

WHAT'S ALL THIS GOT TO DO WITH LEMONS, I HEAR YOU ASK.
NO YOU DIDN'T.

I'M CASUALLY BRINGING UP ONE OF MY FAVORITE EXPERIMENTS. YOU'LL SEE WHY WHEN I'VE DESCRIBED IT.

CASUALLY, SHE SAYS. THIS ISN'T TWITTER, WHERE ONE PRONOUNCEMENT HAS NO REASON TO LINK IN TO WHAT CAME BEFORE.
STOP BEING MEAN. I WANT TO TALK ABOUT THE LEMONS! I PROMISE IT'S RELEVANT.
AH, YOU MEAN THE INTRAVERT/EXTRAVERT* TEST?

WHERE INTRAVERTS SALIVATE MORE THAN EXTRAVERTS AFTER TASTING A DROP OF LEMON JUICE.

*EXTROVERT/INTROVERT SPELLING ALSO ACCEPTED.

THAT TEST IS RATHER OLD HAT, YOU KNOW.

I'M NOT TALKING ABOUT THAT! I MEAN THE LEMON EMPATHY TEST.

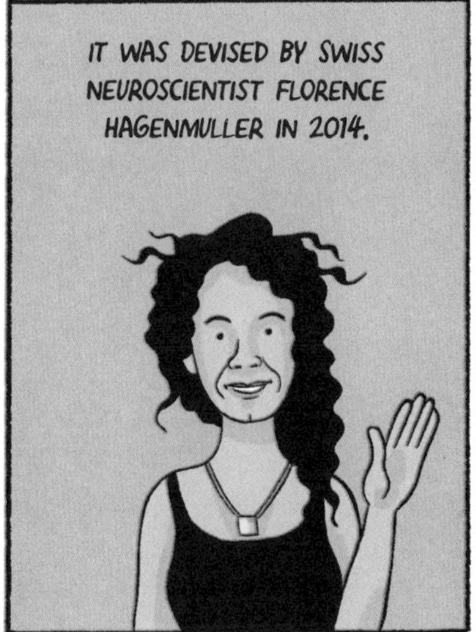
IT WAS DEVISED BY SWISS NEUROSCIENTIST FLORENCE HAGENMULLER IN 2014.

SHE ASKED PEOPLE TO SIT WITH ROLLS OF COTTON WOOL IN THEIR MOUTHS AND WATCH TWO SHORT VIDEOS.
VHS

VIDEO 1: A MAN EATS A LEMON.
VHS

VIDEO 2: A MAN TAKES BALLS OUT OF A BAG.
VHS

YOU WILL NOT BE AT ALL SURPRISED TO LEARN THAT AFTER WATCHING VIDEO 1...

...MOST PEOPLE HAVE PRODUCED A HEALTHY AMOUNT OF SALIVA.

WHILE AFTER VIDEO 2, THEY HAVEN'T.

THE GOOD BIT IS THAT THE AMOUNT OF SALIVA PRODUCED WHILE WATCHING THE LEMON VIDEO CORRELATES TO SCORES PEOPLE GET ON AN EMPATHY QUESTIONNAIRE.
PEOPLE WHO SALIVATED MORE SCORED "HIGHER" ON THE EMPATHY QUESTIONNAIRE – THEY'RE MORE IN TUNE WITH THEIR FELLOW HUMANS.

PLEASE NOTE, "EMPATHY" IN THIS CASE DOESN'T MEAN WHAT IT MEANS IN EVERYDAY CONVERSATION.

WE'RE NOT TALKING ABOUT THE ABILITY TO PUT YOURSELF IN SOMEONE ELSE'S SHOES, OR TO SHARE IN SOMEONE'S SADNESS WHEN THEY HEAR BAD NEWS.

WE'RE TALKING ABOUT AUTOMATIC PHYSIOLOGICAL CONTAGION, THE SAME KIND OF CONTAGION THAT MAKES US LAUGH WITH OTHERS, OR MIMIC THEIR ACTION, AS WE DESCRIBED IN CHAPTER 5.

DID SOMEONE SAY "LAUGH"? THAT'S MY FIELD OF EXPERTISE.

THAT'S MY COLLEAGUE SOPHIE SCOTT. AND IF YOU THINK MOST PSYCHOLOGISTS HAVE A TOUGH TIME PERSUADING PEOPLE THEIR WORK IS WORTHWHILE...

...TRY SECURING FUNDING FOR STUDIES INTO LAUGHTER.

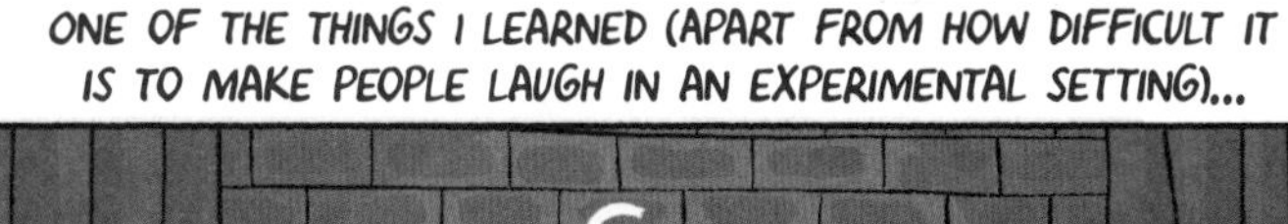
ONE OF THE THINGS I LEARNED (APART FROM HOW DIFFICULT IT IS TO MAKE PEOPLE LAUGH IN AN EXPERIMENTAL SETTING)...

...IS THAT PEOPLE ARE FAR MORE LIKELY (30 TIMES, IN FACT) TO LAUGH IF THEY'RE WITH ANOTHER PERSON.

THEY'RE MORE LIKELY AGAIN TO LAUGH IF THEY'RE WITH A CROWD, ESPECIALLY A CROWD OF SIMILAR PEOPLE.

I ALSO FOUND STRONG EVIDENCE THAT LAUGHTER, EVEN RIOTOUS UPROARING LAUGHTER AT A DAD-JOKE...
HEH-HEH-HEH
WHY DID THE MATH BOOK LOOK SO SAD?

...IS MOSTLY ABOUT FITTING IN SOCIALLY, AND LESS ABOUT HOW FUNNY SOMETHING ACTUALLY IS.
HA-HA-HA-HA-HA
BECAUSE IT WAS FULL OF PROBLEMS.
I'M SUPPOSED TO BE LAUGHING, AREN'T I.

PUTTING PEOPLE IN A SCANNER AND MAKING THEM LAUGH IS NO JOKE.*
*FEEL FREE TO LAUGH AT MY WORDPLAY HERE TO MAKE ME FEEL BETTER, IF NOT BECAUSE IT'S FUNNY.

SO I MOSTLY HAVEN'T. INSTEAD, I'VE SCANNED PEOPLE LISTENING TO LAUGHTER.

TURNS OUT, IT'S EASY TO TELL A FAKE LAUGH FROM A GENUINE ONE.
REAL LAUGH
FAKE LAUGH

UNLIKE REAL LAUGHTER, FAKE LAUGHTER ACTIVATES THE PART OF THE BRAIN THAT RELATES TO "THEORY OF MIND."

MY SUGGESTION IS THAT PEOPLE ARE TRYING TO IMAGINE WHY THE PERSON IS FAKE-LAUGHING.

AND THIS, IN A VERY ROUNDABOUT WAY, BRINGS US BACK TO FREE WILL.

FREE WILL IS STRONGLY ASSOCIATED WITH THE IDEA OF PERSONAL RESPONSIBILITY.
IF YOU DO SOMETHING NAUGHTY, IT'S YOUR FAULT, AND YOU SHOULD TAKE THE BLAME.
SPLAT

IF IT WAS SOMETHING REALLY NAUGHTY, YOU SHOULD BE PUNISHED (MUCH OF SOCIETY SAYS).

BUT IF FREE WILL IS AN ILLUSION, THEN IT CAN NEVER BE YOUR FAULT IF YOU DO SOMETHING BAD.

IT'S JUST THE WAY YOU WERE WIRED UP.

RIGHT?

THE ULTIMATE TEST CASE FOR THIS IS A RARE CONDITION KNOWN AS PSYCHOPATHY.
SOMETHING MY COLLEAGUE ESSI VIDING HAS BEEN STUDYING.

PARTLY BECAUSE OF THE WAYS THEIR BRAINS WORK, PSYCHOPATHS ARE OFTEN UNPLEASANT PEOPLE, WHO MAY HURT OR CON OR OTHERWISE DO HARM.

BUT HOW MUCH IS SUCH BEHAVIOR THEIR FAULT? COULD YOU SAY, PERHAPS, THAT PSYCHOPATHS DO NOT HAVE FREE WILL, THEY CANNOT HELP BUT BEHAVE THE WAY THEY DO?

DIAGNOSING PSYCHOPATHY CURRENTLY DEPENDS ON A DOCTOR ASSESSING PEOPLE AGAINST A CHECKLIST OF TRAITS, SUCH AS: CUNNING, MANIPULATIVENESS, AND A LACK OF EMPATHY.

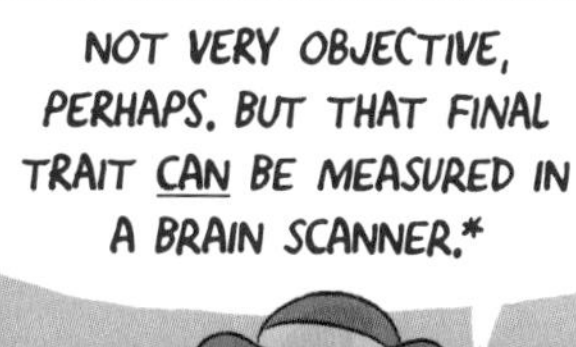
NOT VERY OBJECTIVE, PERHAPS. BUT THAT FINAL TRAIT CAN BE MEASURED IN A BRAIN SCANNER.*

SO WE DID EXACTLY THAT, SCANNING BRAINS WHILE PLAYING VOLUNTEERS AUDIO CLIPS OF PEOPLE LAUGHING.
THE VOLUNTEERS WERE ALL TEENAGE BOYS WITH A HISTORY OF VIOLENCE AND ANTISOCIAL BEHAVIOR.

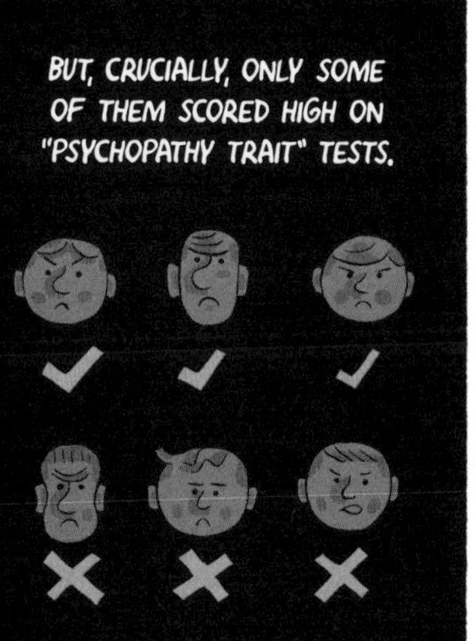
BUT, CRUCIALLY, ONLY SOME OF THEM SCORED HIGH ON "PSYCHOPATHY TRAIT" TESTS.

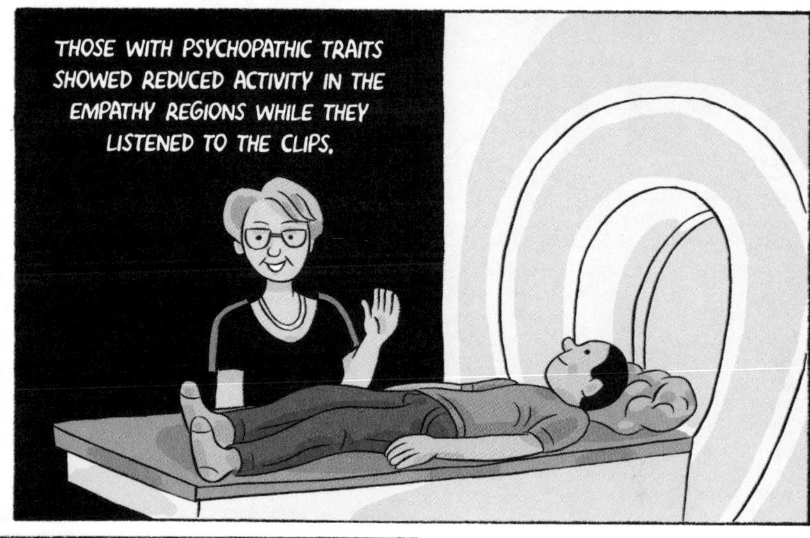
THOSE WITH PSYCHOPATHIC TRAITS SHOWED REDUCED ACTIVITY IN THE EMPATHY REGIONS WHILE THEY LISTENED TO THE CLIPS.

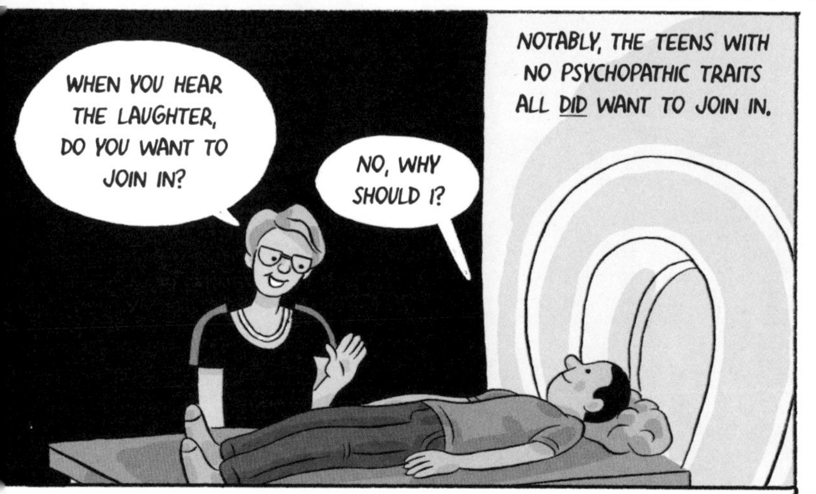
WHEN YOU HEAR THE LAUGHTER, DO YOU WANT TO JOIN IN?
NO, WHY SHOULD I?
NOTABLY, THE TEENS WITH NO PSYCHOPATHIC TRAITS ALL DID WANT TO JOIN IN.

*DO YOU REMEMBER JAMIE WARD TESTING PEOPLE FOR EMPATHY IN CHAPTER 5?

WE HOPED TO BE ABLE TO SAY THAT PSYCHOPATHS ARE RESPONSIBLE FOR THE WICKED THINGS THEY DO – OR, IN OTHER WORDS, THAT THEY HAVE FREE WILL. OUR DATA DOESN'T (AND PERHAPS COULD NEVER) PROVE IT, BUT IT SEEMS PLAUSIBLE.

HOWEVER, WE ALSO CONCLUDED THAT PSYCHOPATHS ARE WORKING WITH A DIFFERENT SET OF MENTAL INPUTS (COMPARED TO NON PSYCHOPATHS) THAT INFORM THEIR DECISIONS.

AFTER EXPLORING PSYCHOPATHY, AND INDEED THE SCIENCE OF LAUGHTER, WE'RE NO CLOSER TO FINDING OUT IF FREE WILL IS AN ILLUSION OR NOT.
BUT WE STILL THINK IT DOESN'T REALLY MATTER.
THE POINT IS, IT'S OUR EXPERIENCE THAT WE ALL LIVE AS IF WE HAVE CONTROL OVER OUR THOUGHTS AND ACTIONS.
WE SUGGEST THAT OUR EXPERIENCE OF VARIOUS KINDS OF REGRET* IS GOOD EVIDENCE OF THIS.
LAUGHTER PROVIDES AN INTERESTING EXAMPLE HERE. AFTER ALL, MOST OF US FIND THAT WE CAN'T HELP LAUGHING AT SOMETHING WE FIND FUNNY – IT SEEMS TO BE AN INVOLUNTARY THING.
AND YET, HAVE YOU EVER LAUGHED AT A JOKE, THEN REGRETTED IT LATER?
I CERTAINLY HAVE.
IF YOU WERE ALIVE DURING THE 20TH CENTURY, YOU QUITE PROBABLY WILL HAVE DONE, TOO.
SO MUCH HUMOR WAS DERIVED FROM PEOPLE PUTTING ON "FUNNY" FOREIGN ACCENTS. BUT NOW, THE MEMORY OF THIS OFTEN BRINGS SHAME.

*PSYCHOPATHS DO EXPRESS FEELINGS OF REGRET, ALTHOUGH THEIR REASONS MAY BE DIFFERENT.

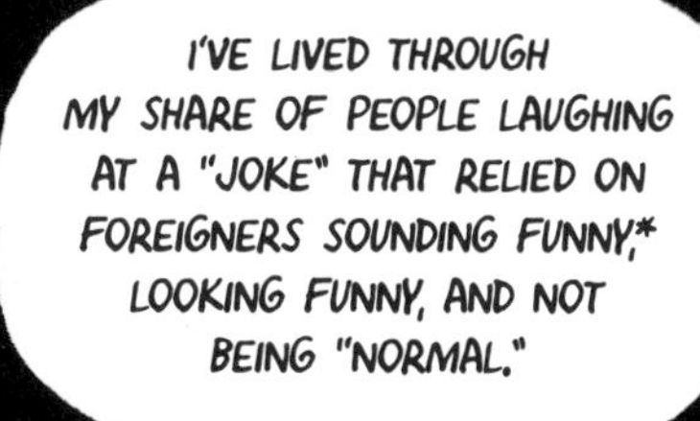
I'VE LIVED THROUGH MY SHARE OF PEOPLE LAUGHING AT A "JOKE" THAT RELIED ON FOREIGNERS SOUNDING FUNNY,* LOOKING FUNNY, AND NOT BEING "NORMAL."
AND WHEN I DIDN'T LAUGH, PEOPLE WOULD SAY "FOREIGNERS JUST DON'T HAVE A SENSE OF HUMOR."
GRRR!
IT'S ALL TO DO WITH THE IDEA OF "IN-GROUPS" AND "OUT-GROUPS," WHICH WE'LL EXPLORE IN THE NEXT CHAPTER.

EARLIER IN THIS CHAPTER, WE DEMONSTRATED HOW A PERSON CAN SPONTANEOUSLY BE "NICER" IF THEY STOP THINKING ABOUT WHAT THEY'RE DOING.
BUT, AS FAR AS WE KNOW, THIS MIGHT ONLY BE TRUE IF WE'RE TALKING ABOUT SITUATIONS INVOLVING PEOPLE WITHIN THE SAME IN-GROUP.

IT TAKES EFFORT TO OVERCOME THE BRAIN'S TENDENCY TO DISDAIN PEOPLE FROM AN OUT-GROUP.

BUT MAYBE THERE ARE SOME WAYS WE CAN TRICK OUR BRAINS, SO WE CAN OVERCOME THOSE TENDENCIES.
ON THAT NOTE OF HOPE, READ ON!

*EVEN MY OWN SONS USED TO LAUGH AT THE WAY I PRONOUNCED THE WORD "CATALOGUE."

Chapter 12

In-groups & out-groups.

PEOPLE OFTEN ASK HOW WE MET.*
BUT THAT'S NOT QUITE THE RIGHT QUESTION.
*LIKE MANY COUPLES, ESPECIALLY IN ACADEMIA, WE MET AT UNIVERSITY. (SEE CHAPTER 2)

THE REAL QUESTION IS, HOW DID WE DISCOVER THAT WE LIKED EACH OTHER ENOUGH TO LIVE AND WORK TOGETHER FOR 60 YEARS?

OUR MUTUAL ATTRACTION GREW, IN LARGE PART, FROM SHARING TASTE IN ALL SORTS OF THINGS.

WHAT ENGLISH I KNEW BEFORE I ARRIVED IN LONDON I HAD LEARNED FROM AGATHA CHRISTIE AND DOROTHY L. SAYERS.

HERCULE POIROT

I HAD LONG BEEN A DEVOURER OF DETECTIVE FICTION MYSELF.
SHERLOCK HOLMES

PERHAPS AS A RESULT OF THIS, MY LANGUAGE AND MAYBE MY GENERAL DEPORTMENT FIT INTO UTA'S EXPECTATION OF WHAT ENGLAND SHOULD BE LIKE.

AT THE VERY LEAST, I HAD A GREAT-AUNT WHO LIVED IN A BIG HOUSE IN THE COUNTRY, ALTHOUGH AS FAR AS I KNOW SHE NEVER SOLVED ANY LOCAL MURDERS.
SHERLOCK HOLMES

MY GREAT-AUNT WAS CONCERNED ABOUT MY CHOICE OF PARTNER. BEING A FOREIGN "ALIEN" WAS ACCEPTABLE, BUT WAS UTA THE RIGHT SORT OF ALIEN?

(OPEN DISCUSSION OF "ALIENS" WAS ALL THE RAGE IN THE 1960S, THANKS TO A TV SHOW AND REPRINT OF A HIT BOOK HOW TO BE AN ALIEN.)

HOW TO BE AN ALIEN
George Mikes

WHAT IF SHE'S A CATHOLIC?

ASKED MY GREAT-AUNT ONE DAY.

WHAT A QUESTION! IN FACT, MY FAMILY WERE STAUNCH LUTHERANS. LUCKILY, THIS WAS SUFFICIENT TO ALLOW HER TO CLASSIFY ME AS A PROTESTANT.

MEANWHILE, I WAS GENERALLY IGNORANT OF GERMAN CULTURE, BUT EAGER TO LEARN! I HAD BEGUN READING THE BEST SORT OF GERMAN – IN TRANSLATION – BERTOLT BRECHT...

A MUST-READ FOR ANY PSYCHOLOGY STUDENT!
THREEPENNY OPERA
Bertolt Brecht
MOTHER COURAGE AND HER CHILDREN
Bertolt Brecht
Tales from the Calendar

LIKE EVERYONE IN THE '60S, I HAD ALREADY READ AUTHOR HERMANN HESSE'S STEPPENWOLF AND THE GLASS BEAD GAME, BUT UTA TOLD ME HE WROTE KITSCH.*

*KITSCH IS GERMAN FOR "LOWBROW TRASH," OR AT LEAST THAT'S HOW I'VE COME TO UNDERSTAND IT.

WITH OUR SHARED LOVE OF READING, WE BOTH DEVELOPED A PASSIONATE INTEREST IN SCIENCE FICTION. AT AROUND THIS TIME, WE EVEN TOOK EVENING CLASSES IN THE FORM.*

*ONE OF UTA'S VERY FIRST PUBLISHED WORKS WAS IN FACT A SCIENCE FICTION SHORT STORY. LOOK FOR IT IN ANTIGRAV (1974).

ALONGSIDE TIME SPENT BOTH PLAYING AND LISTENING TO CLASSICAL MUSIC,
AND TAKING UP WINE TASTING BEFORE IT WAS FASHIONABLE.

IT'S NOT ENTIRELY SURPRISING THAT UTA STILL CLINGS TO THE PORTRAIT OF ENGLISH LIFE AS PAINTED BY CHRISTIE AND SAYERS.

WE LIKE TO JOKE THAT CHRIS TAUGHT ME ENGLISH...
...BUT WITHIN WEEKS UTA WAS ABLE TO CORRECT MY SPELLING.

WE WERE MARRIED AT A SMALL CEREMONY IN YORK IN THE SUMMER OF 1966.

OUR FATHERS GOT ON VERY WELL DESPITE HAVING FOUGHT ON OPPOSITE SIDES DURING WW2.* BOTH WERE TEACHERS.
*NOT THE SAME BATTLES, BUT AT ONE POINT BOTH WERE ON THE ITALIAN FRONT.

WE HAD THE SHARED EXPERIENCE OF CONSCIOUSLY MEETING OUR FATHERS FOR THE FIRST TIME, AFTER THE WAR, AND WONDERING:
WHO IS THIS STRANGE MAN WHO'S SUDDENLY LIVING IN MY HOUSE?

BEING A GERMAN LIVING IN ENGLAND HAS BEEN A STRUGGLE AT TIMES.

YOU LEARN TO COPE WITH A CERTAIN AMOUNT OF STEREOTYPING ON TV, AND CLOSE YOUR EARS WHEN IT HAPPENS IN REAL LIFE.

NEVERTHELESS, THERE ARE TIMES LIVING IN A FOREIGN COUNTRY WHEN I'M MADE TO FEEL THAT I DON'T QUITE BELONG.

AND THIS LEADS TO A MAJOR PART OF SOCIAL COGNITION – THE WAY OUR BRAINS PUT PEOPLE INTO IN-GROUPS AND OUT-GROUPS.

IN BRITAIN – AND WE HAVE TO ASSUME, MOST OF THE WORLD – PEOPLE FIT INTO ALL SORTS OF "IN-GROUP" BOXES, SOME OF THEIR OWN CHOOSING, OTHERS NOT. OFTEN, PEOPLE ARE PROUD OF THE GROUPS THEY'RE IN. HERE ARE A FEW OF THE BOXES WE FIT INTO, SOME APART AND SOME TOGETHER. WE ARE AWARE SOME OF THIS WILL COME ACROSS AS BOASTFUL – AND CERTAINLY, THERE'S A FAIR BIT OF PRIVILEGE ON DISPLAY.

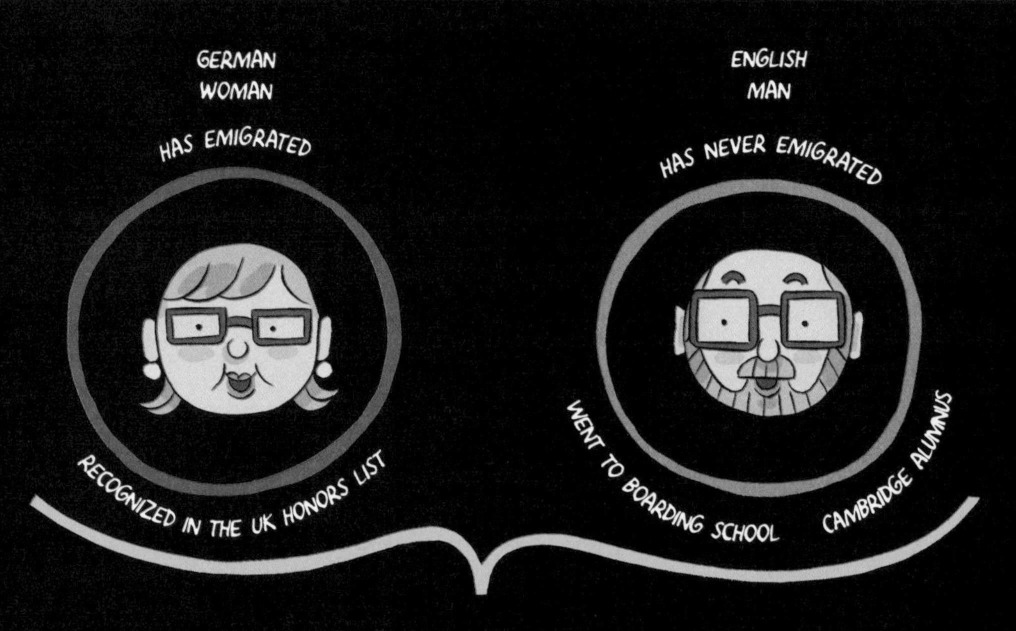
GERMAN WOMAN
HAS EMIGRATED
RECOGNIZED IN THE UK HONORS LIST
ENGLISH MAN
HAS NEVER EMIGRATED
WENT TO BOARDING SCHOOL
CAMBRIDGE ALUMNUS
EUROPEAN
EMERITUS PROFESSOR AT UNIVERSITY COLLEGE LONDON
SCIENTIST AND RESEARCHER (BUT NOT DOCTOR OR CLINICIAN)
FELLOW OF THE ROYAL SOCIETY

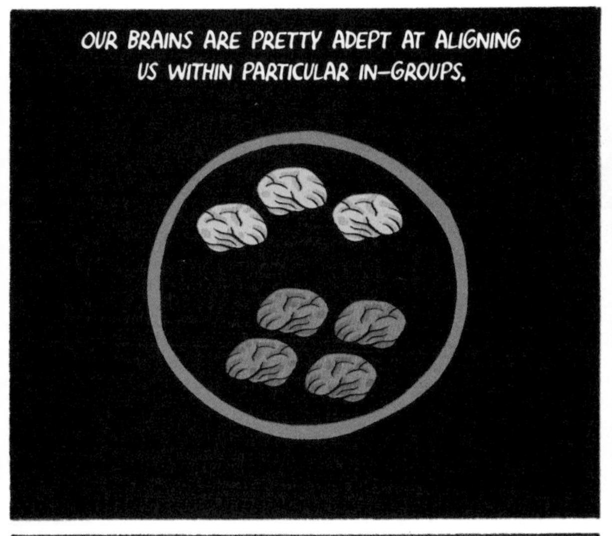
OUR BRAINS ARE PRETTY ADEPT AT ALIGNING US WITHIN PARTICULAR IN-GROUPS.

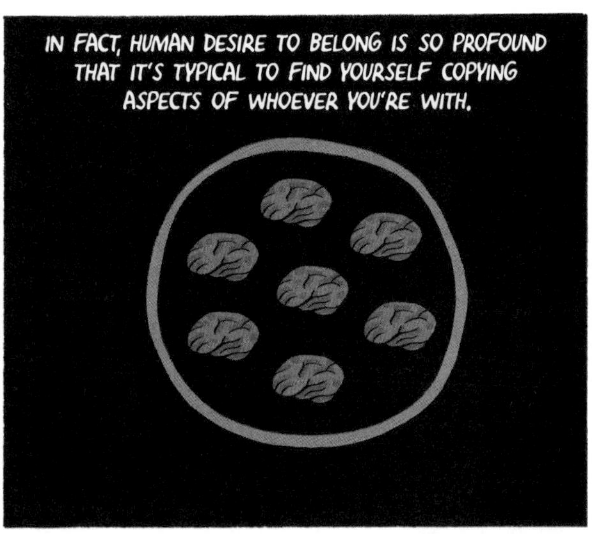
IN FACT, HUMAN DESIRE TO BELONG IS SO PROFOUND THAT IT'S TYPICAL TO FIND YOURSELF COPYING ASPECTS OF WHOEVER YOU'RE WITH.

LIKE THE TIME I DEVELOPED A TEMPORARY LIMP AS I WALKED AND TALKED WITH EMINENT FRENCH NEUROPHYSIOLOGIST MARC JEANNEROD.

OR THE TIME WE BOTH STARTED WHISPERING WHEN WE VISITED A FRIEND IN THE HOSPITAL WHO'D JUST HAD A THROAT OPERATION.
STOP MAKING FUN!
SORRY. WE CAN'T HELP IT.

OF COURSE, THESE ARE JUST TWO ANECDOTES. AND HOPEFULLY BY NOW YOU'RE ASSURED THAT ANECDOTES ARE NOT SCIENCE!
HERE'S A PROPER, REPEATABLE EXPERIMENT ON COPYING.
SPECIFICALLY, HOW HUMANS LINK THE ACT OF COPYING TO THE IDEA OF IN-GROUPS AND OUT-GROUPS.

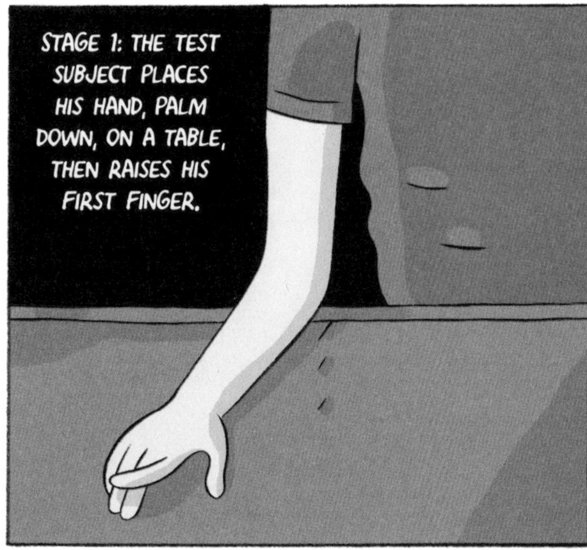
STAGE 1: THE TEST SUBJECT PLACES HIS HAND, PALM DOWN, ON A TABLE, THEN RAISES HIS FIRST FINGER.

STAGE 2: SAME AGAIN, ONLY THIS TIME THE TEST SUBJECT IS WATCHING SOMEONE ELSE DO THAT SAME THING.

OBSERVATION – THE TEST SUBJECT RAISES HIS FINGER MEASURABLY FASTER IN STAGE 2 THAN IN STAGE 1.

STAGE 3: JUST TO RAM THE POINT HOME, THE TEST SUBJECT WATCHES SOMEONE ELSE...
...WHO IS RAISING HER SECOND FINGER.

THIS TIME, IT TAKES HIM EVEN LONGER THAN NORMAL TO RAISE HIS FIRST FINGER.

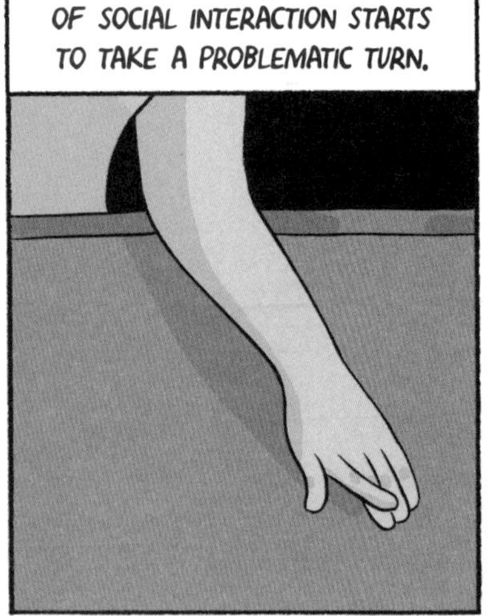
AND THIS IS WHERE THE STORY OF SOCIAL INTERACTION STARTS TO TAKE A PROBLEMATIC TURN.

HERE'S NATALIE SEBANZ TO TALK US THROUGH A 2011 STUDY SHE RAN IN THE NETHERLANDS.

I WANTED TO SEE IF THERE ARE LIMITS TO MIMICKING BEHAVIOR. WILL PEOPLE MIMIC JUST ANYONE?

MY TEST SUBJECTS WERE ALL DUTCH. MORE SPECIFICALLY, WHITE DUTCH.
EACH WAS SHOWN A SET OF VIDEOS SHOWING ONLY ANOTHER PERSON'S HANDS – DARK-SKINNED HANDS.
AS THEY WATCHED, I EXPLAINED WHO THE HANDS BELONGED TO. SOME WERE FROM SURINAMESE PEOPLE, OTHERS WERE FROM MOROCCANS.

WHY THESE TWO COUNTRIES? SURINAME IS A FORMER DUTCH COLONY; MOROCCO IS NOT. PUT ANOTHER WAY, MOST WHITE DUTCH PEOPLE ARE USED TO MEETING SURINAMESE IMMIGRANTS, LESS SO MOROCCAN IMMIGRANTS.

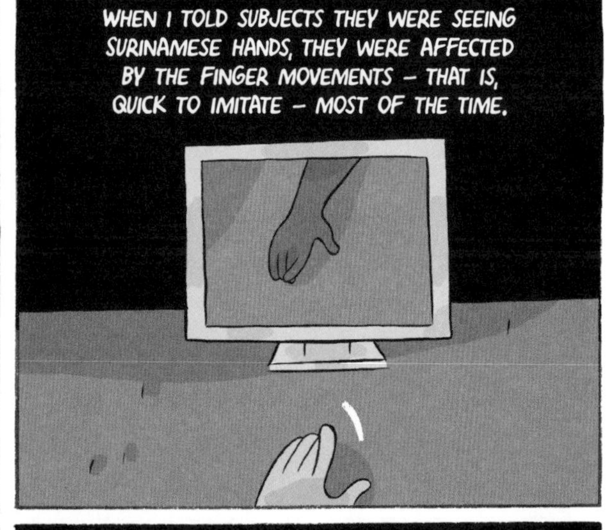
WHEN I TOLD SUBJECTS THEY WERE SEEING SURINAMESE HANDS, THEY WERE AFFECTED BY THE FINGER MOVEMENTS – THAT IS, QUICK TO IMITATE – MOST OF THE TIME.

BUT THEY WERE NOT AFFECTED WHEN I SAID THE HANDS BELONGED TO MOROCCANS.
IN FACT, I HAD BEEN LYING TO THE SUBJECTS.
THE HANDS WERE THE SAME EACH TIME.

THIS SERVES TO CONFIRM THAT THE SURINAMESE ARE SEEN AS A DUTCH IN-GROUP, WHILE MOROCCANS ARE A DEFINITE OUT-GROUP. AND SUPPORTS MY HYPOTHESIS THAT PEOPLE DO NOT SPONTANEOUSLY MIMIC JUST ANYONE.
WE TWO ARE ONE.
YOU ARE "OTHER."

DEFINING WHO COUNTS AS PART OF ANY GIVEN IN-GROUP IS NOT ALWAYS OBVIOUS.

AND, SAD TO SAY, OFTEN THE MOST PRACTICAL DEFINITIONS ARE ABOUT IDENTIFYING WHO IS IN AN OUT-GROUP.

YOU'RE DEFINITELY "OUT."
BUT I'M JUST LIKE YOU!
ARE YOU THOUGH?

IT MAY NOT MATTER HOW MUCH YOU THINK YOU OUGHT TO FIT INTO A GROUP – IF YOU HAVE OBVIOUS OUT-GROUP MARKERS, YOU'LL NEVER BE FULLY WELCOME.
AND BEING "IN" IS NOT A QUESTION OF BELONGING TO A MAJORITY, AS ANY WOMAN CAN TELL YOU!

IN 2014, I BECAME AN HONORARY DAME.* I AM NOW PART OF A VERY EXCLUSIVE BRITISH IN-GROUP. BUT EVEN WITHIN THIS SELECT IN-GROUP, I'M SET APART BY TWO OUT-GROUPINGS.

*ONE OF THE TOP HONORS BESTOWED IN BRITAIN.

OUT-GROUP 1: I'M A WOMAN, SO I CANNOT BE DESCRIBED AS A "KNIGHT."
OUT-GROUP 2: I'M A FOREIGNER, SO I DON'T GET TO BE CALLED "DAME UTA."

IT ALSO MEANT THE HONOR WAS CONFERRED BY A GOVERNMENT MINISTER, NOT A MEMBER OF THE ROYAL FAMILY.
LET ME STRESS – I AM NOT COMPLAINING! I AM VERY PROUD TO HAVE RECEIVED THIS HONOR.

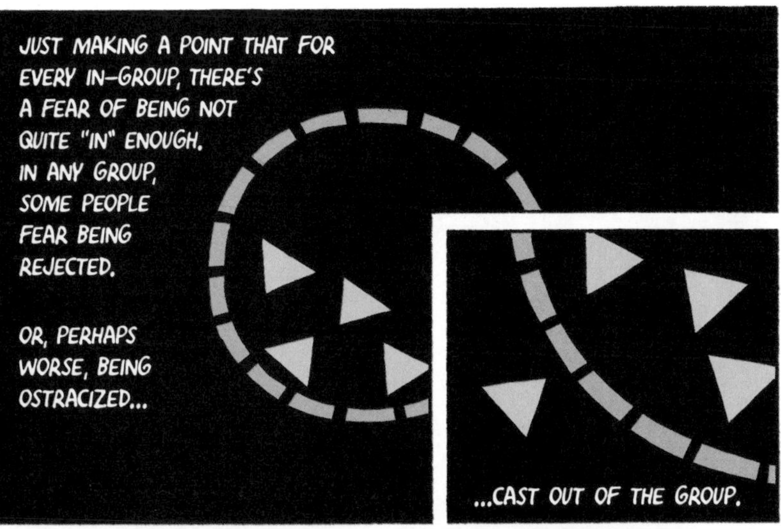
JUST MAKING A POINT THAT FOR EVERY IN-GROUP, THERE'S A FEAR OF BEING NOT QUITE "IN" ENOUGH. IN ANY GROUP, SOME PEOPLE FEAR BEING REJECTED.
OR, PERHAPS WORSE, BEING OSTRACIZED...
...CAST OUT OF THE GROUP.

HOPEFULLY, YOU'VE JUST READ THAT LAST PANEL WITHOUT HAVING TO WONDER WHAT ON EARTH YOU'RE LOOKING AT.

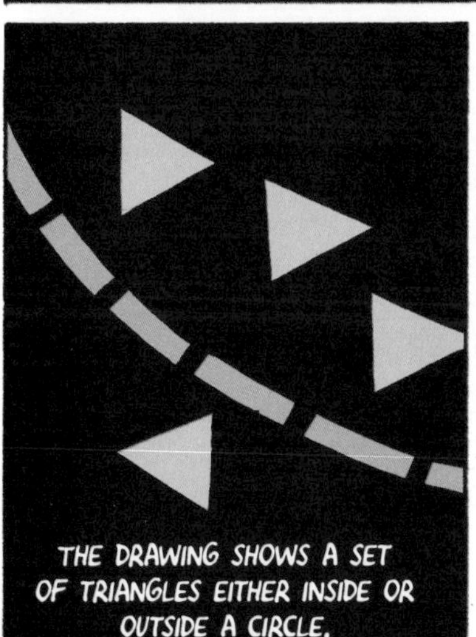
THE DRAWING SHOWS A SET OF TRIANGLES EITHER INSIDE OR OUTSIDE A CIRCLE.

IT'S A PURELY ABSTRACT IMAGE...

...BUT MOST PEOPLE CAN'T HELP THINKING OF THE TRIANGLES AS PEOPLE.

QUICK ASIDE: I DEVELOPED THE IDEA OF TELLING STORIES WITH TRIANGLES WITH MY COLLEAGUE FRANCESCA HAPPÉ.

WE DEVISED A SERIES OF TRIANGLE-BASED ANIMATIONS DURING ONE LONG TRAIN JOURNEY FROM LONDON TO DURHAM.

USING TRIANGLES AS A STAND-IN FOR PEOPLE AND THEIR INTERACTIONS TURNED OUT TO BE SURPRISINGLY VERSATILE, AND PROVED POPULAR AS TESTS OF "THEORY OF MIND." SOME YEARS LATER, IN 2009, HARRIET OVER AND MALINDA CARPENTER USED SIMILAR VIDEOS TO DEVELOP A STUDY ABOUT OSTRACISM...

A SHORT CLIP SHOWS ONE SHAPE BEING OSTRACIZED BY THE MAIN GROUP.

ANOTHER CLIP SHOWED ALL THE SHAPES GETTING ALONG WITH EACH OTHER.

IMMEDIATELY AFTER WATCHING EITHER VIDEO, CHILDREN WERE GIVEN A SET OF OBJECTS.

THEY WERE THEN ASKED TO COPY WHAT THEY SAW SOMEONE ELSE DOING WITH THE OBJECTS.

THOSE WHO HAD SEEN THE OSTRACIZING VIDEO IMITATED THE ACTION MORE CLOSELY THAN THOSE WHO HADN'T.

A SEPARATE STUDY ASKED CHILDREN TO PLAY A GAME CALLED CYBERBALL THROUGH A VR HEADSET. THE CHILD PLAYS CATCH WITH A SET OF PLAYERS WHO HAVE SIMILAR FEATURES TO THE CHILD – AN "IN-GROUP."
AFTER A FEW GOES, THE PLAYERS ALL STOP THROWING THE BALL TO THE CHILD.

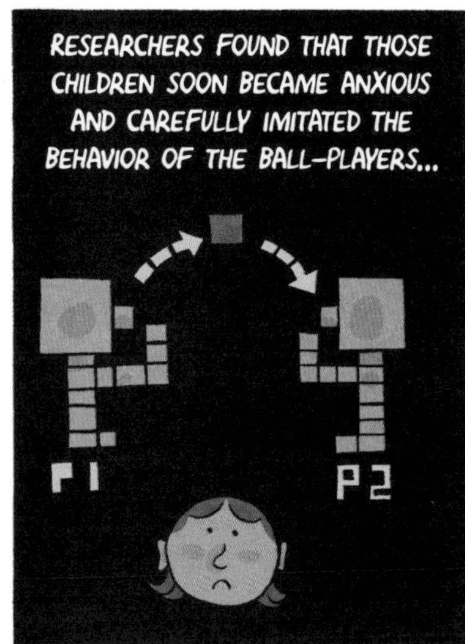
RESEARCHERS FOUND THAT THOSE CHILDREN SOON BECAME ANXIOUS AND CAREFULLY IMITATED THE BEHAVIOR OF THE BALL-PLAYERS...
P1
P2

...PRESUMABLY, BECAUSE THEY HOPED TO BE READMITTED INTO THE GROUP. DON'T WORRY, THIS WAS ALL EXPLAINED AS A DELIBERATE EXPERIMENT!*
*AND THEY DID GET ETHICAL PERMISSION FOR THIS, WE PROMISE!

OTHER CHILDREN PLAYED CYBERBALL WITH PLAYERS WHO WERE PART OF AN OUT-GROUP. WHEN THESE CHILDREN WERE REJECTED, THEY DIDN'T IMITATE ACTIONS MORE CLOSELY.
IT'S AS IF THEY DIDN'T REALLY MIND BEING REJECTED.

TO BE SCIENTIFIC ABOUT IT – AND DISPASSIONATE – STUDIES SHOW THAT PEOPLE SHUNNED BY OUT-GROUPS DO NOT IMITATE THEIR BEHAVIOR.

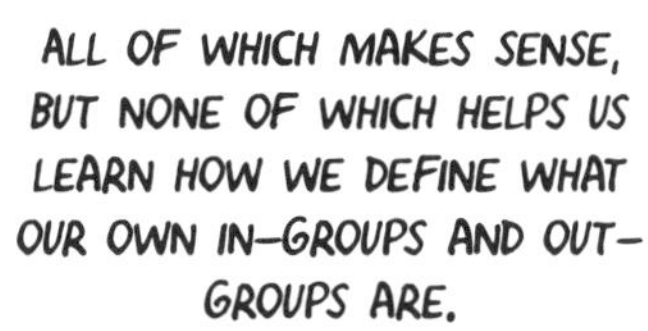
ALL OF WHICH MAKES SENSE, BUT NONE OF WHICH HELPS US LEARN HOW WE DEFINE WHAT OUR OWN IN-GROUPS AND OUT-GROUPS ARE.

OR, MORE IMPORTANTLY, LEARN HOW TO TREAT MEMBERS OF OUT-GROUPS AS IF THEY WERE IN-GROUPS.

YOU COULD ARGUE THAT SOLVING THIS PROBLEM IS ONE OF SOCIETY'S BIGGEST GOALS. WE'RE NOT GOING TO SOLVE IT IN THIS BOOK, BUT WE CAN PROMOTE SOME FINDINGS BASED ON SIMPLE STUDIES.

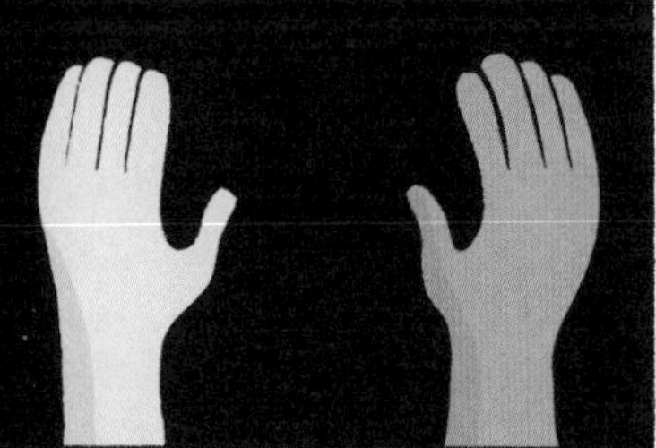

ONE TRICK SEEMS TO BE LEARNING TO SHIFT YOUR PERSPECTIVE...

...TO MAKE EVEN A MINIMAL EFFORT TO SEE THINGS FROM ANOTHER POINT OF VIEW.

THIS IS A TRICK YOU MAY WELL THINK YOU'VE LEARNED FROM FICTION – AND YOU'D BE RIGHT – BUT DOES HARD SCIENCE BACK IT UP? THANKFULLY, THE ANSWER SEEMS TO BE YES.

BELOVED
A NOVEL
TONI MORRISON
Wild Swans
NOUGHTS & CROSSES

THIS PHENOMENON WAS STUDIED USING NO LESS A CULTURAL ICON THAN HARRY POTTER.

GROUPS OF READERS – FROM ELEMENTARY SCHOOL, HIGH SCHOOL, AND EVEN UNIVERSITY – READ SOME OF THE BOOKS IN THE SERIES.

ONE OF THE RECURRING THEMES IS IN-GROUPS OF WIZARDS LORDING IT OVER OUT-GROUPS OF NON-WIZARDS CALLED "MUGGLES."

(AND, TO A LESSER EXTENT, IN-GROUPS DEFINED BY FOUR SCHOOL "HOUSES.")

ALL READERS WERE FOUND TO HAVE AN IMPROVED ATTITUDE TOWARDS MARGINALIZED GROUPS. SPECIFICALLY: IMMIGRANTS, GAY PEOPLE, AND REFUGEES.

NATALIE, ALONG WITH OUR COLLEAGUE MANOS TSAKIRIS (AND HIS TEAM), LOOKED MORE SPECIFICALLY INTO WAYS TO IMPROVE PEOPLE'S ATTITUDES WHEN IT COMES TO SKIN COLOR.

DO YOU REMEMBER THE RUBBER-HAND ILLUSION? THE ONE WHERE SOMEONE TICKLES A RUBBER HAND, BUT MAKES YOU FEEL THAT YOUR OWN HAND IS BEING TICKLED.

MANOS'S TEAM USED IT TO MAKE FAIR-SKINNED TEST SUBJECTS EXPERIENCE THE ILLUSION THAT A DARK-SKINNED RUBBER HAND WAS THEIR OWN.

POKE

AFTER PRIMING PEOPLE WITH THIS EXPERIENCE, THE TEAM CARRIED OUT SOMETHING CALLED AN IAT – AN IMPLICIT ASSOCIATION TEST.

FILTH JOY

TORTURE PLEASURE

VOMIT PEACE

THE TEST EXAMINES HOW QUICKLY PEOPLE ASSOCIATE LIGHT AND/OR DARK FACES WITH GOOD AND/OR BAD WORDS.

SAD TO SAY, MOST LIGHT-SKINNED PEOPLE ARE QUICKER TO ASSOCIATE DARK-SKINNED FACES WITH BAD WORDS.

FILTH JOY

TORTURE PLEASURE

VOMIT PEACE

BUT PEOPLE WHO WERE PRIMED WITH THE RUBBER-HAND ILLUSION FIRST SHOW LESS OF THIS ASSOCIATION.

IF YOU DON'T HAVE A RUBBER HAND, OR THE PATIENCE TO READ A BOOK, ONE POSSIBLE SHORTCUT TO GET THE SAME EFFECT IS TO WATCH A FILM.

NIGHT
of the
LIVING
DEAD

IDEALLY A GOOD-QUALITY FILM...

...BUT IT COULD BE ANY FILM THAT HAS A LEAD CHARACTER NOT OF YOUR SKIN COLOR.*

*NIGHT OF THE LIVING DEAD IS AN EXAMPLE THAT FITS THE BILL NICELY FOR WHITE PEOPLE; COUNTER-EXAMPLES FOR PEOPLE OF COLOR ARE TOO NUMEROUS TO LIST...

EVEN THIS MINIMAL LEVEL OF PRIMING CAN BE ENOUGH TO SHIFT A PERSON'S BIAS.

WHAT DOES BRAIN IMAGING MAKE OF ALL THIS?

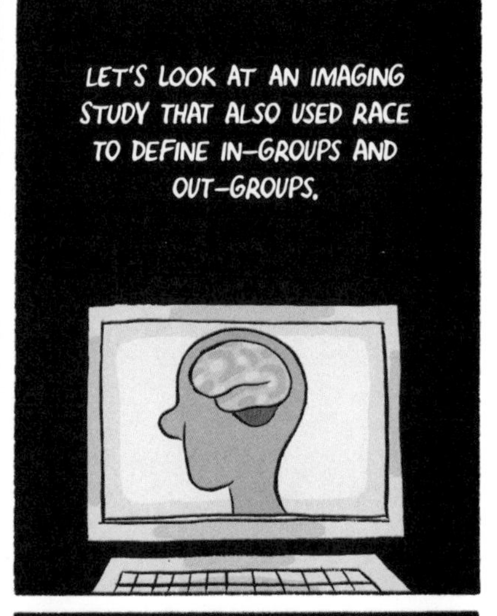
LET'S LOOK AT AN IMAGING STUDY THAT ALSO USED RACE TO DEFINE IN-GROUPS AND OUT-GROUPS.

IF YOU SCAN ONE PERSON WHILE THEY'RE WATCHING A VIDEO OF SOMEONE ELSE IN PAIN...
...PART OF THE BRAIN WILL RESPOND IN A STRAIGHTFORWARDLY EMPATHETIC WAY.

SO LONG AS THE PERSON THEY'RE WATCHING BELONGS TO THEIR RACIAL IN-GROUP.

WHITE EUROPEAN SUBJECTS WERE SHOWN VIDEOS OF A CHINESE PERSON IN TWO DIFFERENT SCENARIOS:
1) POKED WITH A NEEDLE
2) POKED WITH A COTTON BUD
UNDER A SCANNER, THE BRAINS SHOWED THE SAME RESPONSE BOTH TIMES.*
THIS GOES AGAINST THE DISCOVERIES MADE ABOUT MIRROR NEURONS, WHICH, AS WE SAW EARLIER, NORMALLY ACTIVATE OUT OF EMPATHY.
*THE SAME HAPPENED IN REVERSE, WHEN CHINESE SUBJECTS WATCHED VIDEOS OF A WHITE EUROPEAN.

DOES THIS MEAN RACISM IS HARDWIRED INTO OUR BRAINS?

WELL, IN FACT, THERE ARE STRATEGIES THAT CAN FAIRLY QUICKLY CHANGE THIS REACTION.

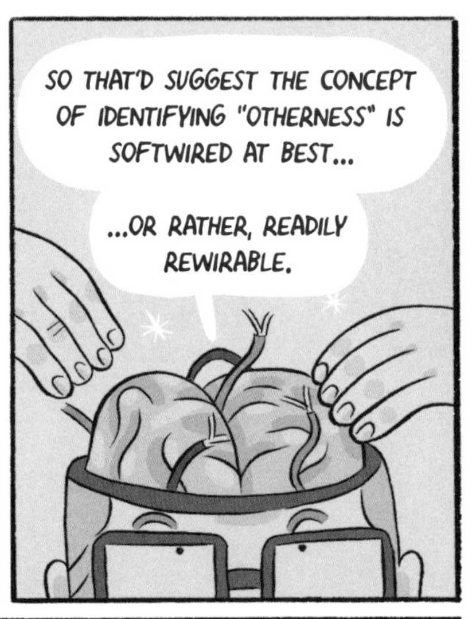
SO THAT'D SUGGEST THE CONCEPT OF IDENTIFYING "OTHERNESS" IS SOFTWIRED AT BEST...
...OR RATHER, READILY REWIRABLE.

THE STUDY TRIED TWO DIFFERENT TRICKS TO OVERCOME THE RACE-BASED EMPATHY RESPONSE.

TRICK 1: GET THE TEST SUBJECT TO DESCRIBE THE SPECIFIC PAIN THAT THE PERSON IN THE VIDEO IS FEELING. SEEING ANYONE AS AN INDIVIDUAL, RATHER THAN AS A GENERIC STAND-IN FOR AN OUT-GROUP, IS ENOUGH TO GET THOSE MIRROR NEURONS WORKING PROPERLY SO TRUE EMPATHY KICKS IN.

TRICK 2: PUT THE TEST SUBJECT INTO A TEAM WITH A SELECTION OF OUT-GROUP MEMBERS...

...AND GET THAT TEAM TO DO SOME SORT OF COMPETITIVE TASK, SUCH AS PLAYING A GAME AGAINST ANOTHER TEAM.

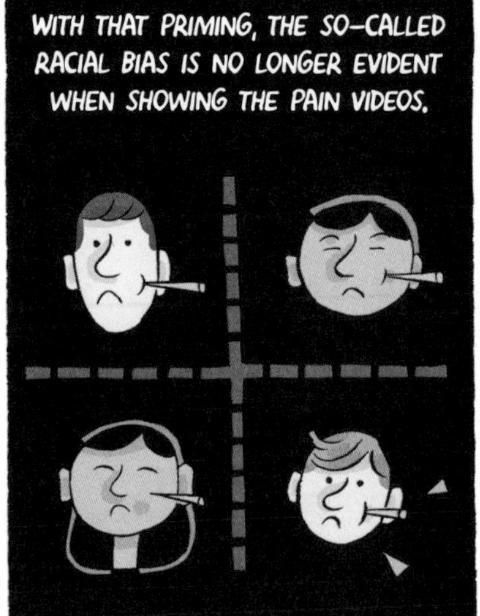
WITH THAT PRIMING, THE SO-CALLED RACIAL BIAS IS NO LONGER EVIDENT WHEN SHOWING THE PAIN VIDEOS.

IT'S COMFORTING, IF NOT GROUNDBREAKING, TO KNOW THAT PEOPLE CAN AND DO OVERCOME THEIR OWN RACIAL BIASES, EVEN ON A TEMPORARY BASIS.

BUT THAT'S NOT THE ONLY DISCOVERY ABOUT HOW IN-GROUPS AND OUT-GROUPS WORK.

OUR "TWO HEADS" RESEARCH IS ABOUT PEOPLE WORKING IN PAIRS.

BUT WHAT ABOUT LARGER GROUPS?

HERE'S A STUDY THAT WE COULDN'T RESIST SHARING, IN WHICH GROUPS OF 4 WERE ASKED TO SOLVE A MURDER MYSTERY BY REVIEWING THE DOCUMENTS IN A CASE FILE.

IN THE STUDY, EACH GROUP STARTED OFF AS JUST THREE PEOPLE, WHO DISCUSSED THE CASE TOGETHER.

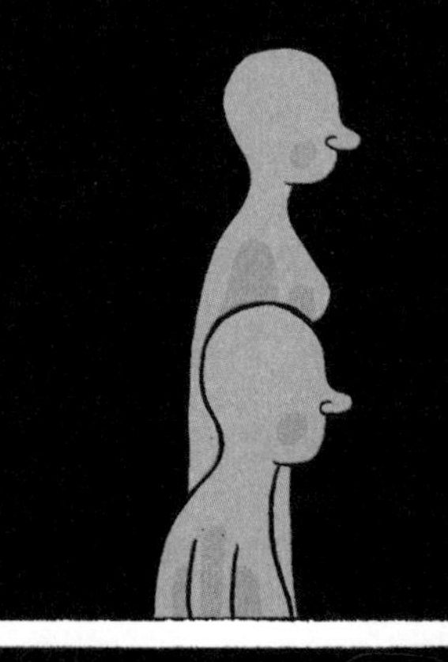
THEN A FOURTH PERSON WAS INTRODUCED TO THE GROUP – HALF OF THE TIME, AN IN-GROUP MEMBER, AND HALF OF THE TIME, AN OUT-GROUP MEMBER.

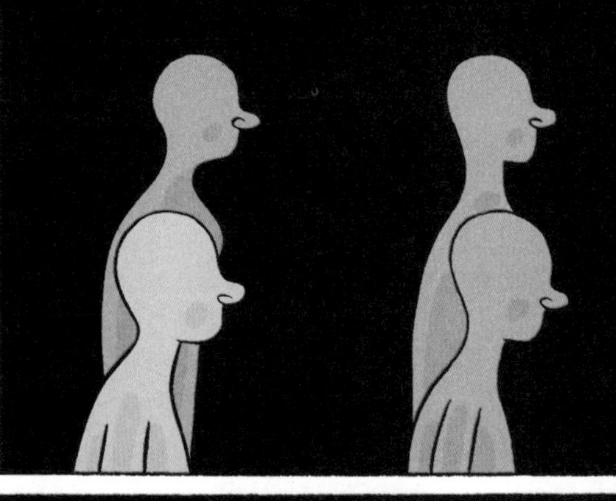
(IN THE ACTUAL STUDY, IN-GROUP VS. OUT-GROUP WAS DETERMINED BY MEMBERSHIP OF US COLLEGE FRATERNITIES AND SORORITIES.)

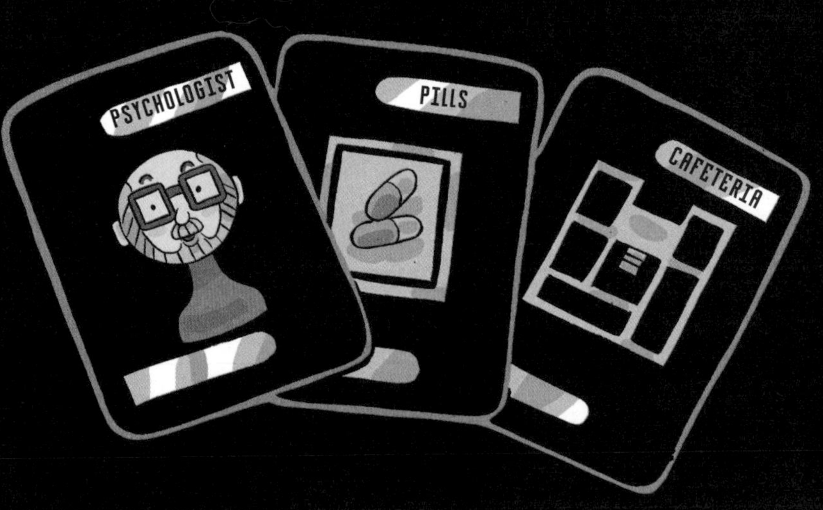
EACH GROUP EXPLAINED THE CASE TO THE NEWCOMER AND GAVE THEM A CHANCE TO REVIEW THE FILE. THEN TOGETHER THEY HAD TO DECIDE WHO WAS THE MURDERER, AND WHY.
PSYCHOLOGIST
PILLS
CAFETERIA

TEAMS WITH AN IN-GROUP NEWCOMER REPORTED A PLEASANT DISCUSSION, AND HIGH CONFIDENCE IN THEIR SOLUTION TO THE MYSTERY.

TEAMS WITH AN OUT-GROUP NEWCOMER REPORTED THE EXACT OPPOSITE – UNHAPPY DISCUSSION, AND LOW CONFIDENCE IN THEIR SOLUTION.

BUT IN FACT, THE DIVERSE GROUPS OBTAINED THE CORRECT SOLUTION MORE CONSISTENTLY.

THE REASONS WHY HAVEN'T BEEN CONCLUSIVELY PROVED, BUT THERE ARE SOME LIKELY THEORIES.
DETECTIVE
FICTION
SHERLOCK HOLMES

A NEWCOMER FROM AN IN-GROUP IS PRIMED TO GO ALONG WITH THE GROUP. LIKE WATSON MEETING HOLMES, THIS NEWCOMER IS UNLIKELY TO QUESTION OR CHALLENGE DECISIONS MADE BY THE GROUP, EVEN IF THEY ARE SKEPTICAL – BECAUSE FITTING IN IS IMPORTANT.
YOUR IDEAS ARE BIZARRE BUT I SHALL AGREE WITH EVERYTHING YOU SAY!

ON THE OTHER HAND, A NEWCOMER FROM AN OUT-GROUP (SAY, BELGIUM) STIMULATES A VERY DIFFERENT RESPONSE.
I, POIROT, TELL YOU: YOU ARE WRONG!

NOT ONLY IS THE NEWCOMER MORE LIKELY TO CHALLENGE THE FINDINGS...
A. CONAN DOYLE

...BUT ALSO THE OTHER GROUP MEMBERS FEEL THE NEED TO EXPLAIN THEIR DECISIONS MORE CAREFULLY, WHICH FORCES THEM TO REASSESS EVERYTHING.
AGATHA CHRISTIE

CLEARLY, WORKING WITH DIVERSE GROUPS IS BENEFICIAL.
WHICH MAKES US WONDER ABOUT TWO COMMON FEATURES OF SOCIAL COGNITION.

WHY DO HUMAN BRAINS FIND IT SO EASY TO CREATE IN-GROUPS AND OUT-GROUPS, EVEN WHEN WE'RE NOT THINKING ABOUT IT?

AND WHY IS IT THAT THE SAME BRAINS CAN THEN REDEFINE WHO COUNTS AS AN IN-GROUP WHEN CIRCUMSTANCES CHANGE?
THE ENEMY OF MY ENEMY IS MY FRIEND!

THIS IS GETTING A LITTLE BEYOND NEUROSCIENCE AND INTO EVOLUTIONARY BIOLOGY...
...OR SOMETHING LIKE THAT.
BUT HERE'S ONE ANSWER:

WHEN PEOPLE ARE COMPETING AS INDIVIDUALS, IT'S BETTER TO BE SELFISH —MEANING, SEEING EVERYONE WHO ISN'T YOU AS AN OUT-GROUP WILL GIVE YOU AN ADVANTAGE.

I'LL CRUSH ALL OF YOU!
WHACK A MOLE

WHACK!

WHACK!

BUT WHEN PEOPLE COMPETE AS PART OF LARGE GROUPS...

...THOSE GROUPS MADE UP OF COOPERATIVE INDIVIDUALS DO BEST.
IF WE WORK AS ONE, NO ONE CAN DEFEAT US!

WHAT'S IMPORTANT IS THE PEOPLE IN THE SAME GROUP TRUST EACH OTHER.
PSST! DO YOU KNOW THAT GUY?
!

DID ROMAN SOLDIERS – RECRUITED FROM ACROSS THE ENTIRE EMPIRE – REALLY TRUST EACH OTHER?
I DON'T TRUST YOU.

AS FAR AS HISTORY RECORDS, THEY DID. OR AT LEAST, THE TRAINING WAS ENOUGH TO OVERCOME ANY TRUST ISSUES.
I TRUST YOU IMPLICITLY.

OF COURSE, JUST BECAUSE ONE PERSON IS PART OF AN IN-GROUP DOESN'T MEAN YOU OUGHT TO ACTUALLY TRUST THEM.

IN TIME, EVERYONE DEVELOPS A REPUTATION, AND COMMUNICATES THIS TO OTHER PEOPLE.
BRAVE

SOMETIMES THE TWO DON'T LINE UP. YOU CAN PRETEND TO BE ONE THING BUT ACTUALLY BE SOMETHING ELSE.
BRAVE?
COWARD

YOU CAN'T GET MUCH MORE OF AN IN-GROUP THAN A MARRIED COUPLE.

BUT THIS DOESN'T STOP US WORRYING ABOUT THE INSIDES OF EACH OTHER'S HEADS.

WHAT DO YOU THINK OF ME?
YOU ARE A VERY WONDERFUL PERSON.
WHAT DOES SHE REALLY THINK?

WHAT DO I THINK SHE THINKS OF ME?
WHAT'S HE THINKING??
LITTLE THINGS I DO ANNOY HER TERRIBLY.

SHE MUST BASICALLY LIKE ME OR SHE WOULDN'T HAVE STAYED WITH ME FOR MORE THAN 50 YEARS.
HE MUST BASICALLY LIKE ME OR HE WOULDN'T HAVE STAYED WITH ME FOR MORE THAN 50 YEARS.

WE CELEBRATED OUR GOLDEN WEDDING ANNIVERSARY IN 2016.
THE SECRET TO A LASTING MARRIAGE?
WELL, WE DID READ A PAPER RECENTLY THAT UNCOVERED A BRAIN-BASIS FOR MARRIAGE.

IT'S TO DO WITH OUR EXPECTATION OF WHAT WE GET OUT OF STAYING MARRIED.
OUR BRAINS PERFORM A SIMPLE CALCULATION: CAN WE EXPECT TO GET MORE BENEFIT FROM STAYING TOGETHER, OR LESS?
IN THEORY, OUR BRAINS ARE CONSTANTLY MAKING THIS CALCULATION.

ALTHOUGH WE PREFER ANOTHER PAPER, WHICH FOUND A CORRELATION BETWEEN HIGH INTELLIGENCE IN A COUPLE, AND LENGTH OF MARRIAGE.*

*(ALTHOUGH THAT PAPER WAS SPECIFICALLY ABOUT DUTCH PEOPLE BORN IN THE LATE 1940S.)

EITHER WAY, THE TRUTH IS THAT WE PUT LOTS OF CONSCIOUS EFFORT INTO WORRYING ABOUT WHAT WE THINK OF EACH OTHER (AND OURSELVES).
COME TO THAT, WHAT DO THE READERS THINK OF US??

LOGICAL (AND ILLOGICAL) THOUGHTS ARE JUST PART OF HOW WE COPE WITH AND MANAGE THE IDEA OF OUR OWN REPUTATIONS.
WE'LL GET INTO THIS IN OUR NEXT AND FINAL CHAPTER.

Chapter 13

YOU MIGHT WONDER HOW ON EARTH THIS COMIC GOT PUBLISHED. IT'S PRETTY NICHE, AS COMICS GO.

REPUTATION MATTERS.
WE'D LIKE TO FLATTER OURSELVES THAT IT'S BECAUSE OF ITS ORIGINALITY OF SUBJECT MATTER.
ROPOSAL FOR A GRAPHIC NOVEL
AND I THINK IT'S BECAUSE THE WORLD IS FINALLY READY TO NOTICE THAT NONFICTION IS MADE EASIER TO UNDERSTAND – AND MORE FUN TO READ – AS COMICS!

HEY, IT'D ALL BE FOR NOTHING WITHOUT THE ART!

UTA
CHRIS
BUT YOU CAN BE CERTAIN THAT IT ONLY CAME TO BE IN THE FIRST PLACE...
...AND ONLY STOOD A CHANCE OF FINDING AN AUDIENCE...
...BY TRADING ON THE REPUTATIONS OF TWO SCIENTISTS.

UTA FRITH
Date of PhD: 1969
Publications: 219
Impact Factor: 116
Reputation: 99
CHRIS FRITH
Date of PhD: 1969
Publications: 565
Impact Factor: 231
Reputation: 99
EXCUSE ME FOR A MOMENT WHILE I BOAST YET MORE ABOUT MY PARENTS' ACCOLADES. I'M USING THEM TO MAKE A POINT, I PROMISE.

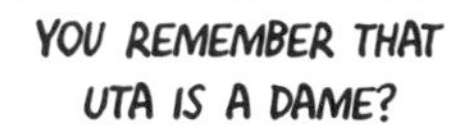
YOU REMEMBER THAT UTA IS A DAME?

SHE HAS A "DECORATION" THAT DISPLAYS THIS FACT – AND ALSO HER REPUTATION – TO THE WORLD.

CHRIS DOESN'T HAVE ANY SHINY DECORATIONS – BUT HE DOES HAVE AN INCREASINGLY LARGE COLLECTION OF ACADEMIC GOWNS.

AND THEY BOTH HAVE A SMALL COLLECTION OF HONORARY DEGREES.

CHRIS IS ESPECIALLY PROUD OF HIS BRIEF TIME AT ALL SOULS OXFORD...

...THE COLLEGE SO DEVOTED TO ACADEMIA THAT IT HAS NO TIME FOR UNDERGRADUATES AND TEACHING.

(BUT PLENTY OF TIME FOR FINE DINING)

WITHIN THE SCIENTIFIC COMMUNITY IN THE UK, THE ULTIMATE ACCOLADE IS PERHAPS BEING ELECTED AS A FELLOW OF THE ROYAL SOCIETY.
(THEY'RE BOTH FELLOWS, SO THEY WOULD SAY THIS...)

WHILE TO THE WORLD AT LARGE, OCCASIONAL APPEARANCES ON RADIO AND TV HELP TO BOOST THEIR REPUTATIONS AS EXPERTS.
BBC

THE POINT IS, HAVING A NAME THAT PEOPLE RECOGNIZE MAKES A DIFFERENCE.
WOULD YOU BUY THE BOOK I'M HOLDING?
I MIGHT. BUT I'D WANT TO KNOW, WHAT DOES THIS AURNHAMMER PERSON KNOW ABOUT AUTISM?
WELL, SHE'S ME!* AND I HAVE A STRONG REPUTATION AS AN AUTISM EXPERT.
Autism explained
by U. Aurnhammer
*MY MAIDEN NAME IS AURNHAMMER.

History of science
LET'S TAKE A SMALL DETOUR INTO THE HISTORY OF SCIENCE...

IN PARTICULAR, A STORY FROM THE WORLD OF MEDICINE. IT STARTS IN THE 2ND CENTURY...

...WITH THE SCHOLAR GALEN, WHO WROTE AND DREW BOOKS ON HUMAN ANATOMY. HIS REPUTATION WAS SO STRONG THAT DOCTORS ACROSS THE ROMAN AND ARAB WORLDS DIDN'T BOTHER TO DO THEIR OWN RESEARCH.

I AM THE LAST WORD IN ANATOMY!

THIS BEGAN TO CHANGE IN THE 15TH CENTURY, WHEN DUTCH SCHOLAR VESALIUS GAINED ACCESS TO SOME HUMAN BODIES AND MADE HIS OWN DRAWINGS. HE WAS ABLE TO CORRECT VARIOUS MISTAKES – MISTAKES THAT EMERGED BECAUSE, IT TURNED OUT, GALEN'S WORK WAS DERIVED IN LARGE PART FROM DISSECTIONS OF MONKEYS AND PIGS.

MEDIEVAL VERSION OF A SKELETON, DRAWN BY ARAB SCHOLARS LEARNING FROM GALEN

VESALIUS'S VERSION OF A SKELETON, DRAWN FROM AN ACTUAL SKELETON

VESALIUS WAS PART OF A WAVE OF THINKERS WHO KICK-STARTED MODERN SCIENCE BY FINDING THINGS OUT FOR THEMSELVES, AND VERIFYING THEM BY EXPERIMENTS – A WAVE BEGUN BY THE SAME ARAB SCHOLARS* WHO HAD KEPT GALEN IN PRINT. PLEASE NOTE, THE POINT HERE IS NOT TO LAUGH AT THE MISTAKES GALEN MADE, OR EVEN TO CRITICIZE THE PEOPLE WHO WERE SEEMINGLY SLAVISH TO HIS WORK.

*IBN SINA, IBN AL-HAYTHAM, JABIR IBN HAYYAN, IBN AL-NAFIS, ABU BAKR AL-RAZI, TO NAME JUST A FEW.

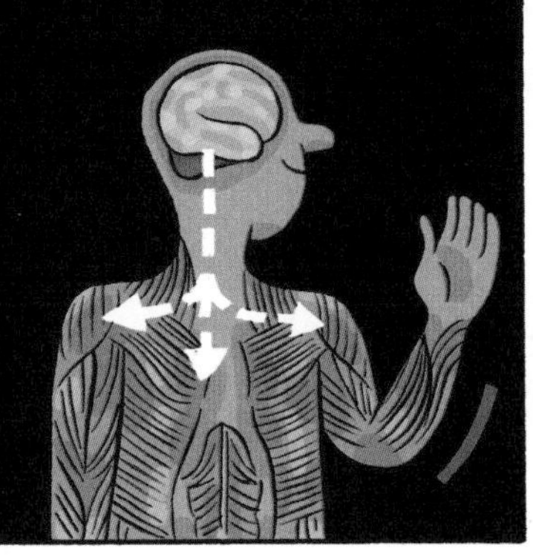
IN FACT, GALEN'S WORK WAS CORRECT IN A LOT OF PRACTICAL DETAILS – FOR EXAMPLE, THAT THE BRAIN CONTROLS MUSCLES.

LIKEWISE, CHOOSING TO LEARN FROM A RESPECTED AUTHORITY SERVED MANY DOCTORS, ESPECIALLY GIVEN THAT THEIR OWN OPTIONS FOR STUDY WERE LIMITED BY WIDELY HELD TABOOS AGAINST CUTTING UP CADAVERS.
AVICENNA! THESE ETHICS COMMITTEES – TSCH.
I HEAR YOU, RHAZES.

MODERN SCIENCE, OF COURSE, IS ENTIRELY BASED ON A MIXTURE OF THOROUGH LITERATURE REVIEWS, ORIGINAL THOUGHTS, AND REPEATED EXPERIMENTAL VERIFICATION.

THIS RECALLS THE MOTTO OF THE ROYAL SOCIETY: NULLIUS IN VERBA, "TAKE NO ONE'S WORD FOR IT."
NULLIUS
IN
VERBA

IT HAS NO BASIS IN BLINDLY FOLLOWING WORK BY RESPECTED AUTHORITY FIGURES.

WE'RE REPEATING THE SILENT, DEADPAN PANEL HERE TO ASSURE YOU THAT WE ARE INDEED BEING SARCASTIC.*
HOPEFULLY, SCIENTISTS WITH GOOD REPUTATIONS HAVE EARNED THOSE REPUTATIONS THROUGH THEIR WORK.
*SURPRISINGLY HARD TO CONVEY IN COMICS.

THERE'S EVEN AN OBJECTIVE WAY TO MEASURE THIS LEVEL OF REPUTATION AMONG SCIENTISTS.
COLD
HOT
REPUTATION METER

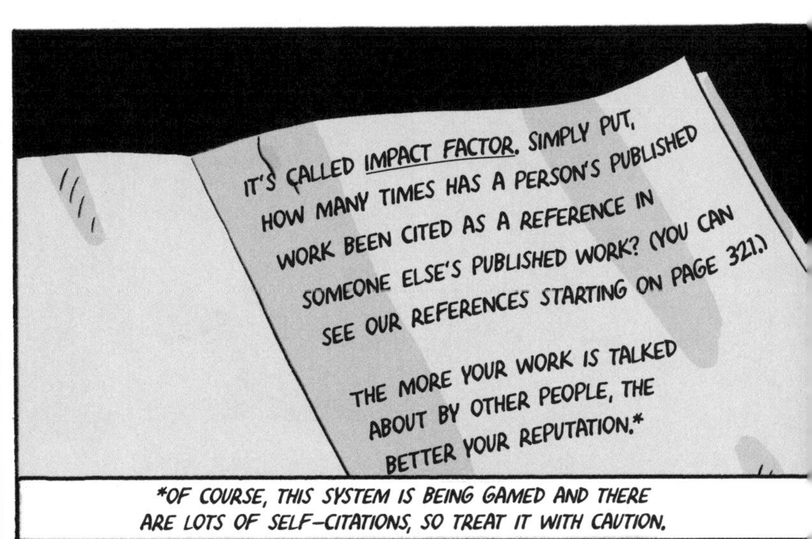
IT'S CALLED IMPACT FACTOR. SIMPLY PUT, HOW MANY TIMES HAS A PERSON'S PUBLISHED WORK BEEN CITED AS A REFERENCE IN SOMEONE ELSE'S PUBLISHED WORK? (YOU CAN SEE OUR REFERENCES STARTING ON PAGE 321.)
THE MORE YOUR WORK IS TALKED ABOUT BY OTHER PEOPLE, THE BETTER YOUR REPUTATION.*

*OF COURSE, THIS SYSTEM IS BEING GAMED AND THERE ARE LOTS OF SELF-CITATIONS, SO TREAT IT WITH CAUTION.

LET'S GO BACK TO TALKING ABOUT BRAINS.

BACK IN THE CONVERSATION ABOUT IN-GROUPS AND OUT-GROUPS, COMPETITION WAS KEY.

A SUCCESSFUL GROUP DEPENDS ON ITS MEMBERS COOPERATING WHEN COMPETING WITH A RIVAL GROUP.

BUT WHEN INDIVIDUALS WITHIN A GROUP ARE COMPETING, COOPERATION IS NOT WANTED. IT'S WINNER TAKES ALL!

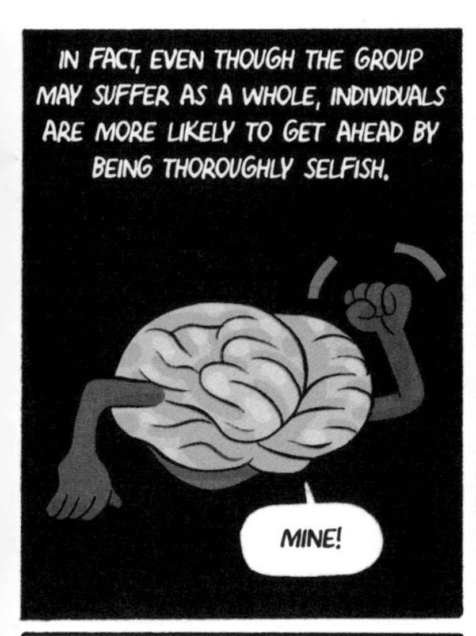
IN FACT, EVEN THOUGH THE GROUP MAY SUFFER AS A WHOLE, INDIVIDUALS ARE MORE LIKELY TO GET AHEAD BY BEING THOROUGHLY SELFISH.
MINE!

AND THE ONES WHO REALLY GET AHEAD ARE PEOPLE WHO CAN MAKE OTHERS THINK THEY ARE COOPERATIVE...
YOURS!

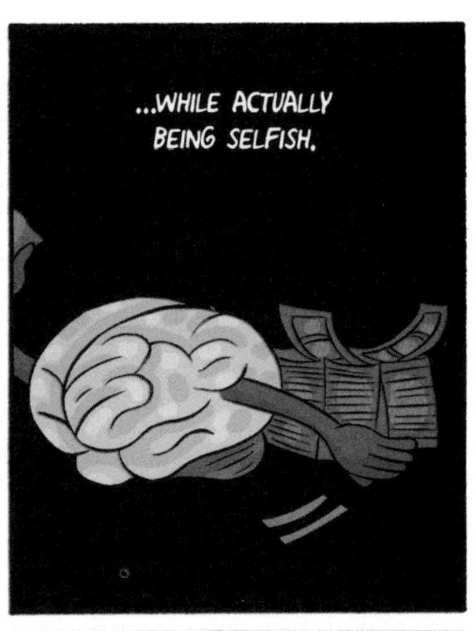
...WHILE ACTUALLY BEING SELFISH.

SUCH PEOPLE ARE DESCRIBED (IN ACADEMIC PAPERS, ANYWAY) AS FREE RIDERS.

THESE ARE THE PEOPLE WHO CONSISTENTLY "FORGET" TO PUT MONEY INTO THE OFFICE TEA/BISCUIT COLLECTION...

...BUT STILL CONSUME PLENTY OF TEA AND BISCUITS.
A PROBLEM, INCIDENTALLY, THAT CAN PARTLY BE SOLVED...

...BY STICKING A PICTURE OF A PAIR OF EYES ABOVE THE COLLECTION JAR.
(JUST THE IDEA OF BEING WATCHED IS ENOUGH TO MAKE MOST PEOPLE PROTECT THEIR REPUTATION BY CONFORMING.)*
*SO MANY PSYCHOLOGISTS KNOW ABOUT THIS EXPERIMENT THAT IT PROBABLY DOESN'T WORK ANYMORE. (IN PSYCHOLOGY DEPARTMENT BREAK ROOMS, ANYWAY.)

THE HOWS AND WHYS OF FREE RIDERS HAVE BEEN STUDIED MORE OR LESS EXTENSIVELY BY TWO OTHER ACADEMIC DISCIPLINES, FAR FROM THE WORLD OF NEUROSCIENCE:

ZOOLOGY AND ECONOMICS.

ONE OF THE MORE STRAIGHTFORWARD ZOOLOGY EXAMPLES CONCERNS FISH.
(BASED ON RESEARCH BY BEHAVIORAL ECOLOGIST REDOUAN BSHARY.)

THIS IS A LABROIDES DIMIDIATUS, OR BLUESTREAK CLEANER WRASSE, A SMALL FISH FOUND IN REEFS.

IN EXCHANGE FOR A MEASURE OF PROTECTION, IT AGREES TO CLEAN OFF LITTLE PARASITES (WHICH IS FOOD FOR THE WRASSE, ANYWAY) FROM THE BACKS OF CLIENTS – VARIOUS BIG FISH.

SOUNDS GOOD FOR THE CLIENT, RIGHT? BUT IT'S NOT THAT SIMPLE.

BECAUSE ALTHOUGH THE WRASSE WILL EAT THE PARASITES, WHAT THEY REALLY LIKE IS EATING THE MUCUS SECRETED BY THE SKIN OF THEIR CLIENTS.

AND TO GET AT THAT MUCUS, THEY HAVE TO BITE THE FISH...

...WHICH IS GOING TO GET THEM THROWN OFF THE CLIENT PRETTY QUICK, AND MAYBE EVEN HUNTED DOWN.

SO, THE WRASSE HAVE TO DEVELOP A REPUTATION AS TRUSTWORTHY CLEANERS.
FLUTTER

WAVE

IDEALLY, A CLIENT WANTS TO SEE A WRASSE FINISHING A PREVIOUS CLEANING JOB, MAKING IT OBVIOUS IF THE CURRENT CLIENT WAS HAPPY OR NOT.

OTHERWISE, CLIENTS WATCH WRASSES PERFORMING A SORT OF DANCE, AN INTERSPECIES ADVERTISEMENT OF REPUTATION.

ARGUABLY, THE MOST SUCCESSFUL WRASSES ARE THE ONES THAT CAN SELL A GOOD REPUTATION...
MARK OF A DAME AMONG WRASSE

...WHILE ACTUALLY PROCEEDING TO BITE THEIR CLIENTS TO GET SOME OF THAT PRECIOUS MUCUS.
THESE ARE THE FREE RIDERS.

IF THERE ARE TOO MANY FREE RIDERS IN A SCHOOL OF WRASSE, EVENTUALLY THE SCHOOL AS A WHOLE WILL FIND IT HARD TO GET CLIENTS.

WHAT ABOUT HUMANS?
LET'S PLAY SOME MORE ECONOMICS GAMES.

IT'S POSSIBLE TO SCAN PAIRS OF PEOPLE WHILE THEY PLAY SO-CALLED COOPERATION GAMES.
GAMES THAT INVOLVE EACH PERSON TRYING TO GUESS WHAT THE OTHER IS GOING TO DO.

ONE EXAMPLE IS CALLED THE TRUST GAME, DEVELOPED BY ACCOUNTING PROFESSOR JOYCE BERG.

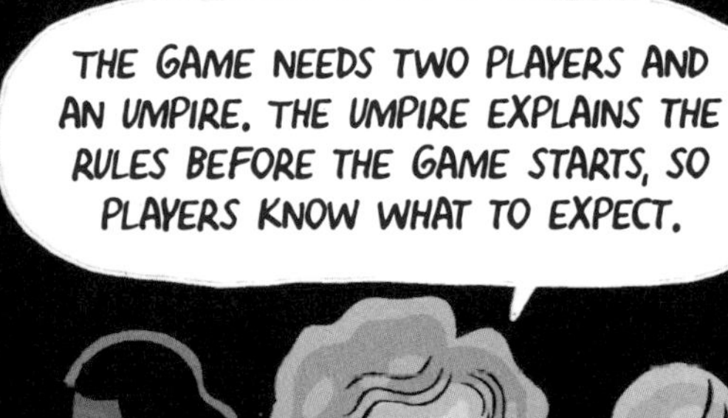
THE GAME NEEDS TWO PLAYERS AND AN UMPIRE. THE UMPIRE EXPLAINS THE RULES BEFORE THE GAME STARTS, SO PLAYERS KNOW WHAT TO EXPECT.

PLAYER ONE, I WILL GIVE YOU A POT OF MONEY. YOU CAN GIVE AS MUCH (OR AS LITTLE) AS YOU LIKE TO ME...
...THANK YOU.

RIGHT, NOW I'M GOING TO TRIPLE THE AMOUNT YOU GAVE ME AND GIVE IT TO PLAYER TWO.*
*BOTH PLAYERS KNOW IN ADVANCE THAT THIS IS WHAT WILL HAPPEN.

NOW, PLAYER TWO, YOU CAN GIVE AS MUCH (OR AS LITTLE) FROM YOUR POT BACK TO ME...

...AND THIS TIME I'LL TRIPLE THE AMOUNT AND GIVE IT TO PLAYER ONE.

CLASSIC ECONOMIC THEORY PREDICTED THAT PLAYER ONE WOULD GIVE NO MONEY, AND WOULD GET NO MONEY BACK.

BUT BERG AND OTHERS HAVE FOUND ONLY ABOUT 11% OF PLAYER ONES GIVE NO MONEY. IN FACT, MOST PLAYER ONES GIVE MORE THAN HALF OF WHAT THEY HAVE.

AND MOST PLAYER TWOS ARE EVEN MORE GENEROUS. IN FACT, THE GAME DOESN'T HAVE TO STOP AT ONE ROUND. SOME EXPERIMENTS USE MULTIPLE PLAYERS, MIXING UP THE PAIRS EACH TIME. IN THIS VERSION, PLAYERS LEARN WHO WILL BE GENEROUS, AND WHO WON'T BE.

NEUROSCIENTISTS HAVE TRIED THE GAME OUT ON PEOPLE WHILE SCANNING THEIR BRAINS.

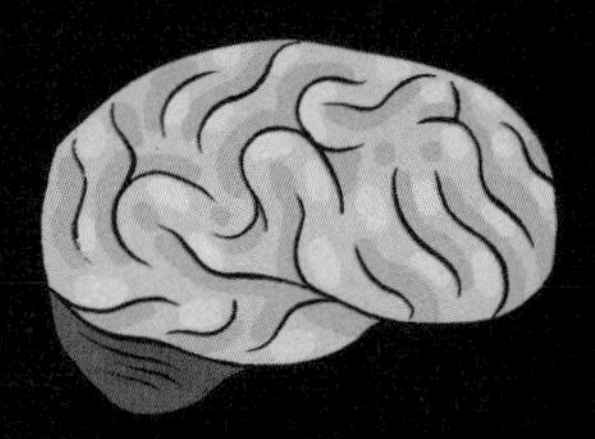

TWO CONTRASTING AREAS SHOWED THE MOST ACTIVITY:

THE CAUDATE NUCLEUS...

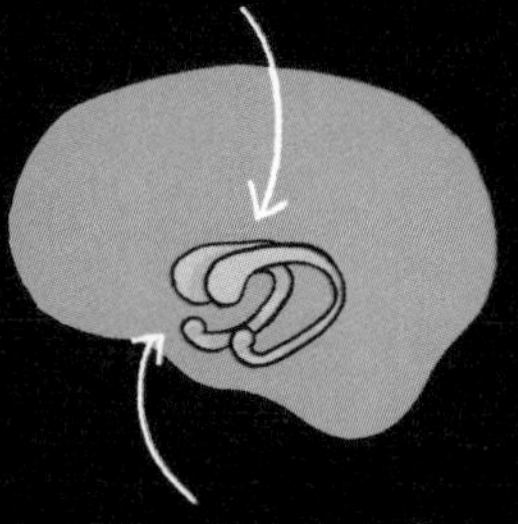

...AND THE AMYGDALA.

THE CAUDATE NUCLEUS IS PART OF THE BRAIN'S REWARD/LEARNING MECHANISM. AFTER EVERY ITERATION OF THE GAME, IF YOUR OPPOSITE NUMBER IS MORE GENEROUS THAN YOU EXPECTED, THE NUCLEUS ACTIVATES.

IN OTHER WORDS, YOU GET A SIGNAL TELLING YOU THIS PERSON IS A GOOD AND TRUSTWORTHY PERSON.
THEY EARN A GOOD REPUTATION.

MEANWHILE, THE AMYGDALA ACTIVATES WHEN YOU ENCOUNTER SOMETHING YOU SHOULD AVOID – SUCH AS A PERSON WHO CONTINUALLY GIVES NO MONEY. THEY EARN A BAD REPUTATION.

ECONOMISTS MAY BE SURPRISED THAT MOST PEOPLE ARE INSTINCTIVELY NICE.
PSYCHOLOGISTS AREN'T SURPRISED.

LEARNING TO WIN GAMES LIKE THIS IS A MATTER OF BUILDING UP YOUR OWN, AND TESTING EACH OTHER'S, REPUTATIONS.
IF YOU HAVE A GOOD REPUTATION, YOU'LL GET MORE MONEY.

YOU ONLY HAVE TO LOOK AT THE EXAMPLE OF CHILDREN. THEY WORRY ABOUT THEIR REPUTATIONS TOO, BUT FAR, FAR LESS.

ONE STUDY INVOLVED HUNDREDS OF CHILDREN, AGED EITHER 3–4 OR 11–12.

EACH CHILD WAS GIVEN THE CHOICE OF DOING TWO ACTIVITIES:

SINGING AND DANCING...OR SITTING QUIETLY AND DOING SOME COLORING.

159 OF THE 315 YOUNGER CHILDREN ELECTED TO SING AND DANCE.

NOT A SINGLE ONE OF THE 11- OR 12-YEAR-OLDS MADE THAT CHOICE.

AND YES, THE REASON SEEMS TO BE THE OBVIOUS ONE:
FEAR OF EMBARRASSMENT.

I ENJOY SINGING AND DANCING VERY MUCH.
I'M QUITE HAPPY TO BE SEEN DANCING IN A DRAWING...

CENSORED
...BUT YOU'LL NEVER SEE ME DO IT IN REAL LIFE – A PRIVILEGE ONLY ENJOYED (THESE DAYS) BY MY GRANDCHILDREN.

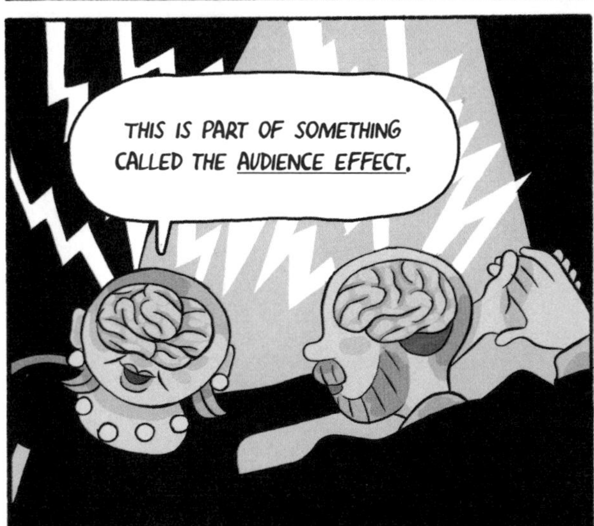
THIS IS PART OF SOMETHING CALLED THE AUDIENCE EFFECT.

CHILDREN AS YOUNG AS 10 MONTHS CAN BE SEEN ADJUSTING THEIR BEHAVIOR WHEN THEY KNOW THERE'S AN AUDIENCE WATCHING.

FOR EXAMPLE BY SMILING, OR BY SMILING IN A DIFFERENT WAY.

WE ALL KNOW WE BEHAVE DIFFERENTLY WHEN THERE'S AN AUDIENCE, BUT THERE'S A LOT WE DON'T KNOW.

JUST HOW EARLY DOES THE AUDIENCE EFFECT START TO APPEAR IN YOUNG CHILDREN, AND HOW DO THEY LEARN IT?

IS IT SOMETHING THAT DOESN'T AFFECT CERTAIN PEOPLE IN THE SAME WAY, FOR EXAMPLE THOSE WITH AUTISM?

HOW DOES THE AUDIENCE EFFECT WORK WHEN WE CAN'T DIRECTLY SEE OUR AUDIENCES?
YES, WE'RE TALKING ABOUT YOU!*
*AND YES, WE'RE ASSUMING YOU SEE US AS PART OF YOUR IN-GROUP.

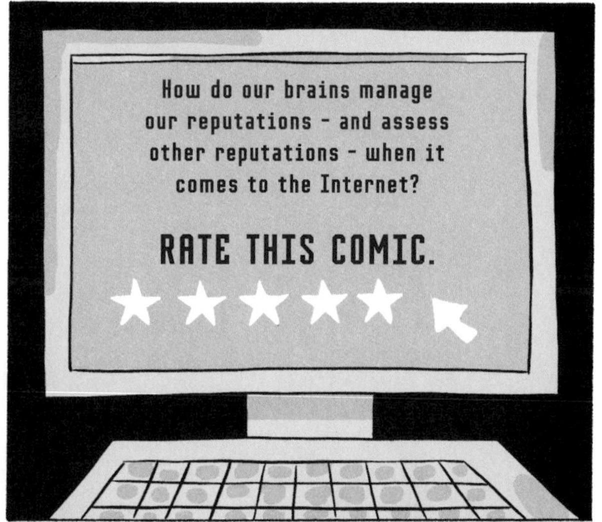
How do our brains manage our reputations - and assess other reputations - when it comes to the Internet?
RATE THIS COMIC.

TWITTER, AS WITH MOST FORMS OF SOCIAL MEDIA, HAS A WEIRDLY ADDICTIVE QUALITY THAT IS BASED IN REPUTATION MANAGEMENT.

TWITTER IS ALSO WELL-PLACED TO TRICK OUR BAYESIAN BRAINS, BECAUSE IT CAN MAKE IT APPEAR THAT INFORMATION HAS COME FROM MULTIPLE SOURCES – WHEN IN FACT IT'S JUST ONE IDEA RETWEETED MULTIPLE TIMES.

HOW MANY FOLLOWERS DO I HAVE NOW?

HOW MANY TIMES HAS MY LATEST POST BEEN RETWEETED?

GOD FORBID ANY ATTEMPTS TO BE FUNNY ON TWITTER.
WHY HAVEN'T ALL MY FOLLOWERS IMMEDIATELY LIKED MY HILARIOUS POST?*
*ALTHOUGH, SINCE I'M A WOMAN AND A GERMAN, THE ENGLISH-SPEAKING WORLD DOESN'T BELIEVE I'M CAPABLE OF HUMOR...

SOCIAL MEDIA SITES SUCH AS TWITTER ARE, IN THEORY, A SIMPLE AND DIRECT SOURCE OF NEWS AND INFORMATION.

BUT OF COURSE THEY'RE ONLY AS RELIABLE AS THE PEOPLE POSTING THE INFORMATION.

IT'S MORE SIMILAR TO GOSSIP, SOMETHING THAT HAS ALWAYS DOMINATED HUMAN INTERACTION.

AS TWO COLLEAGUES WHO WORK IN THE SAME FIELD...
...BUT RARELY IN THE SAME TEAMS OR EVEN BUILDINGS...

...WE'VE HAD PLENTY OF FUEL FOR GOSSIP ABOUT PEOPLE WE BOTH KNOW.
AND WE DID GOSSIP...

...BUT THE TRUTH IS THAT ALL THE PEOPLE WE'VE EVER WORKED WITH ARE PERFECTLY LOVELY.*
WINK
WINK
*SORRY IF THAT'S DISAPPOINTING TO HEAR!

NO TELL-ALL BIOGRAPHY ABOUT VILLAINS AND CHARLATANS IN THE REALM OF CONTEMPORARY NEUROSCIENCE EXISTS TO BE SHARED.

THESE TWO FACTS MADE OUR LIFE EASIER, BUT IT'S ALSO TRUE THAT BEING ABLE TO GOSSIP WITH EACH OTHER HAS HELPED OUR MARRIAGE TO ENDURE AND EVEN THRIVE.

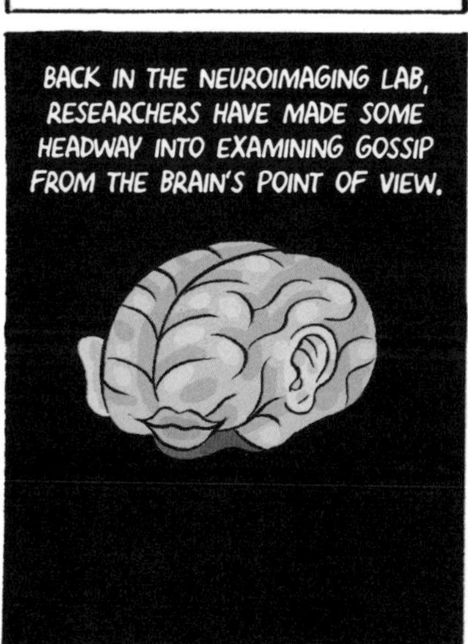
BACK IN THE NEUROIMAGING LAB, RESEARCHERS HAVE MADE SOME HEADWAY INTO EXAMINING GOSSIP FROM THE BRAIN'S POINT OF VIEW.

IT TURNS OUT THAT SIGNALS WE GET FROM OTHERS, NOT LEAST GOSSIP ABOUT A PERSON, WILL SWAY OUR BRAINS MORE...
...THAN OUR OWN OBSERVATIONS OF THAT SAME PERSON.

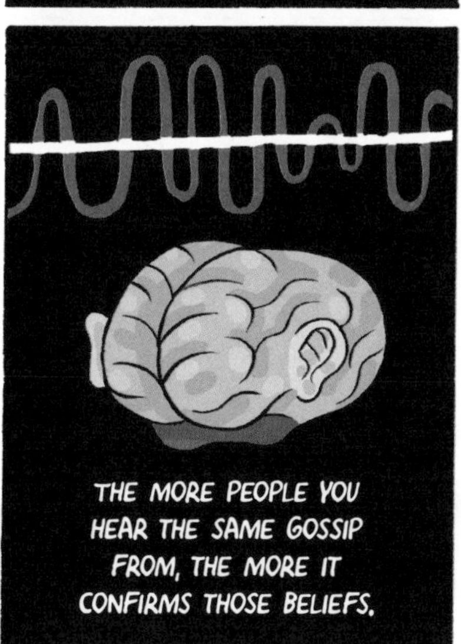
THE MORE PEOPLE YOU HEAR THE SAME GOSSIP FROM, THE MORE IT CONFIRMS THOSE BELIEFS.

ON THE ONE HAND, THIS SEEMS OUTRAGEOUS AND UNFAIR.

IF I SEE A PERSON BEHAVING WELL, WHY SHOULD I BELIEVE OTHER PEOPLE WHO TELL ME THEY'RE BAD?
WELL...PERHAPS THEY KNOW SOMETHING I DON'T KNOW?

ON THE OTHER HAND, MOST GOSSIP ENDS UP COMING FROM MANY PEOPLE.
SO, ARGUABLY, YOUR BRAIN IS JUDGING THAT INFORMATION GLEANED FROM SEVERAL SOURCES IS MORE RELIABLE THAN INFORMATION FROM A SINGLE SOURCE (IN THIS CASE, YOURSELF).

DO YOU REMEMBER THE CAUDATE NUCLEUS? THIS IS PART OF YOUR BRAIN THAT NOTICES AND REMEMBERS WHEN A PERSON IS DOING SOMETHING NICE FOR YOU.

OR AT LEAST, IT DOES WHEN YOU'RE PLAYING THE TRUST GAME.
A 2005 EXPERIMENT TRIED A VARIATION ON THE GAME, ADDING AN ELEMENT OF GOSSIP.

PEOPLE TAKING PART WERE GIVEN INFORMATION SHEETS ABOUT THEIR OPPOSITE NUMBER BEFORE PLAYING.
TOP TRUMPS
RECIPROCATION 2
3
TRUSTWORTHY 2
1

THE INFORMATION SHEETS EITHER PUSHED PEOPLE AS VERY TRUSTWORTHY, OR AS UNTRUSTWORTHY.
TOP TRUMPS
RECIPROCATION 4
5
TRUSTWORTHY 8
REPUTATION 7

NOT SURPRISINGLY, THIS AFFECTED HOW PEOPLE PLAYED THE GAME: THEY WERE MORE GENEROUS WITH TRUSTWORTHY PEOPLE.

WHAT WAS MORE INTERESTING WAS THAT THE BRAIN ITSELF BEHAVED DIFFERENTLY.

THE CAUDATE NUCLEUS SHOWED FAR LESS ACTIVITY – IT SEEMS PEOPLE WERE RELYING ON THE GOSSIP, RATHER THAN ON BUILDING UP THEIR OWN PICTURE DURING THE GAME.

WHEN YOU ENCOUNTER PEOPLE WHO TELL YOU THEY NEVER PAY ATTENTION TO GOSSIP, THEY MAY MEAN WELL...
...BUT THEY'RE LYING TO THEMSELVES.

A STUDY BY A JOINT TEAM OF MATHEMATICIANS AND EVOLUTIONARY BIOLOGISTS LOOKED INTO WHY GOSSIP IS SO IMPORTANT.

COULD IT BE THAT GOSSIP, USUALLY SEEN AS SOMETHING NAUGHTY, MAY ACTUALLY BE VITAL TO SUCCESSFUL GROUP COOPERATION?

THE STUDY INVOLVED MULTIPLE GROUPS OF PEOPLE, WHO STARTED BY WATCHING EACH OTHER PLAYING A TYPE OF TRUST GAME.

AFTER WATCHING SEVERAL ROUNDS, THEY GOSSIPED WITH EACH OTHER ABOUT THE WAY THE PLAYERS BEHAVED DURING THE GAME.
HE'S STINGY
SHE'S GENEROUS

THE EXPERIMENTERS* THEN OBSERVED THE GROUP AS A WHOLE PLAYING ROUNDS OF THE TRUST GAME WITH EACH OTHER. THEY TOOK CAREFUL NOTE OF HOW EACH PERSON MADE DECISIONS, BASED ON WHAT THE GOSSIP HAD SAID ABOUT THEIR OPPONENTS.
*THEY DON'T ACTUALLY WEAR LAB COATS, IT'S JUST A COMICS SHORTHAND FOR "SCIENTIST."

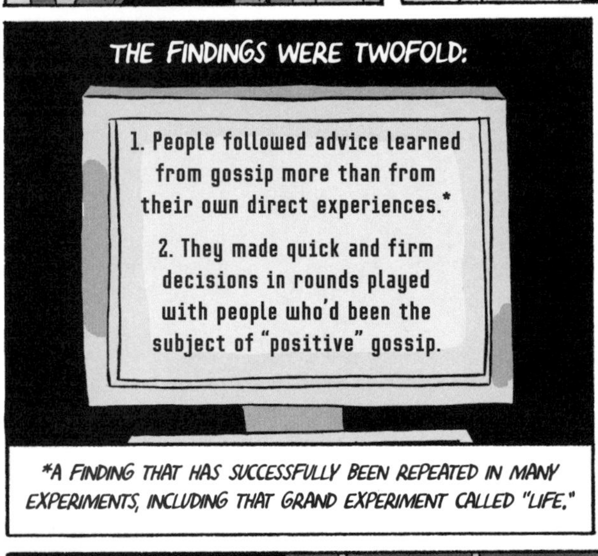
THE FINDINGS WERE TWOFOLD:
1. People followed advice learned from gossip more than from their own direct experiences.*
2. They made quick and firm decisions in rounds played with people who'd been the subject of "positive" gossip.
*A FINDING THAT HAS SUCCESSFULLY BEEN REPEATED IN MANY EXPERIMENTS, INCLUDING THAT GRAND EXPERIMENT CALLED "LIFE."

POINT 2 SUGGESTS THAT WE FIND IT EASY TO REINFORCE A GOOD OPINION OF A PERSON, WHICH IS NICE.

THIS IS THE SORT OF FINDING THAT IS NOT ENTIRELY SURPRISING, BUT YOU COULDN'T BE SURE OF WITHOUT DOING THE SCIENCE!

IT SEEMS THAT, TO GET PEOPLE TO TRUST YOU, YOU HAVE TO PAY VARIOUS COSTS.
THE BANK OF REPUTATION

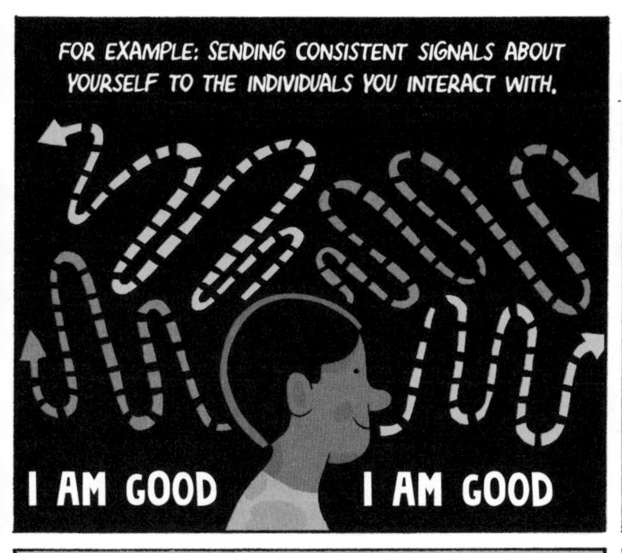
FOR EXAMPLE: SENDING CONSISTENT SIGNALS ABOUT YOURSELF TO THE INDIVIDUALS YOU INTERACT WITH.
I AM GOOD
I AM GOOD

AND, ABOVE ALL, MAKING SURE THAT THESE SIGNALS ARE SO CONSISTENT THAT WHEN PEOPLE TALK ABOUT YOU BEHIND YOUR BACK, THEY'RE GOING TO SAY NICE THINGS.

HE IS GOOD HE IS GOOD

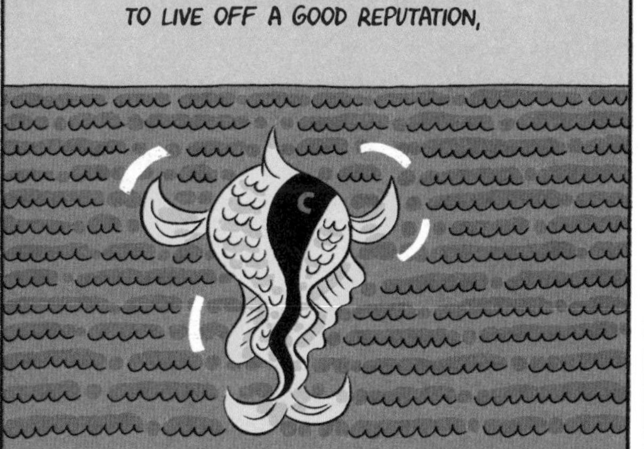
IT'S PROBABLY ALSO TRUE IN NATURE: MAKING IT EASIER FOR CLEANER WRASSE TO LIVE OFF A GOOD REPUTATION,

AND FOR SCOUT BEES TO PERSUADE OTHERS WITH THEIR DANCES.

BUT EVEN WITH ALL THAT EFFORT, IT'S NOT POSSIBLE TO COMPLETELY CONTROL YOUR OWN REPUTATION.
FOR INSTANCE, CONSIDER THE POWER OF THE PRESS.
IF YOU SEE SOMETHING IN PRINT, YOU TRUST IT MORE.*
IT'S AN UNDEREXPLORED AREA; WE AWAIT FURTHER RESEARCH!

*DON'T YOU?

LET'S GO ALL THE WAY BACK TO CHAPTER 1, AND OUR NEURONS.

ARGUABLY, EVEN OUR OWN NEURONS WORK A LITTLE LIKE REPUTATION ASSESSORS.
THEY HAVE TO CHANNEL SIGNALS COMING FROM SURROUNDING NEURONS, AND DETERMINE WHICH ONES TO PASS ALONG, BASED ON RELIABILITY.

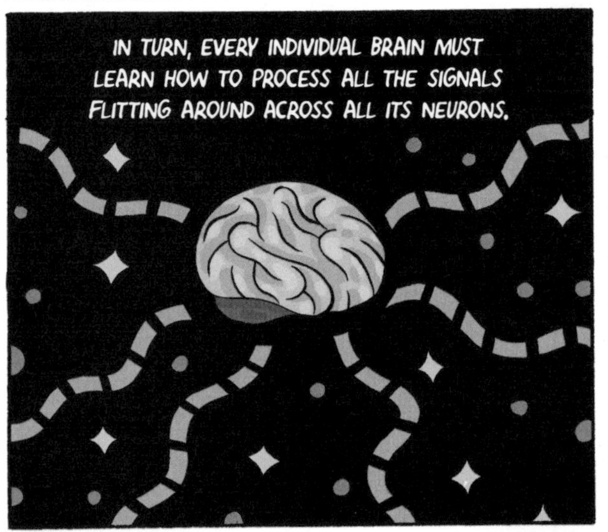
IN TURN, EVERY INDIVIDUAL BRAIN MUST LEARN HOW TO PROCESS ALL THE SIGNALS FLITTING AROUND ACROSS ALL ITS NEURONS.

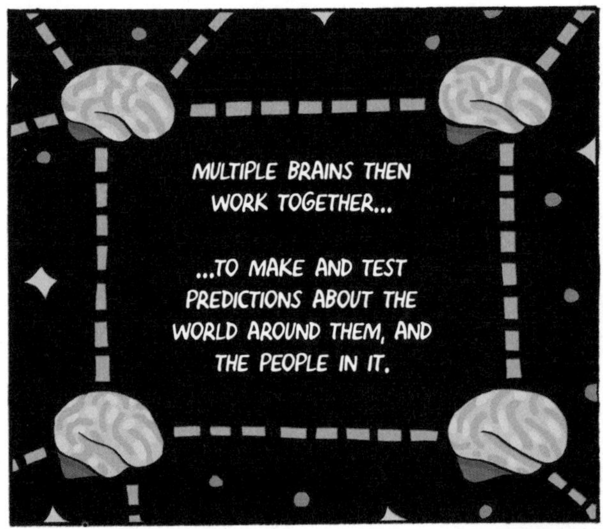
MULTIPLE BRAINS THEN WORK TOGETHER...
...TO MAKE AND TEST PREDICTIONS ABOUT THE WORLD AROUND THEM, AND THE PEOPLE IN IT.

SOME OF THIS IS DONE WITH FULL, CONSCIOUS AWARENESS.
MUCH OF IT HAPPENS BY AUTOMATIC INSTINCT.
IT'S ALL PART OF SOCIAL COGNITION. WE THINK COOPERATION WITH OTHERS IS ONE OF THE CENTRAL THINGS OUR BRAINS ARE DESIGNED TO DO.
ARGUABLY IT'S MORE IMPORTANT THAN THE JOB OF SENSING AND MANAGING WHAT OUR BODIES ARE DOING – IF YOU'RE TALKING ABOUT WHAT GIVES LIFE MEANING.

I HOPE BY NOW WE'VE BUILT UP A COMPELLING PICTURE OF HOW YOUR BRAIN LIVES AND THRIVES IN A WORLD OF OTHER BRAINS.
WE'D LIKE TO THINK WE'VE HELPED PUSH AGAINST ANY DUALIST INSTINCTS – YOUR MIND AND YOUR BRAIN ARE NOT SEPARATE THINGS. AND "YOU" ARE NOT SEPARATE FROM EITHER OF THEM!
WE MAY HAVE HELPED YOU TO UNDERSTAND WHY YOU BEHAVE SLIGHTLY DIFFERENTLY AROUND DIFFERENT PEOPLE.
THAT'S YOUR BRAIN ALIGNING YOU TO THE OTHER BRAINS AROUND YOU. THEY'RE NOT ALL THE SAME!
MAYBE WE'VE ENCOURAGED YOU TO THINK MORE ABOUT WHAT YOU'RE THINKING ABOUT...
...AND, EVEN MORE, TO THINK ABOUT WHAT OTHER PEOPLE ARE THINKING ABOUT.
(BUT NOT TOO MUCH – WE DON'T WANT YOU TO GET TRAPPED IN A CYCLE OF WORRYING!)

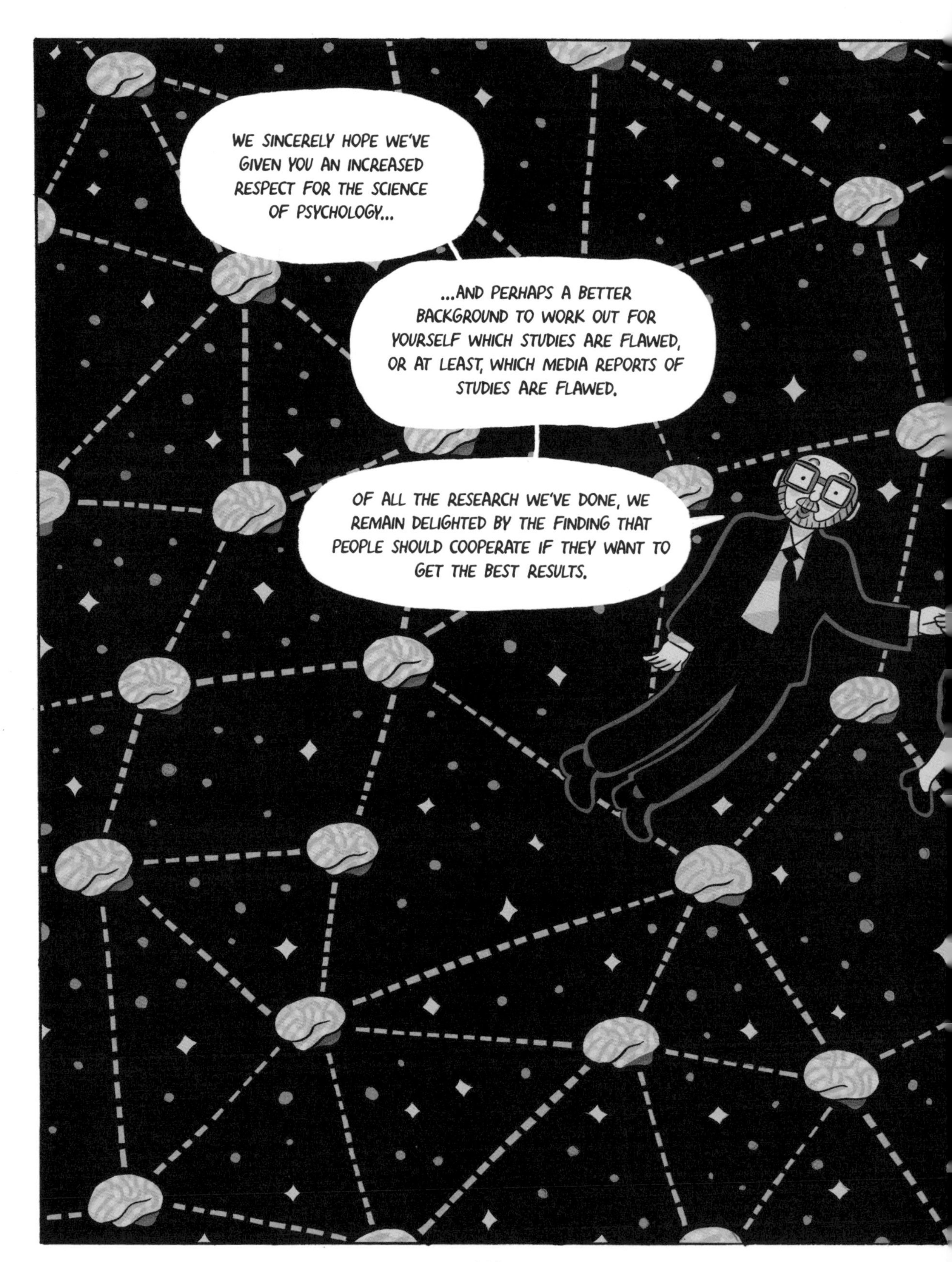
WE SINCERELY HOPE WE'VE GIVEN YOU AN INCREASED RESPECT FOR THE SCIENCE OF PSYCHOLOGY...
...AND PERHAPS A BETTER BACKGROUND TO WORK OUT FOR YOURSELF WHICH STUDIES ARE FLAWED, OR AT LEAST, WHICH MEDIA REPORTS OF STUDIES ARE FLAWED.
OF ALL THE RESEARCH WE'VE DONE, WE REMAIN DELIGHTED BY THE FINDING THAT PEOPLE SHOULD COOPERATE IF THEY WANT TO GET THE BEST RESULTS.

AND THAT THE ULTIMATE EXPRESSION OF COOPERATION IS TO COME TOGETHER IN GROUPS THAT REFLECT THE DIVERSITY OF THE WORLD.
YES, THIS MEANS DIVERSITY OF GENDER, CULTURE, FAMILY HISTORY, EDUCATION, AND SO ON –
BUT WE'RE REALLY TALKING ABOUT DIVERSITY OF HOW OUR MINDS WORK (WHICH IS OF COURSE SHAPED BY ALL THOSE OTHER THINGS).
TO PROGRESS HUMAN KNOWLEDGE, WE NEED TO HARNESS THE MIXTURES OF EXPLOITERS AND EXPLORERS, OF RISK TAKERS AND DATA ANALYSTS, OF INTROVERTS AND EXTROVERTS, OF BIG THINKERS AND CURIOUS READERS.

Epilogue

In which the Friths host a party.

FOR THE LAST 30 YEARS, WE'VE HOSTED A PARTY AT OUR HOUSE ON NEW YEAR'S DAY.
THE GUESTLIST CHANGES, BUT ONE THING'S ALWAYS THE SAME – THERE'S AN ABUNDANCE OF HOME-BAKED GERMAN CAKES!

OF COURSE, IT'S MOSTLY AN EXCUSE TO CATCH UP WITH OLD FRIENDS.
BUT IT ALWAYS PUTS OUR SOCIAL BRAINS TO THE TEST.
CAN WE REMEMBER EVERYONE'S PARTNERS' NAMES?
HAVE WE INTRODUCED THE RIGHT PEOPLE TO EACH OTHER?
CAN WE KEEP AN EYE ON OUR PERSONAL SPACE, SO AS NOT TO BUMP INTO ANYONE...
...BUT ALSO KEEP OUR "WE MODE" PRIMED, SO WE DON'T FORGET TO OFFER PEOPLE FOOD AND DRINK THAT'S OUT OF THEIR REACH?
BAYES

CAN WE KEEP A DIVERSE CROWD OF PEOPLE EQUALLY AT EASE...
...BUT ALSO REVEL IN THE FACT THAT THEY APPROACH THEIR LIFE AND WORK IN THEIR OWN WAYS, AND EACH HAS VERY DIFFERENT STORIES TO TELL?
YES, 75% OF THEM ARE IN PSYCHOLOGY, AND MOST IF NOT ALL ARE WEIRD, BUT THEY HAIL FROM 14 DIFFERENT COUNTRIES, AND AGES RANGE FROM 0–100 IN A SURPRISINGLY EVEN SPREAD, MOST YEARS.
F
B

IT'S NEVER POSSIBLE TO GET ALL OF OUR FRIENDS TO THESE PARTIES.

AND THERE ARE TOO MANY PEOPLE MISSING FROM THIS BOOK, TOO.

WE SINCERELY HOPE THAT IT'S OBVIOUS BY THIS POINT THAT ALL OF OUR ACHIEVEMENTS IN OUR LIFE AND WORK ARE THE RESULT OF COLLABORATION.

LOOK, HERE'S THE THING: EVERYONE WANTS TO DO THINGS IN LIFE WELL.
AND WE NOW HAVE SCIENTIFIC PROOF THAT PEOPLE ACHIEVE THE BEST RESULTS BY WORKING TOGETHER.

WE'RE A LONG WAY FROM UNDERSTANDING ALL THE DETAILS, BUT HERE'S WHAT WE DO KNOW:

PEOPLE IN PAIRS WORK BETTER THAN PEOPLE WORKING ALONE.
OF COURSE, THERE ARE CAVEATS.

TO ACHIEVE THE BEST RESULTS, PAIRS OF PEOPLE SHOULD BE ROUGHLY EQUALLY MATCHED IN THEIR ABILITY TO DO A TASK...
...AND IN THEIR ABILITY TO ASSESS AND DESCRIBE THEIR OWN CONFIDENCE IN HOW GOOD THEY ARE AT IT.

IN THEORY, WE'RE TALKING ABOUT ANY KIND OF TASK. ALTHOUGH WE COULD BE TALKING ABOUT OUR OWN MARRIAGE!
BUT AS FAR AS THIS BOOK IS CONCERNED, WHAT WE'RE REALLY TALKING ABOUT IS THE TASK OF LEARNING MORE ABOUT NEUROSCIENCE.
WE'VE LEARNED A LOT ABOUT NEUROSCIENCE IN OUR LIFETIMES, INCLUDING ALL SORTS OF THINGS THAT NOBODY KNEW BEFORE.

BUT, AS WITH ANY ACADEMIC DISCIPLINE, THE MORE WE LEARN, THE MORE NEW QUESTIONS ARISE.
WE LIKE TO THINK WE'VE OPENED UP ONE OR TWO DOORS THAT SHED NEW LIGHT ON HOW THE BRAIN WORKS...

...BUT BEHIND EACH DOOR, THERE'S AN ENORMOUS SPACE SCATTERED WITH NEW DOORS JUST WAITING TO BE OPENED.

THE END

Acknowledgments

Special thanks to Hannah Chater and Emily Frith for providing an extra layer of cooperation in the creation of this book, shepherding us to a better result.

Chris and Uta would like to thank:

All those superheroes who provided crucial discussion and help in developing our ideas. Many of them appear in the book, but here we mention just some of those we are particularly sad not to see in a picture:
Frederique de Vignemont (who instigated the idea of this book);
Pierre Jacob (who critiqued on our ideas);
Dan Sperber (who made us think about communication);
Apostolos Doxiadis (whose Logicomix inspired the whole idea of this book);
John Morton (who created the intellectual environment for studying cognition);
Cecilia Heyes (who held us to account on matters of cultural evolution);
Paul Fletcher (who gave advice on schizophrenia and use of ECT);
Dorothy Bishop (who gave advice on the reproducibility crisis);
Richard Frackowiak (who enabled us to do brain imaging);
Dick Passingham (who taught us brain anatomy);
Gergely Csibra (who taught us about infant development);
Stephen Fleming (who penetrated the basis of confidence monitoring);
Tim Behrens (who taught us about computational modeling);
Claudio Tennie (who introduced us to reputation management);
Micah Allen (who taught us about the body and the brain);
Barry Smith (who provided a stimulating ambience for discussions).

Alex would like to thank:

Jenny Tyler and Usborne Publishing for permission to pursue this project, and Jane Chisholm in particular for making me a better writer.
Ben and Rowan for not getting annoyed with me being on the computer all the time.
And Holly, for being born in the middle of all this, somehow.

Dan would like to thank:

Alex, a generous collaborator, who during the making of this book has become my friend.
Uta and Chris, for their brilliant brains, support and feedback, and their hospitality and kindness.

The hivemind team would like to thank:

Patrick Walsh for his enthusiasm when the book was only half-written and still in black-and-white, and to John Ash and Margaret Halton at PEW Literary for championing the book.

Alex Reuben, who helped us find a thread connecting chapter to chapter.

Sarah Goldberg and Daniel Loedel at Scribner, for making the book more coherent and fun to read. And to Kathryn Belden and Rebekah Jett, for their steady hands steering the book through its final obstacles.

Alexis Kirschbaum and Jasmine Horsey at Bloomsbury.

References

It might not look like one, but this is intended to be a proper academic work. All the studies described are real, and all the findings described have been published in various papers, journals, and books. We don't think any are from graphic novels, sadly. Here they are, listed by chapter and page number. In small print, so we don't run out of pages.

Prologue

4: No one understands how the brain works: For an excellent justification of this statement, see Matthew Cobb, *The Idea of the Brain: The Past and Future of Neuroscience* (Basic Books, 2020).

11: The mind is affected by the body: S. Gallagher, *How the Body Shapes the Mind* (Oxford: Clarendon Press, 2006).

13: Descartes: "And accordingly, it is certain that I am really distinct from my body, and can exist without it." René Descartes, *Meditations on First Philosophy*, Meditation VI, 9 (1641).

13: The dualist heresy comes naturally to all of us: A.I. Jack, A.J. Dawson, K.L. Begany, R.L. Leckie, K.P. Barry, A.H. Ciccia, and A.Z. Snyder, "fMRI reveals reciprocal inhibition between social and physical cognitive domains," *NeuroImage* (2013), 66: 385-401.

16: Energy used by brain: M.E. Raichle and D.A. Gusnard, "Appraising the brain's energy budget," *Proc Natl Acad Sci USA* (2002), 99(16): 10237-39.

17: Hormone release in anticipation of stress: J.W. Mason, L.H. Hartley, T.A. Kotchen, E.H. Mougey, P.T. Ricketts, and L.G. Jones, "Plasma cortisol and norepinephrine responses in anticipation of muscular exercise," *Psychosom Med* (1973), 35(5): 406-14.

Chapter 1

20: Galvani: See Wikipedia.

21: Learning as refining connections in the brain and neural pruning: S.-J. Blakemore and U. Frith, *The Learning Brain: Lessons for Education* (Oxford: Wiley-Blackwell, 2005).

21: Improving skills via imagination: D.L. Feltz and D.M. Landers, "The effects of mental practice on motor skill learning and performance: A meta-analysis," *J Sport Psychol* (1983), 5: 27-57.

23: Decision-making in bees and similarity to how neurons do it:
T.D. Seeley and P.K. Visscher, "Group decision making in nest-site selection by honey bees," *Apidologie* (2004), 35(2): 101-16.

T.D. Seeley, P.K. Visscher, T. Schlegel, P.M. Hogan, N.R. Franks, and J.A.R. Marshall, "Stop Signals Provide Cross Inhibition in Collective Decision-Making by Honeybee Swarms," *Science* (2012), 335(6064): 108-11.
"For neural models of decision-making, cross inhibition between integrating populations is crucial for effective decision-making and has been shown to allow optimal decisions under some circumstances. As we have shown here, cross inhibition between integrating populations is also present in honeybee swarms and is very important for their success when making decisions. It is tempting to think that the ability to implement a highly reliable strategy of decision-making is what underlies the astonishing convergence in the functional organization of these two distinct forms of decision-making system: a brain built of neurons and a swarm built of bees."

25: Vision and expectation: C.D. Frith, *Making Up the Mind: How the Brain Creates Our Mental World* (Oxford: Blackwell, 2007).

26: Shape from shading: V.S. Ramachandran, "Perception of shape from shading," *Nature* (1988), 331: 163.

28: Bayes's paper: T. Bayes and R. Price, "An Essay towards Solving a Problem in the Doctrine of Chances. By the Late Rev. Mr. Bayes, F.R.S. Communicated by Mr. Price, in a Letter to John Canton, A.M.F.R.S.," *Philosophical Transactions of the Royal Society of London* (1763), 53: 370-418.
Reprinted: T. Bayes, "Studies in the History of Probability and Statistics: IX. Thomas Bayes' Essay Towards Solving a Problem in the Doctrine of Chances," *Biometrika* (1763/1958), 45(3 & 4): 296-315.

29: K. Friston, "The free-energy principle: a unified brain theory?" *Nat Rev Neurosci* (2010), 11(2): 127-38.
Gregory Huang, "Is This a Unified Theory of the Brain?" *New Scientist*, May 23, 2008.
J. Hohwy, *The Predictive Mind* (Oxford: Oxford University Press, 2013).

Chapter 2

34: Innate priors: B.J. Scholl, "Innateness and (Bayesian) visual perception: Reconciling nativism and development," in *The Innate Mind: Structure and Contents*, ed. P. Carruthers, S. Laurence, and S. Stich (Oxford: Oxford University Press, 2005), 34-52.

36: Taxi drivers' brains: E.A. Maguire, D.G. Gadian, I.S. Johnsrude, C.D. Good, J. Ashburner, R.S. Frackowiak, and C.D. Frith, "Navigation-related structural change in the hippocampi of taxi drivers," *Proc Natl Acad Sci USA* (2000), 97(8): 4398-403.
Effects of retirement: K. Woollett, H.J. Spiers, and E.A. Maguire, "Talent in the taxi: a model system for exploring expertise," *Philos Trans R Soc Lond B Biol Sci* (2009), 364(1522): 1407-16.

38: Sensory substitution: P. Bach-y-Rita, C.C. Collins, F.A. Saunders, B. White, and L. Scadden, "Vision Substitution by Tactile Image Projection," *Nature* (1969), 221: 963.
N. Twilley, "Seeing with Your Tongue," *New Yorker*, 15 May 2017.

42: E. Claparède, "Reconnaissance et moitié [Recognition and "me-ness"]," *Archives de Psychologie Genève*, II (1911): 79-90.

43: Examples of symptoms: C.S. Mellor, "First rank symptoms of schizophrenia," *British Journal of Psychiatry* (1970), 117:15-23.

Chapter 3

48: Different kinds of learning (association learning): There seems to be a need for someone to write a short review of this topic (Chapter 12 in *What Makes Us Social?* Frith & Frith, MIT Press, forthcoming).

48: Start-up kits: U. Frith, "Are there innate mechanisms that make us social beings?" in *Neurosciences and the Human Person: New Perspectives on Human Activities*, Pontifical Academy of Sciences, *Scripta Varia 121* (Vatican City, 2013).

49: Early blurred vision helps the development of face perception: L. Vogelsang, S. Gilad-Gutnick, E. Ehrenberg, A. Yonas, S. Diamond, R. Held, and P. Sinha, "Potential downside of high initial visual acuity," *Proceedings of the National Academy of Sciences* (2018), 115: 11333-38.

49: Newborns identify human faces: M.H. Johnson, S. Dziurawiec, H. Ellis, and J. Morton, "Newborns' preferential tracking of face-like stimuli and its subsequent decline," *Cognition* (1991), 40(1-2): 1-19.

49: Fusiform face area: N. Kanwisher and G. Yovel, "The fusiform face area: a cortical region specialized for the perception of faces," *Philos Trans R Soc Lond B Biol Sci* (2006), 361(1476): 2109-28.

50: Copying because we want to be liked: T.L. Chartrand and J.A. Bargh, "The chameleon effect: the perception-behavior link and social interaction," *J Pers Soc Psychol* (1999), 76(6): 893-910.

51: Computer learning to play Atari game: V. Mnih, K. Kavukcuoglu, D. Silver, A.A. Rusu, J. Veness, M.G. Bellemare, A. Graves, M. Riedmiller, A.K. Fidjeland, G. Ostrovski, et al., "Human-level control through deep reinforcement learning," *Nature* (2015), 518, 529.

52: Advantages of copying in man and machine: L. Rendell, R. Boyd, D. Cownden, M. Enquist, K. Eriksson, M.W. Feldman, L. Fogarty, S. Ghirlanda, T. Lillicrap, and K.N. Laland, "Why copy others? Insights from the social learning strategies tournament," *Science* (2010), 328: 208-213.

54: Copying in fruit flies: F. Mery, S.A. Varela, E. Danchin, S. Blanchet, D. Parejo, I. Coolen, and R.H. Wagner, "Public versus personal information for mate copying in an invertebrate," *Curr Biol* (2009), 19: 730-34.

55: Detection of agents: B.J. Scholl and P.D. Tremoulet, "Perceptual causality and animacy," *Trends Cogn Sci* (2000), 4: 299-309.

56: Gaze following: K. Zuberbühler, "Gaze following," *Curr Biol* (2008), 18: R453-55.

57: Whites of the eye in humans: H. Kobayashi and S. Kohshima, "Unique morphology of the human eye," *Nature* (1997), 387: 767-68.

58: Gaze following: T. Farroni, S. Massaccesi, D. Pividori, and M.H. Johnson, "Gaze following in newborns," *Infancy* (2004), 5: 39-60.

59: Pouring sake: The most basic etiquette rule of serving sake is known as o-shaku. The main tenet of o-shaku is that it is considered most polite to pour sake for others but never directly for yourself.

62: Free improvisation example: *Improvised Music New York 1981* by Bill Laswell, Sonny Sharrock, Derek Bailey, Fred Frith, and John Zorn (MuWorks Records), https://www.youtube.com/watch?v=h8qex2ODYnM.

63: George Lewis's Voyager program:
Synthesizer: *Rainbow Family* (telefilm, IRCAM, Paris, 1984), https://www.anthropocene-curriculum.org/contribution/5-rainbow-family.
Piano: "Interactive Trio," Geri Allen, piano; George Lewis, trombone; interactive computer pianist, http://leccap.engin.umich.edu/leccap/view/7td666n16oht47labax/15205.

64: Listening to laughter: G.A. Bryant, D.M.T. Fessler, R. Fusaroli, et al. "Detecting affiliation in co-laughter across 24 societies," *Proceedings of the National Academy of Sciences* (2016), 113: 4682-87.

66: Laughter as social glue: S.K. Scott, N. Lavan, S. Chen, and C. McGettigan, "The social life of laughter," *Trends Cogn Sci* (2014), 18: 618-20.

68: Mind in the eyes test: S. Baron-Cohen, S. Wheelwright, J. Hill, Y. Raste, and I. Plumb, "The 'Reading the Mind in the Eyes' Test Revised Version: A Study with Normal Adults, and Adults with Asperger Syndrome or High-functioning Autism," *Journal of Child Psychology and Psychiatry and Allied Disciplines* (2001), 42: 241-51.

Chapter 4

72: Uniquely human teaching: M.A. Kline, "How to learn about teaching: An evolutionary framework for the study of teaching behavior in humans and other animals," *Behavioral and Brain Sciences* (2015), 38, e31.

72: Naming in dolphins: S.L. King and V.M. Janik, "Bottlenose dolphins can use learned vocal labels to address each other," *Proceedings of the National Academy of Sciences of the United States of America* (2013), 110: 13216-21.

73: Learning through imitation in chimpanzees: A. Whiten, "Imitation of the sequential structure of actions by chimpanzees (Pan troglodytes)," *J Comp Psychol* (1998), 112: 270-81.

73: Teaching in meerkats: A. Thornton and K. McAuliffe, "Teaching in wild meerkats," *Science* (2006), 313: 227-29.

74: Overimitation in children: D.E. Lyons, A.G. Young, and F.C. Keil, "The hidden structure of overimitation," *Proc Natl Acad Sci USA* (2007), 104: 19751-56.

75: Overimitation in adults: N. McGuigan, J. Makinson, and A. Whiten, "From over-imitation to super-copying: Adults imitate causally irrelevant aspects of tool use with higher fidelity than young children," *Br J Psychol* (2011), 102, 1-18.

75: Overimitation in chimpanzees: V. Horner and A. Whiten, "Causal knowledge and imitation/emulation switching in chimpanzees (Pan troglodytes) and children (Homo sapiens)," *Anim Cogn* (2005), 8: 164-81.

76: M. McDougall, "Imitation, Play, and Habit," Chapter 15 in *An Introduction to Social Psychology*, rev. ed. (Boston: John W. Luce, 1926), 332-58.

76: Overimitation in autism: L. Marsh, A. Pearson, D. Ropar, and A. Hamilton, "Children with autism do not overimitate," *Curr Biol* (2013), 23: R266-68.

77: Asperger syndrome: U. Frith, "Asperger and his syndrome," *Autism and Asperger Syndrome* (1991), 14: 1-36.

77: Characterization of autism: U. Frith, *Autism: Explaining the Enigma* (Oxford: Blackwell, 1989).

78: L. Wing, "The autistic spectrum," *The Lancet* (1997), 350: 1761-66.

81: B. Hermelin and N. O'Connor, *Psychological Experiments with Autistic Children* (Oxford: Pergamon Press, 1970).

82: Pretend play: A.M. Leslie, "Pretence and representation: The origins of 'theory of mind,'" *Psychological Review* (1987), 94: 412-26.

82: Autistic children don't use pretend play: L. Wing, J. Gould, S.R. Yeates, and L.M. Brierley, "Symbolic Play in Severely Mentally Retarded and in Autistic Children," *Journal of Child Psychology and Psychiatry* (1977), 18: 167-78.

84: H. Wimmer and J. Perner, "Beliefs About Beliefs—Representation and Constraining Function of Wrong Beliefs in Young Children's Understanding of Deception," *Cognition* (1983), 13: 103-28.

85: Baron-Cohen: S. Baron-Cohen, A.M. Leslie, and U. Frith, "Does the autistic child have a 'theory of mind'?" *Cognition* (1985), 21: 37-46.

87: Ágnes Kovács: Á.M. Kovács, E. Téglás, and A.D. Endress, "The social sense: susceptibility to others' beliefs in human infants and adults," *Science* (2010), 330: 1830-34.

88: Implicit theory of mind and autism: A. Senju, V. Southgate, S. White, and U. Frith, "Mindblind eyes: an absence of spontaneous theory of mind in Asperger syndrome," *Science* (2009), 325: 883-85.

89: "Empathy" for unseen fear: P.J. Whalen, S.L. Rauch, N.L., Etcoff, S.C. McInerney, M.B. Lee, and M.A. Jenike, "Masked presentations of emotional facial expressions modulate amygdala activity without explicit knowledge," *J Neurosci* (1998), 18: 411-18.

89: "Empathy" for disgust: B. Wicker, C. Keysers, J. Plailly, J.P. Royet, V. Gallese, and G. Rizzolatti, "Both of us disgusted in My insula: the common neural basis of seeing and feeling disgust," *Neuron* (2003), 40: 655-64.

89: Instinctive (automatic) gaze following: A.P. Bayliss and S.P. Tipper, "Predictive gaze cues and personality judgments: Should eye trust you?" *Psychol Sci* (2006), 17: 514-20.

89: Emotion and empathy in autism: A.P. Jones, F.G.E. Happé, F. Gilbert, S. Burnett, and E. Viding, "Feeling, caring, knowing: different types of empathy deficit in boys with psychopathic tendencies and autism spectrum disorder," *Journal of Child Psychology and Psychiatry* (2010), 51: 1188-97.

90: Affordance with others: P. Cardellicchio, C. Sinigaglia, and M. Costantini, "Grasping affordances with the other's hand: A TMS study," *Social Cognitive and Affective Neuroscience* (2013), 8: 455-59.

91: We mode: M. Gallotti and C.D. Frith, "Social cognition in the we-mode," *Trends Cogn Sci* (2013), 17: 160-65.

91: A. Iriki, M. Tanaka, and Y. Iwamura, "Coding of modified body schema during tool use by macaque postcentral neurones," *Neuroreport* (1996), 7: 2325-30.

Chapter 5

96: Brain basis for empathy: B.C. Bernhardt and T. Singer, "The neural basis of empathy," *Annu Rev Neurosci* (2012), 35: 1-23.

98: Rizzolatti, mirror neurons: G. Rizzolatti and L. Craighero, "The mirror-neuron system," *Annu Rev Neurosci* (2004), 27: 169-92.

98: Jamie Ward, mirror-touch synesthesia: S.-J. Blakemore, D. Bristow, G. Bird, C. Frith, and J. Ward, "Somatosensory activations during the observation of touch and a case of vision-touch synaesthesia," Brain (2005), 128: 1571-83. Read his book *The Student's Guide to Cognitive Neuroscience* (Hove, UK: Psychology Press, 2015).

100: Switching off mirror neurons: K.A. Cross, S. Torrisi, E.A.R. Losin, and M. Iacoboni, "Controlling automatic imitative tendencies: Interactions between mirror neuron and cognitive control systems," *Neuroimage* (2013), 83: 493-504.

101: Sensation in phantom limbs: V.S. Ramachandran and W. Hirstein, "The perception of phantom limbs. The D. O. Hebb lecture," *Brain* (1998), 121: 1603-30.

101: Rubber hand illusion: M. Botvinick and J. Cohen, "Rubber hands 'feel' touch that eyes see," *Nature* (1998), 391: 756.

103: Johannes Müller, "Law of specific nerve energies," *Handbuch der Physiologie des Menschen für Vorlesungen.* (Coblenz: J. Hölscher, 1840).

103: Seeing light from pressure on the eyeball: C.W. Tyler, "Some new entoptic phenomena," *Vision Res* (1978), 18: 1633-39.
[According to our exploration of the Internet, it was Helmholtz who first wrote about the phenomenon of pressure phosphenes.]

104: Speed of nerve conduction: H. Helmholtz, "Vorläufiger Bericht über die Fortpflanzungs-Geschwindigkeit der Nervenreizung," in *Archiv für Anatomie, Physiologie und wissenschaftliche Medicin. Jg. 1850* (Berlin: Veit & Comp., 1850), 71-73.
(Good Wikipedia entry for Helmholtz.)

104: Wilhelm Wundt: Wikipedia entry.

104: Pavlov: Wikipedia entries for I.P. Pavlov and for classical conditioning.

106: Golgi: Wikipedia entry.

106: Ramón y Cajal: Wikipedia entry.
E.A. Newman, A. Araque, and J.M. Dubinsky, *The Beautiful Brain: The Drawings of Santiago Ramón y Cajal* (New York: Abrams, 2017).
Golgi's Nobel lecture attacking the neuron doctrine of Ramón y Cajal: https://www.nobelprize.org/prizes/medicine/1906/golgi/lecture/.

108: Broca: N.F. Dronkers, O. Plaisant, M.T. Iba-Zizen, and E.A. Cabanis, "Paul Broca's historic cases: high resolution MR imaging of the brains of Leborgne and Lelong," *Brain* (2007), 130: 1432-41.

109: Inouye: M. Glickstein and D. Whitteridge, "Tatsuji Inouye and the mapping of the visual fields on the human cerebral cortex," *Trends in Neurosciences* (1987), 10: 350-53.

110: I.H. Hyde, "A micro-electrode and unicellar stimulation," *Biol Bull* (1921), 40: 130-33.

110: Elizabeth Warrington: R.A. McCarthy and E.K. Warrington, *Cognitive Neuropsychology* (London: Academic Press, 1990).

111: Brenda Milner: B. Milner, S. Corkin, and H.L. Teuber, "Further analysis of the hippocampal amnesic syndrome: 14-year follow-up study of H.M.," *Neuropsychologia* (1968), 6: 215-34.

112: B. Milner, "Some cognitive effects of frontal-lobe lesions in man," *Philos Trans R Soc Lond B Biol Sci* (1982), 298: 211-26.

112: Norbert Wiener: N. Wiener, "Cybernetics," *Scientific American* (1948), 179: 14-19.

112: Claude Shannon: C.E. Shannon and W. Weaver, *The Mathematical Theory of Communication* (Urbana: University of Illinois Press, 1949).

113: The brain as an information processing device: W.S. McCulloch and W.H. Pitts, "A logical calculus of the ideas immanent in nervous activity," *Bulletin of Mathematical Biophysics* (1943), 5: 115-33.

114: Alan Turing: A.M. Turing, "Computing machinery and intelligence," *Mind* (1950), 59: 433-60.

114: John von Neumann: J. von Neumann, *The Computer and the Brain* (New Haven/London: Yale University Press, 1958).

116: Mirror neurons, innate or learned?: C. Heyes, "Where do mirror neurons come from?" *Neuroscience & Biobehavioral Reviews* (2010), 34: 575-83.

117: Mirror neurons and the problem of who is acting: N. Georgieff and M. Jeannerod, "Beyond Consciousness of External Reality: A 'Who' System for Consciousness of Action and Self-Consciousness," *Consciousness and Cognition* (1998), 7: 465-77.

Chapter 6

120: *Fooling the brain with robot arms:* S.-J. Blakemore, C.D. Frith, and D.M. Wolpert, "Spatio-temporal prediction modulates the perception of self-produced stimuli," *J Cogn Neurosci* (1999), 11: 551-59. Read Sarah-Jayne's book, *Inventing Ourselves: The Secret Life of the Teenage Brain* (Doubleday, 2018).

120: *Why you can't tickle yourself:* L. Weiskrantz, J. Elliott, and C. Darlington, "Preliminary observations on tickling oneself," *Nature* (1971), 230: 598-99.

121: C.D. Frith and E.C. Johnstone, *Schizophrenia: A Very Short Introduction* (Oxford: Oxford University Press, 2003).

122: *Examples of delusions and hallucinations:* J. Chapman, "The Early Symptoms of Schizophrenia," *British Journal of Psychiatry* (1966), 112: 225-51.

123: *Animal models:* C.A. Jones, D.J.G. Watson, and K.C.F. Fone, "Animal models of schizophrenia," *British Journal of Pharmacology* (2012), 164: 1162-94.

123: *Hallucinations in the healthy:* I.E. Sommer, K. Daalman, T. Rietkerk, K.M. Diederen, S. Bakker, J. Wijkstra, and M.P. Boks, "Healthy individuals with auditory verbal hallucinations; who are they? Psychiatric assessments of a selected sample of 103 subjects," *Schizophr Bull* (2010), 36: 633-41.

124: *Nature and treatment of schizophrenia:* C.D. Frith and E.C. Johnstone, *Schizophrenia: A Very Short Introduction* (Oxford: Oxford University Press, 2003).

126: *Measuring postmortem brains:* C.J. Bruton, T.J. Crow, C.D. Frith, E.C. Johnstone, D.G. Owens, and G.W. Roberts, "Schizophrenia and the brain: a prospective clinico-neuropathological study," *Psychol Med* (1990), 20: 285-304.

126: *Drugs that block dopamine:* E.C. Johnstone, T.J. Crow, C.D. Frith, M.W. Carney, and J.S. Price, "Mechanism of the antipsychotic effect in the treatment of acute schizophrenia," *Lancet* (1978), 1: 848-51.

127: *Hearing voices:* T.H. Nayani and A.S. David, "The auditory hallucination: A phenomenological survey," *Psychological Medicine* (1996), 26: 177-89.

127: C.D. Frith, *The Cognitive Neuropsychology of Schizophrenia* (Classic Edition) (Hove, UK: Psychology Press, 1992/2015).

128: K.E. Stephan, K.J. Friston, and C.D. Frith, "Dysconnection in schizophrenia: from abnormal synaptic plasticity to failures of self-monitoring," *Schizophr Bull* (2009), 35: 509-27.

129: *Thought insertion:* I. Feinberg, "Efference copy and corollary discharge: implications for thinking and its disorders," *Schizophr Bull* (1978), 4: 636-40.

131: *Victims of violence:* H. Khalifeh, S. Johnson, L.M. Howard, R. Borschmann, D. Osborn, K. Dean, C. Hart, J. Hogg, and P. Moran, "Violent and non-violent crime against adults with severe mental illness," *Br J Psychiatry* (2015), 206: 275-82, doi:10.1192/bjp.bp.114.147843.

132: Guilty, but insane (M'Naghten rule): https://en.wikipedia.org/wiki/M%27Naghten_rules.

133: (Some) patients with schizophrenia can tickle themselves: S.-J. Blakemore, J. Smith, R. Steel, E.C. Johnstone, and C.D. Frith, "The perception of self-produced sensory stimuli in patients with auditory hallucinations and passivity experiences: evidence for a breakdown in self-monitoring," *Psychol Med* (2000), 30: 1131-39.

134: Libet experiment: B. Libet, C.A. Gleason, E.W. Wright, and D.K. Pearl, "Time of conscious intention to act in relation to onset of cerebral activity (readiness-potential). The unconscious initiation of a freely voluntary act," *Brain* (1983), 106 (Pt 3): 623-42.

136: The stir caused by Libet's experiment: https://en.wikipedia.org/wiki/Benjamin_Libet.

137: Libet revisited: C.D. Frith and P. Haggard, "Volition and the Brain–Revisiting a Classic Experimental Study," *Trends in Neurosciences* (2018), 41: 405-7.

139: Schizophrenia and predictions: P.C. Fletcher and C.D. Frith, "Perceiving is believing: a Bayesian approach to explaining the positive symptoms of schizophrenia," *Nat Rev Neurosci* (2009), 10: 48-58.

Interlude

142: Libet experiment with fMRI: C.S. Soon, M. Brass, H.J. Heinze, and J.D. Haynes, "Unconscious determinants of free decisions in the human brain," *Nat Neurosci* (2008), 11: 543-45.

144: The MNI brain, used in many brain imaging packages, is the brain of Colin Holmes, scanned 27 times to create a very-high-detail MRI image of one brain: http://www.bic.mni.mcgill.ca/ServicesAtlases/Colin27.

145: Libet experiment: B. Libet, C.A. Gleason, E.W. Wright, and D.K. Pearl, "Time of conscious intention to act in relation to onset of cerebral activity (readiness-potential). The unconscious initiation of a freely voluntary act," *Brain* (1983), 106 (Pt 3): 623-42.

148: Finding significant results that aren't really there: https://en.wikipedia.org/wiki/Data_dredging.

150: Replication crisis: https://en.wikipedia.org/wiki/Replication_crisis.
–In psychology research: Leslie K. John, George Loewenstein, and Drazen Prelec, "Measuring the Prevalence of Questionable Research Practices with Incentives for Truth Telling," *Psychological Science* (2012), 23 (5): 524-32.
–In cancer research: C.G. Begley and M.E. Lee, "Drug Development: Raise Standards for Preclinical Cancer Research," *Nature* (2012), 483: 531-33.

150: Thinking about old age and walking slowly: J.A. Bargh, M. Chen, and L. Burrows, "Automaticity of social behavior: Direct effects of trait construct and stereotype activation on action," *Journal of Personality and Social Psychology* (1996), 71: 230-44.

151: Priming the experimenters: S. Doyen, O. Klein, C.-L. Pichon and A. Cleeremans, "Behavioral Priming: It's All in the Mind, but Whose Mind?" *PLoS ONE* (2012), 7: e29081.
But has the stereotype that the elderly move more slowly dissipated since 1996?

151: **WEIRD**: J. Henrich, S.J. Heine, and A. Norenzayan, "The weirdest people in the world?" *Behav Brain Sci* (2010), 33: 61-83, discussion 83-135.

153: Correlation is not causation: https://en.wikipedia.org/wiki/Correlation_does_not_imply_causation.

154: Coffee consumption and anxiety: W.W. Eaton and J. McLeod, "Consumption of coffee or tea and symptoms of anxiety," *American Journal of Public Health* (1984), 74: 66-68.

155: Individual differences in coffee drinking: A. Steptoe and J. Wardle, "Mood and drinking: a naturalistic diary study of alcohol, coffee and tea," *Psychopharmacology* (1999), 141: 315-21.

156: Coffee and schizophrenia: J.D. Mann and E.H. Labrosse, "Urinary excretion of phenolic acids by normal and schizophrenic male patients," *AMA Archives of General Psychiatry* (1959), 1: 547-51.

Chapter 7

159: Metacognition in nonhuman animals: J.D. Crystal and A.L. Foote, "Metacognition in animals," *Comparative Cognition & Behavior Reviews* (2009), 4: 1-16.

161: Theory of mind in nonhuman animals: C. Heyes, "Animal mindreading: what's the problem?" *Psychonomic Bulletin & Review* (2015), 22(2): 313-27.

163: Recursion in the songs of starlings: G.F. Marcus, "Startling starlings," *Nature* (2006), 440: 1117.

164: Beauty contest game: "It is not a case of choosing those [faces] that, to the best of one's judgment, are really the prettiest, nor even those that average opinion genuinely thinks the prettiest. We have reached the third degree where we devote our intelligences to anticipating what average opinion expects the average opinion to be. And there are some, I believe, who practice the fourth, fifth, and higher degrees." (John Maynard Keynes, *General Theory of Employment, Interest, and Money*, 1936, chapter 12).

165: P-beauty contest game, limited recursion in people: C.F. Camerer, T.-H. Ho, and J.-K. Chong, "A Cognitive Hierarchy Model of Games," *Quarterly Journal of Economics* (2004), 119: 861-98.

165: Had to guess 2/3 of average: Astrid Schou, "Gæt-et-tal konkurrence afslører at vi er irrationelle," *Politiken* (in Danish), 22 September 2005, retrieved 29 August 2017. Includes a histogram of the guesses. Note that some of the players guessed close to 100. A large number of players guessed 33.3 (i.e., 2/3 of 50), indicating an assumption that players would guess randomly (one level of recursion). A smaller but significant number of players guessed 22.2 (i.e., 2/3 of 33.3, two levels of recursion). The final number of 21.6 was slightly below this peak, implying that on average each player had 1.07 levels of recursion.

166: L. Goupil and S. Kouider, "Developing a Reflective Mind: From Core Metacognition to Explicit Self-Reflection," *Current Directions in Psychological Science* (2019), 28: 403-8.
In preverbal infants: L. Goupil and S. Kouider, "Behavioral and Neural Indices of Metacognitive Sensitivity in Preverbal Infants," *Current Biology* (2016), 26: 3038-45.

170: Don't know button/nonhuman animals:
J.D. Smith, W.E. Shields, and D.A. Washburn, "The comparative psychology of uncertainty monitoring and metacognition," *Behavioral and Brain Sciences* (2003), 26: 317-39, discussion 340-73.
A.L. Foote and J.D. Crystal, "Metacognition in the rat," *Current Biology* (2007), 17: 551-55.

Chapter 8

182: Results of ECT trial: E.C. Johnstone, J.F. Deakin, P. Lawler, C.D. Frith, M. Stevens, K. McPherson, and T.J. Crow, "The Northwick Park electroconvulsive therapy trial," Lancet (1980), 2: 1317-20.

183: Positron Emission Tomography: https://en.wikipedia.org/wiki/Brain_positron_emission_tomography.

184: fMRI signals caused by movement: J.V. Hajnal, R. Myers, A. Oatridge, J.E. Schwieso, I.R. Young, and G.M. Bydder, "Artifacts due to stimulus correlated motion in functional imaging of the brain," *Magnetic Resonance in Medicine* (1994), 31: 283-91.

185: John Bohannon, "A computer program just ranked the most influential brain scientists of the modern era," *Science*, 11 November 2016, 9:45 a.m.

187: The Lab (St. John's House): A. Roepstorff, "Transforming subjects into objectivity: an 'ethnography of knowledge' in a brain imaging laboratory," *J Dan Ethnogr Soc* (2002), 44: 145-70.

190: Need to study two brains at once: L. Schilbach, B. Timmermans, V. Reddy, A. Costall, G. Bente, T. Schlicht, and K. Vogeley, "Toward a second-person neuroscience," *Behav Brain Sci* (2013), 36: 393-414.

191: The problem of instructing the participant: A. Roepstorff and C. Frith, "What's at the top in the top-down control of action? Script-sharing and 'top-top' control of action in cognitive experiments," *Psychol Res* (2004), 68: 189-98.

Chapter 9

196: Effects of prayer on pain: E.-M. Elmholdt, J. Skewes, M. Dietz, A. Møller, M.S. Jensen, A. Roepstorff, K. Wiech, and T.S. Jensen, "Reduced Pain Sensation and Reduced BOLD Signal in Parietofrontal Networks during Religious Prayer," *Frontiers in Human Neuroscience* (2017), 11: 337.

196: Group interaction using LEGO®: R. Fusaroli, J.S. Bjørndahl, A. Roepstorff, and K. Tylén, "A heart for interaction: Shared physiological dynamics and behavioral coordination in a collective, creative construction task," *Journal of Experimental Psychology: Human Perception and Performance* (2016), 42: 1297-1310.

198: Psychophysics of two heads better than one:
Dot counting: B. Bahrami, D. Didino, C. Frith, B. Butterworth, and G. Rees, "Collective enumeration," *J Exp Psychol Hum Percept Perform* (2013), 39, 338-47.
Two-interval, oddball, contrast detection task: B. Bahrami, K. Olsen, P.E. Latham, A. Roepstorff, G. Rees, and C.D. Frith, "Optimally interacting minds," *Science* (2010), 329: 1081-85.

200: Talking about confidence: R. Fusaroli, B. Bahrami, K. Olsen, A. Roepstorff, G. Rees, C. Frith, and K. Tylén, "Coming to terms: quantifying the benefits of linguistic coordination," *Psychol Sci* (2012), 23: 931-39.

207: CIA committee: S. Kent, *Sherman Kent and the Board of National Estimates: Collected Essays* (Washington, DC: History Staff, Center for the Study of Intelligence, Central Intelligence Agency; University of Michigan Library, 1994).

208: Calibrating confidence: D. Bang, L. Aitchison, R. Moran, S. Herce Castanon, B. Rafiee, A. Mahmoodi, J.Y.F. Lau, P.E. Latham, B. Bahrami, and C. Summerfield, "Confidence matching in group decision-making," *Nature Human Behaviour* (2017), 1: 0117.

210: World rankings of trust: A cross-country project coordinated by the Institute for Social Research of the University of Michigan, www.worldvaluessurvey.org/.

210: No cultural differences: A. Mahmoodi, D. Bang, K. Olsen, Y.A. Zhao, Z. Shi, K. Broberg, S. Safavi, S. Han, M. Nili Ahmadabadi, C.D. Frith, et al., "Equality bias impairs collective decision-making across cultures," *Proceedings of the National Academy of Sciences* (2015), 112: 3835-40.

212: Better problem-solving in groups (bat and ball problem): E. Trouche, E. Sander, and H. Mercier, "Arguments, more than confidence, explain the good performance of reasoning groups," *J Exp Psychol Gen* (2014), 143: 1958-71.

214: Exploring and exploiting: T.T. Hills, P.M. Todd, D. Lazer, A.D. Redish, I.D. Couzin, and Cognitive Search Research Group, "Exploration versus exploitation in space, mind, and society," *Trends Cogn Sci* (2015), 19: 46-54.

214: Proportion of bees who are scouts: T.D. Seeley, *Honeybee Democracy* (Princeton, NJ: Princeton University Press, 2010).

217: M. Dewey, "Living with Asperger's syndrome," in *Autism and Asperger Syndrome*, ed. U. Frith (Cambridge: Cambridge University Press, 1991), 184-206.

218: Attention to detail in autism: A. Shah and U. Frith, "An islet of ability in autistic children: A research note," *Journal of Child Psychology and Psychiatry* (1983), 24: 613-20.

Chapter 10

222: Solving coordination problems: D.A. Braun, P.A. Ortega, and D.M. Wolpert, "Motor coordination: when two have to act as one," *Exp Brain Res* (2011), 211: 631-41.

226: Third party advice to solve coordination problems: R.J. Aumann, "Correlated equilibrium as an expression of Bayesian rationality," *Econometrica* (1987), 55: 1-18.

227: O. Henry story: http://www.auburn.edu/~vestmon/Gift_of_the_Magi.html.

228: Game theory: https://en.wikipedia.org/wiki/Game_theory.

229: Value of economic games in real life: A.C. Pisor, M.M. Gervais, B.G. Purzycki, and C.T. Ross, "Preferences and constraints: the value of economic games for studying human behaviour," *Royal Society Open Science* (2020), 7: 192090.

229: Unconscious unselfishness: D.G. Rand, J.D. Greene, and M.A. Nowak, "Spontaneous giving and calculated greed," *Nature* (2012), 489: 427-30.

229: Far Pavilions: V.P. Crawford, M.A. Costa-Gomes, and N. Iriberri, "Structural Models of Nonequilibrium Strategic Thinking: Theory, Evidence, and Applications," *Journal of Economic Literature* (2013), 51: 5-62 (part 4).

231: More recursion good in competition, bad in cooperation: M. Devaine, G. Hollard, and J. Daunizeau, "Theory of Mind: Did Evolution Fool Us?" *PLoS ONE* (2014), 9: e87619.

233: Automatic imitation in matching shapes game: M. Belot, V.P. Crawford, and C. Heyes, "Players of Matching Pennies automatically imitate opponents' gestures against strong incentives," *Proceedings of the National Academy of Sciences of the United States of America* (2013), 110: 2763-68.

236: Haunted house experiment: G. Dezecache, J. Grèzes, and C.D. Dahl, "The nature and distribution of affiliative behaviour during exposure to mild threat," *Royal Society Open Science* (2017), 4: 170265.

236: Intuitive desire to cooperate: D.G. Rand, "Cooperation, Fast and Slow: Meta-Analytic Evidence for a Theory of Social Heuristics and Self-Interested Deliberation," *Psychological Science* (2016), 27: 1192-1206.

Chapter 11

242: Epicurus and free will: S. Bobzien, "Moral Responsibility and Moral Development in Epicurus' Philosophy," in *The Virtuous Life in Greek Ethics*, ed. B. Reis (New York: Cambridge University Press, 2006), 206-99.

243: Effects of anticipated regret on behavior: M. Zeelenberg, "Anticipated regret, expected feedback and behavioral decision making," *J Behav Decis Mak* (1997), 12: 93-106.

244: Regret and proactive choice: T. Gilovich and V.H. Medvec, "The experience of regret: what, when, and why," *Psychol Rev* (1995), 102: 379-95.

245: The trolley problem: P. Foot, "The Problem of Abortion and the Doctrine of Double Effect," *Oxford Review* (1967), 5: 5-15.

247: Cash version of the trolley problem: O. FeldmanHall, D. Mobbs, D. Evans, L. Hiscox, L. Navrady, and T. Dalgleish, "What we say and what we do: the relationship between real and hypothetical moral choices," *Cognition* (2012), 123: 434-41.

249: Auctions with anticipated regret: E. Filiz-Ozbay and E.Y. Ozbay, "Auctions with anticipated regret: Theory and experiment," *American Economic Review* (2007), 97: 1407-18.

251: Lemon juice and empathy: F. Hagenmuller, W. Rössler, A. Wittwer, and H. Haker, "Juicy lemons for measuring basic empathic resonance," *Psychiat Res* (2014), 219: 391-96.

253: Empathy and contagion: F. de Vignemont and T. Singer, "The empathic brain: how, when and why?" *Trends Cogn Sci* (2006), 10: 435-41.

253: Laughter: S.K. Scott, N. Lavan, S. Chen, and C. McGettigan, "The social life of laughter," *Trends in Cognitive Sciences* (2014), 18: 618-20.

255: Brain responses to fake laughter: N. Lavan, G. Rankin, N. Lorking, S. Scott, and C. McGettigan, "Neural correlates of the affective properties of spontaneous and volitional laughter types," *Neuropsychologia* (2017), 95: 30-39.

256: Neuroscience, free will and responsibility: D. Talmi, and C.D. Frith, "Neuroscience, Free Will, and Responsibility," in *Conscious Will and Responsibility: A Tribute to Benjamin Libet*, eds. W. Sinnott-Armstrong and L. Nadel (New York: Oxford University Press, 2011), 124-33.

257: Psychopaths don't join in with laughter: E. O'Nions, C.F. Lima, S.K. Scott, R. Roberts, E.J. McCrory, and E. Viding, "Reduced Laughter Contagion in Boys at Risk for Psychopathy," *Current Biology* (2017), 27: 3049-55. e3044.

Chapter 12

267: Action observation affects action: M. Brass, H. Bekkering, and W. Prinz, "Movement observation affects movement execution in a simple response task," *Acta Psychol (Amst)* (2001), 106: 3-22.

269: Natalie Sebanz experiment: Not yet published. See https://www.ceu.edu/article/2013-06-18/good-news-racial-attitudes-can-change.

271: Frith-Happé triangles: F. Abell, F. Happé, and U. Frith, "Do triangles play tricks? Attribution of mental states to animated shapes in normal and abnormal development," *Cognitive Development* (2000), 15: 1-16.

272: Seeing ostracism increases imitation: H. Over and M. Carpenter, "Priming third-party ostracism increases affiliative imitation in children," *Developmental Science* (2009), 12: F1-F8.

272: Experiencing ostracism increases imitation: R.E. Watson-Jones, H. Whitehouse, and C.H. Legare, "In-Group Ostracism Increases High-Fidelity Imitation in Early Childhood," *Psychol Sci* (2016), 27: 34-42.

273: Reading Harry Potter reduces prejudice: L. Vezzali, S. Stathi, D. Giovannini, D. Capozza, and E. Trifiletti, "The greatest magic of Harry Potter: Reducing prejudice," *Journal of Applied Social Psychology* (2014), 45: 105-21.

274: Reducing prejudice: L. Maister, N. Sebanz, G. Knoblich, and M. Tsakiris, "Experiencing ownership over a dark-skinned body reduces implicit racial bias," *Cognition* (2013), 128: 170-78.

275: No empathy for out-group members: X. Xu, X. Zuo, X. Wang, and S. Han, "Do you feel my pain? Racial group membership modulates empathic neural responses," *J Neurosci* (2009), 29: 8525-29.

276: Tricks for reducing racial bias: F. Sheng and S. Han, "Manipulations of cognitive strategies and intergroup relationships reduce the racial bias in empathic neural responses," *Neuroimage* (2012), 61: 786-97.

278: Group solving of murder mysteries: K.W. Phillips, K.A. Liljenquist, and M.A. Neale, "Is the Pain Worth the Gain? The Advantages and Liabilities of Agreeing with Socially Distinct Newcomers," *Pers Soc Psychol B* (2008), 35: 336-50.

280: Individuals do best by being selfish, groups do best by having altruistic individuals: E. Sober and D.S. Wilson, *Unto Others* (Cambridge, MA: Harvard University Press, 1998).

281: Reputation: See Wikipedia.

282: Brain regions involved in love: B.P. Acevedo, A. Aron, H.E. Fisher, and L.L. Brown, "Neural correlates of long-term intense romantic love," *Soc Cogn Affect Neurosci* (2012), 7: 145-59.

283: Intelligence and divorce: J. Dronkers, "Bestaat er een samenhang tussen echtscheiding en intelligentie? [Is there a relation between divorce and intelligence?]" *Mens en Maatschappij* (2002), 77:25-42.

Chapter 13

288: Galen and Vesalius: See Wikipedia.

290: Impact factor, h-index: See Wikipedia.

290: E. Sober and D.S. Wilson, *Unto Others* (Cambridge, MA: Harvard University Press, 1998).

291: Putting eyes over the collection jar: M. Bateson, D. Nettle, and G. Roberts, "Cues of being watched enhance cooperation in a real-world setting," *Biol Lett* (2006), 2: 412-14.

292: Cleaner wrasse like to bite clients: A.S. Grutter, and R. Bshary, "Cleaner wrasse prefer client mucus: support for partner control mechanisms in cleaning interactions," *Proceedings of the Royal Society of London Series B-Biological Sciences* (2003), 270: S242-S244.

292: Cleaner wrasse seek a good reputation: A. Pinto, J. Oates, A. Grutter, and R. Bshary, "Cleaner Wrasses Labroides dimidiatus Are More Cooperative in the Presence of an Audience," *Current Biology* (2011), 21: 1140-44.

293: Cleaner wrasse dance: A.S. Grutter, "Cleaner Fish Use Tactile Dancing Behavior as a Pre-conflict Management Strategy," *Current Biology* (2004), 14: 1080-83.

293: Avoiding schools with too many free riders: R. Bshary and A. D'Souza, in *Animal Communication Networks*, ed. P.K. McGregor (Cambridge University Press, 2005), 521-39.

293: Joyce Berg, trust games: J. Berg, J. Dickhaut, and K. McCabe, "Trust, Reciprocity, and Social History," *Games and Economic Behavior* (1995), 10: 122-42.

295: Trust with caudate and amygdala: J.K. Rilling and A.G. Sanfey, "The Neuroscience of Social Decision-Making," *Annual Review of Psychology* (2010), 62: 23-48.

297: Singing and dancing children: L.N. Chaplin and M.I. Norton, "Why We Think We Can't Dance: Theory of Mind and Children's Desire to Perform," *Child Development* (2014), 86: 651-58.

298: Audience effect in babies: S. Jones, K. Collins, and H. Hong, "An Audience Effect on Smile Production in 10-Month-Old Infants," *Psychological Science* (1991), 2: 45-49.

299: Audience effect in autism: K. Izuma, K. Matsumoto, C.F. Camerer, and R. Adolphs, "Insensitivity to social reputation in autism," *Proceedings of the National Academy of Sciences* (2011), 108: 17302-307.

299: Reputation on the Internet: C. Tennie, U. Frith, and C.D. Frith, "Reputation management in the age of the world-wide web," *Trends Cogn Sci* (2010), 14: 482-88.

300: Gossip: P.M. Spacks, "In Praise of Gossip," *Hudson Review* (1982), 35: 19-38.

303: Gossip as an alternative to direct experience (with MRI): M.R. Delgado, R.H. Frank, and E.A. Phelps, "Perceptions of moral character modulate the neural systems of reward during the trust game," *Nat Neurosci* (2005), 8: 1611-18.

303: Many sources of gossip: R.D. Sommerfeld, H.J. Krambeck, and M. Milinski, "Multiple gossip statements and their effect on reputation and trustworthiness," *Proc Biol Sci* (2008), 275: 2529-36.

303: More gossip: R.D. Sommerfeld, H.J. Krambeck, D. Semmann, and M. Milinski, "Gossip as an alternative for direct observation in games of indirect reciprocity," *Proc Natl Acad Sci USA* (2007), 104: 17435-40.

305: Reputation (or at least confidence) in the brain: A. Clark, *Surfing Uncertainty: Prediction, Action, and the Embodied Mind*, (Oxford University Press, 2015).

309: In praise of collaboration and diversity: D. Bang and C.D. Frith, "Making better decisions in groups," *R Soc Open Sci* (2017), 4, 170193.

Index

About the Authors

Uta Frith, DBE, FBA, AcMedSci, FRS, is emeritus professor of Cognitive Development at the Institute of Cognitive Neuroscience at University College London. She is a member of the Leopoldina, the German National Academy of Sciences, and a Foreign Associate of the NAS. Uta is known for her research on autism and dyslexia, which has resulted in new insights into the links between mind, brain, and behavior in these conditions. She was listed in 2014 as among the two hundred most eminent psychologists of the modern era. She also has a special interest in science communication and public engagement, and has made a number of acclaimed TV documentaries for BBC *Horizon*. She was chair of the Diversity Committee at the Royal Society from 2015 to 2018 and has raised awareness of the value of diversity in making group decisions.

Chris Frith is emeritus professor of Neuropsychology at the Wellcome Centre for Human Neuroimaging at University College London, visiting professor at the Interacting Minds Centre, Aarhus University, and honorary research fellow at the Institute of Philosophy, University of London. Since completing his PhD on experimental psychology in 1969, he was funded by the Medical Research Council and the Wellcome Trust to study the relationship between the mind and the brain. He is a pioneer in the application of brain imaging to the study of mental processes. He has contributed more than five hundred papers to scientific journals and is known especially for his work on agency, social cognition, and understanding the minds of people with mental disorders such as schizophrenia. In 2016, he was listed among the top ten most influential brain scientists of the modern era (*Science*, News, Nov 11, 2016). He is a Fellow of the Royal Society (2000), the American Association for the Advancement of Science (2000), and the British Academy (2008).

Alex Frith has been a children's nonfiction author since 2005. Working exclusively for Usborne Publishing, he has written more than fifty books covering almost any subject you can think of, from the origins of the universe to the meaning behind world religions, from extinct animals to prototype AIs. Two of his books have been shortlisted for the Royal Society Young People's Book Prize: *See Inside Inventions* (2012) and *100 Things to Know about Space* (2017).

Daniel Locke is an artist and graphic novelist. Much of his work has been informed and shaped by the discoveries of contemporary science. His work ranges between commissions for commercial bodies and socially engaged organizations, such as universities and charities, alongside personal work and residencies. He has collaborated with a wide range of researchers and artists on a diverse series of projects, including the Wellcome Trust, Nobrow, the NHS, the National Trust, and Arts Council England.